全球新建筑

Global Architecture Today Ⅳ

博览空间

《设计家》编

天津大学出版社
TIANJIN UNIVERSITY PRESS

图书在版编目（CIP）数据

全球新建筑：全 5 册 /《设计家》编 . — 天津：天津大学出版社，2012.6
ISBN 978-7-5618-4283-6

Ⅰ . ①全 ••• Ⅱ . ①设 ••• Ⅲ . ①建筑设计—作品集—世界—现代 Ⅳ . ① TU206

中国版本图书馆 CIP 数据核字（2012）第 009669 号

主编：许晓东
编辑：李颖 宣文
翻译：魏颖卓 窦慧 张芳 赵如梦 卞鹏霄 季祎鸣 瞿伟兵
校对：李颖 魏颖卓
美术编辑：刘慧平
责任编辑：油俊伟

出版发行 天津大学出版社
出 版 人 杨欢
地　　址 天津市卫津路 92 号天津大学内（邮编：300072）
电　　话 发行部：022—27403647　邮购部：022—27402742
网　　址 publish.tju.edu.cn
印　　刷 上海瑞时印刷有限公司
经　　销 全国各地新华书店
开　　本 240mm×320mm
印　　张 93.75
字　　数 1426 千
版　　次 2012 年 6 月第 1 版
印　　次 2012 年 6 月第 1 次
定　　价 1288.00 元（全 5 册）

序言——全景展现全球建筑最新成果与动态

《全球新建筑》是《设计家》编辑部继成功出版《中国新建筑》之后，及时推出的具有全球视野，关注全球范围内最具创新性、引导性的优秀建筑作品集，旨在为正在快速成长中的中国建筑业提供一个深入了解并感受世界各文化地域近年来最前沿的建筑发展动向及成果的机会，以期为中国的建筑实践提供有价值的参考。

《全球新建筑》共五册，书中汇集了来自全球七十多家著名建筑师事务所在全球范围内完成或在建的最新原创建筑设计作品200多个，涵盖商业、办公、酒店、文化、教育、体育、公共服务、宗教、会议、剧院、博览、美术馆、图书馆、居住等各种建筑类型，是当下国际建筑创作的集中体现，也是城市建筑风貌的一个很好缩影。

I 商业空间：

商业空间以全球最具代表性的商业类建筑事务所的最新优秀作品为主体，如美国凯里森的作品：杭州万象城；美国捷得的作品：日本桥本市 Konoha 购物中心；英国贝诺的作品：St.David 购物中心等。甄选作品时，我们希望兼顾全面性与代表性，既有大型购物中心及旗舰店、商业综合体，亦收录了商业步行街及其它休闲娱乐项目的优秀代表作。

II 办公、酒店：

办公部分由企业总部办公及展示、办公楼、市政办公、办公园区及小型创意办公五个章节组成。清晰详细的分类不仅提供了方便有效的阅读，同时也旨在展示一幅全景式的办公作品盛宴。如果说索尼大崎新楼、米兰杰尼亚总部等作品展现了高层、多层办公及企业展示项目的丰富经验，那么，光之屋、福戈岛艺术工作室等则为创意类工作场所的实践提供了丰富的灵感与想象。

此次酒店部分的作品甄选更加关注作品本身的独特性，因而我们分别从城市五星级酒店、度假型酒店、设计酒店案例中慎重筛选收录了十个最前沿、最经典、最设计、最浪漫的酒店作品，如：新加坡 St. Regis 酒店，克罗地亚 Lone 酒店，及悦榕庄的最新力作。

III ，IV 文教空间：博览空间

文教空间包含文化中心、剧院、音乐厅、会议中心、教育、体育及宗教等方面的作品。博览空间包括：博物馆、美术馆、展览馆、图书馆、公共服务设施等方面的作品。

本套书的征稿过程中，文教及博览类项目给我们带来了连连惊喜。这些事务所及作品大多集中在欧洲、北美，也有部分作品分布在泰国、日本及中国等。它们在为我们展示多层次、多方面的空间形体的同时，也带来了全球各文化地域在建筑材料、技术等方面的丰富经验与创新成果，如 Henning Larsen 建筑事务所在雷克雅未克“哈帕”音乐厅与会议中心的立面设计中运用新材料——“准砖”（quasi-brick）创造出了一种模糊室内与室外，主体与客体，幻象与再现的视觉装置，令人印象深刻。

V：居住及小型独立建筑

居住及小型独立建筑旨在收录全球各文化地域内既具本土特色又显现代设计理念的精品别墅及公寓，如：美国 Preston Scott Cohen 公司设计的林间雅舍、秘鲁 Jarvier Artadi 建筑事务所设计的浪漫的拉斯帕尔梅拉斯海滨别墅，日本富士山建筑事务所的代表作树屋，丹麦 Henning Larsen 事务所设计的滨水地标性公寓——瓦埃勒海浪等。

在中国建筑实践丰富多彩的当下，深入了解全球建筑的最新成果与动态，无疑将是中国建筑融入世界并凸显自我的最佳路径。

《设计家》编辑部

2012 年 5 月

Global Architecture Today is a collection of the architectural works in the worldwide area by Designer & Designing Magazine, instantly after the successful publication of the series books- ***New Architecture In China***. The new collection focuses on the most innovative and leading projects with a global vision, and aims at providing a valuable reference for the fast growing Chinese buildings with the thorough touch of the latest architectural development achievements and trend of the worldwide regions of different cultures.

Global Architecture Today includes five volumes in all, which provides a professional platform for more than 200 works, finished or unfinished, from a global wide of 70 top architects studios. These latest achievements covered a wide range of fields, including commercial areas, offices, hotels, culture, education, sports, public facilities, religion, conference, theatre, exhibition, art museum, library, housing, etc. They showed readers with an overview of the latest international architecture works, and reflections of the images of urban buildings.

The Editorial Department of Designer & Designing Magazine

May, 2012

Contents

目 录

Museum

博物馆

Hoki Museum "Gallery Degree Zero"

Hoki博物馆“零度画廊”

NIKKEN SEKKEI 日建设计

Location: Chiba, Japan
Program: Museum, Gallery
Site Area: 3,862.70m^2
Building Area: 1,602.39m^2
Status: Completed in 2010
Image Courtesy: Harunori Noda

区位：日本，千叶
功能：博物馆，画廊
基地面积：3 862.70 m^2
占地面积：1 602.39 m^2
状态：2010年建成
图片提供：野田東徳

This is a private museum to exhibit and preserve painting and wines that Mr. Hoki has collected.

According to Krzysztof Pomian, the essence of collection is not only just to gather things, but also to reconstruct them, exhibit them and establish new meanings. So, the action Mr. Hoki is performing here through gathering realistic paintings and aiming to establish new meanings can be exactly recognized as "collection".

Once Aby Warburg tried to propose "Mnemosyne Atlas" as a method to exhibit collections. He could not complete it but we felt the concept is still effective. So we thought Hoki museum could be a chance to establish architectural Mnemosyne Atlas of Mr. Hoki's collection.

Main theme of his collection are realistic oil paintings miraculously painted in detail with marvelous technique, so we thought that "gallery" is most suited for these works and should be selected as an archetype of this museum space. Until now, gallery has rarely existed only by itself as an architecture and always has been treated as subsidiary space by architects. So, we tried to generate all architectural elements for this facility only from ensemble of galleries this time.

This architecture is nothing more than a gallery, but also less than a gallery. Our aim was to establish "Gallery Degree Zero".

Inside galleries, people can not only access each paintings in traditional sequential move, but also take a random-access way because galleries are slightly curved and people can recognize positions of all the paintings at a glance. Galleries that extrude longitudinally along the site, measure up to 100m long. The visitors will face to the super-realistic paintings in a seamless gallery without any joints of finish and outstanding existence of lights. At the end of a gallery, they can enjoy natural light and scenery outside. After a short rest, they will move to the next gallery. These are the fundamental system of this museum. All the galleries have the same style but proportions, dimensions and amount of natural light are gradually changed. Through these delicate control, we tried to build up the sequence of spaces, like a bolero, having different densities in the simple system composed only by repetition of galleries.

Outside, we tried to build the presence fit in to the context by stacking galleries. In the western side where the visitors mainly approach to this facility, the galleries are intensively stacked to balance between houses around the site.

In the north and southeast side where this facility faces to a forest, the galleries are discretely composed and shows the program of this facility clearly. Using characteristic structure effectively, the upper gallery is cantilevered and flies toward the forest 30m long.

60 cm high horizontal slit between the galleries works as a route of winds that makes twigs of trees inside courtyard flap.

As just described here, this museum is entirely designed to exhibit the collection at its best condition. At the same time, under the limited condition using only galleries, we tried to reply to Mr. Hoki's request that the architecture also has to constitute a part of his collection.

Site Plan 总图

这是用于展览和贮藏Hoki先生所收藏的画作及葡萄酒的私人博物馆。

在克里斯托夫·波米扬看来，收藏的本质不仅仅在于收集事物本身，也在于对其的重塑、展览及赋予新意。因此，Hoki先生在这里所展示的正是通过收集写实画作，创造能够被精确定义为“收藏”的新意。

艺术史家阿比·沃伯格曾试图提议以“记忆女神摩涅莫辛涅·阿特拉斯”作为展览收藏品的一种方法。尽管他的提议并没有功德圆满，但是我们仍能感觉到这一理念的效用。因此我们萌生出一种想法：Hoki博物馆或许为建造其收藏品建筑的摩涅莫辛涅·阿特拉斯提供了一个契机。

其收藏品的主题即技艺精湛、精细非凡的现实主义油画作品，所以我们认为“画廊”最能实现与这些作品的完美契合，且应当被选作该博物馆空间的建筑原型。直到现在，仍鲜有画廊作为独立的建筑物存在于世，它们通常以建筑物附属空间的角色示人。因而此次我们试图将所有的建筑元素融合于画廊这一整体设施之中。

该建筑仅仅是一个画廊，却又不是完全意义上的画廊。我们的目标是建造“零度画廊”。

在画廊内部，人们不仅可以循着传统的顺序式观摩，依次领略画作之美，也可以随机欣赏。得益于画廊所采用的略微弯曲的设计构造，参观者一眼便可看出所有画作的位置。画廊沿建址纵向延伸，长达100米。完美无缝的画廊中没有任何衔接的痕迹，人们置身其中，在面对着一幅幅超现实主义画作的同时，赞叹着出色的灯光效果。移步至画廊末端，人们又可沐浴在自然光线之下，室外风景可谓美不胜收。小憩过后，人们又将来到下一个画廊。这些便是该博物馆的基本系统。所有的画廊风格相仿，但建造比例、维度及自然采光程度却是逐渐变化着的。我们试图通过把握这些设计的精细之处来创造空间的连续性，就如同波列罗舞一般，让仅由画廊重复构成的简易体系内部也拥有不同的密度。

对于室外设计，我们试图通过叠加画廊来建造与周围环境相适应的存在。参观者主要从建筑物西侧步入该设施，此处画廊密布，从而实现了该建筑与该建址周边房屋间的平衡。

该建筑南北向直面森林，坐落于此处的画廊零散布局，清晰地展示了该设施的规划。通过有效利用特征结构，上部画廊呈悬臂式构造，向森林飞跨30米。

各画廊间60厘米处的水平缝隙作为气流穿行的路线发挥作用，风力涌动，庭院中树木的细枝末梢随之拍打。

正如我们这里所描述的一样，该博物馆完全是为以最好的条件展示收藏品而设计的。与此同时，在有限的条件下仅仅采用画廊的设计，也响应了Hoki先生希望建筑物本身作为其收藏品的一部分而存在的要求。

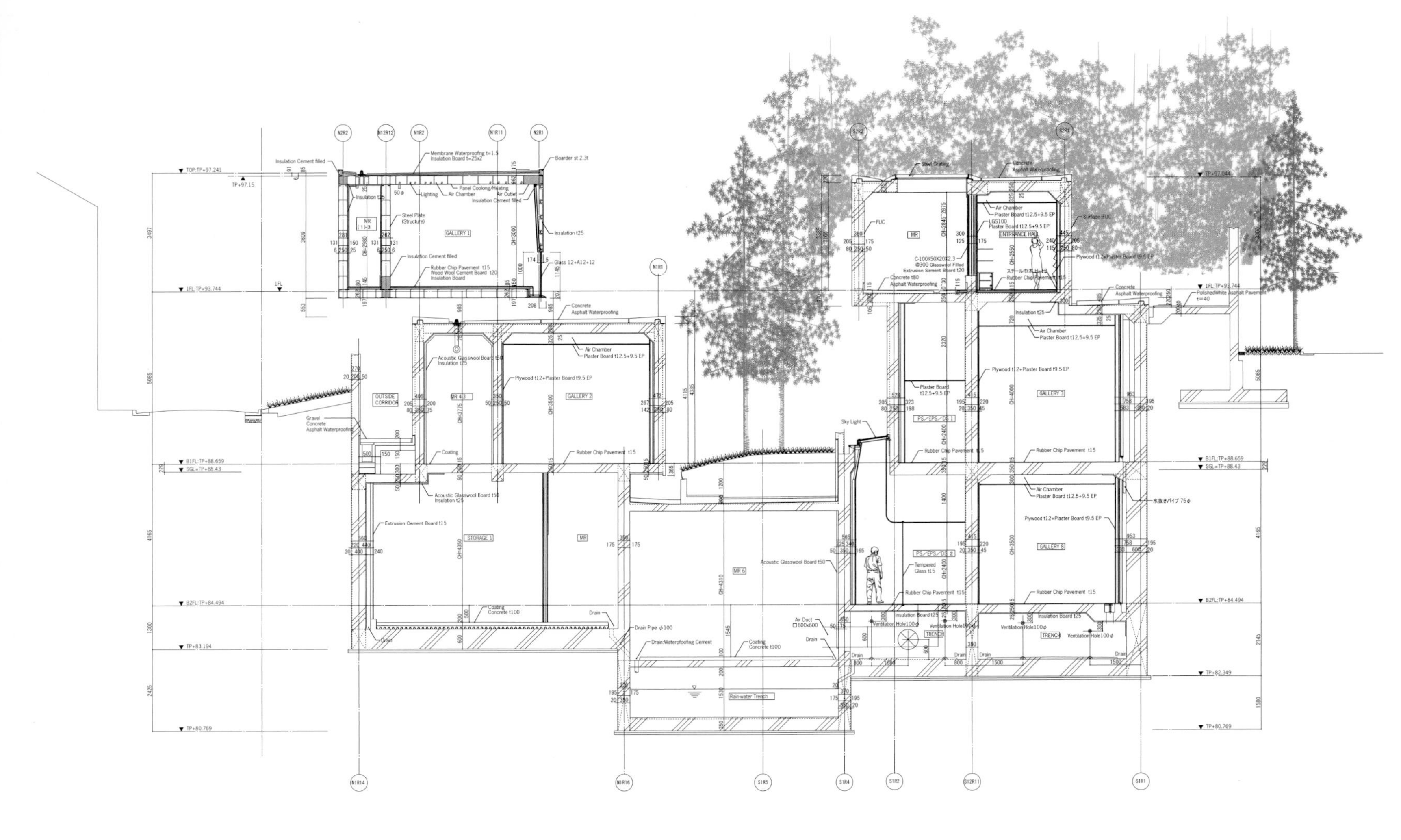

Detailed Section (North-South) S=1/100

Section Details 剖面详图

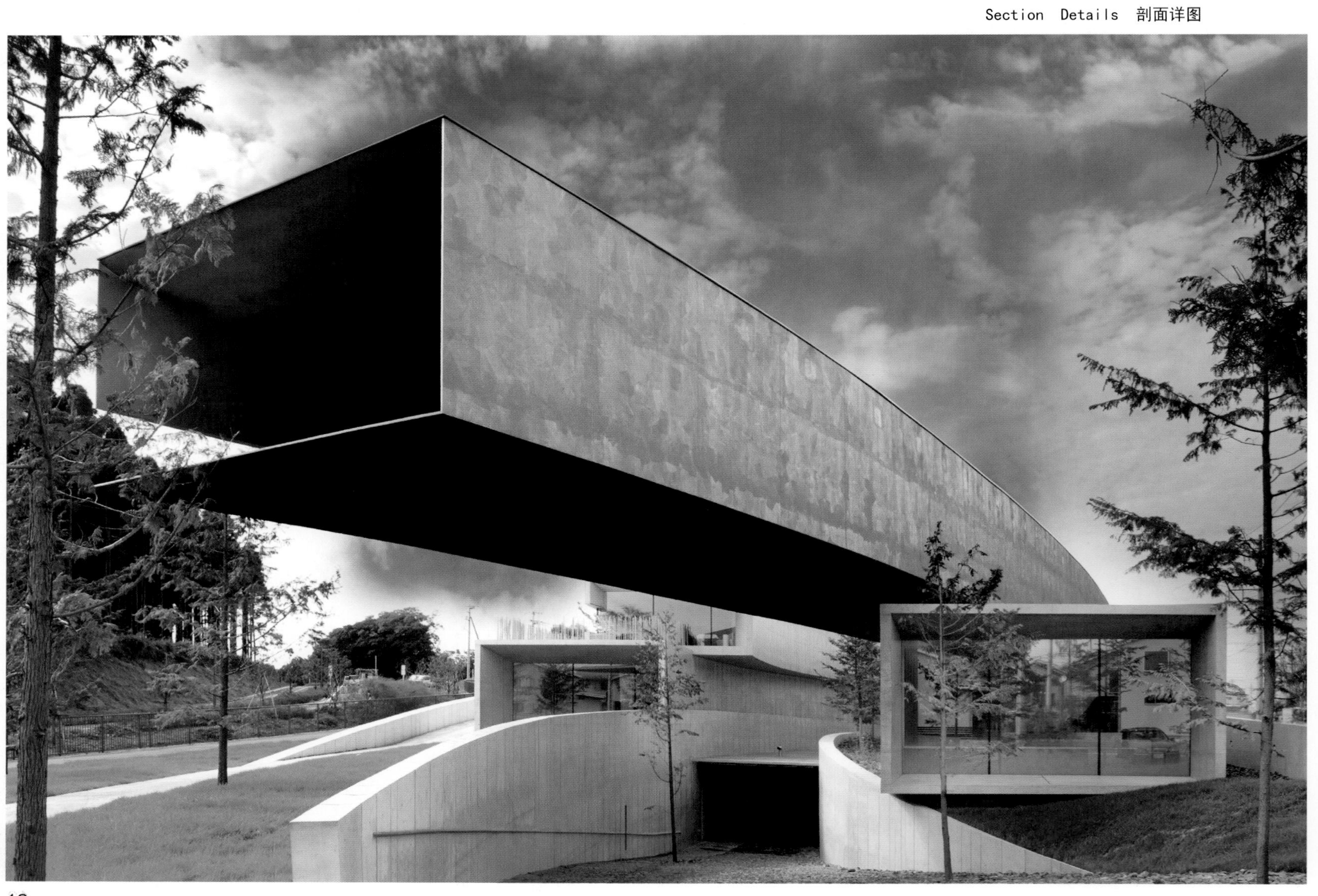

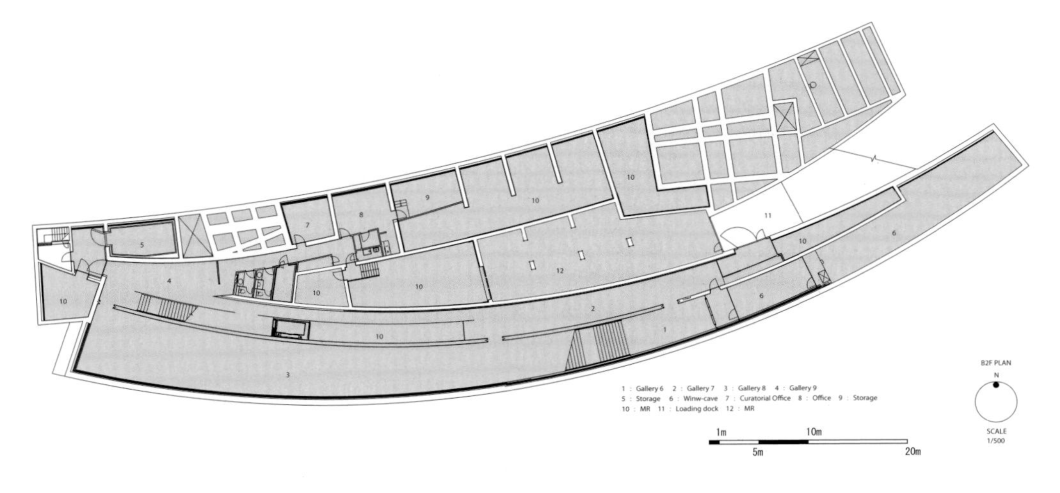

Plan Level -2 地下二层平面

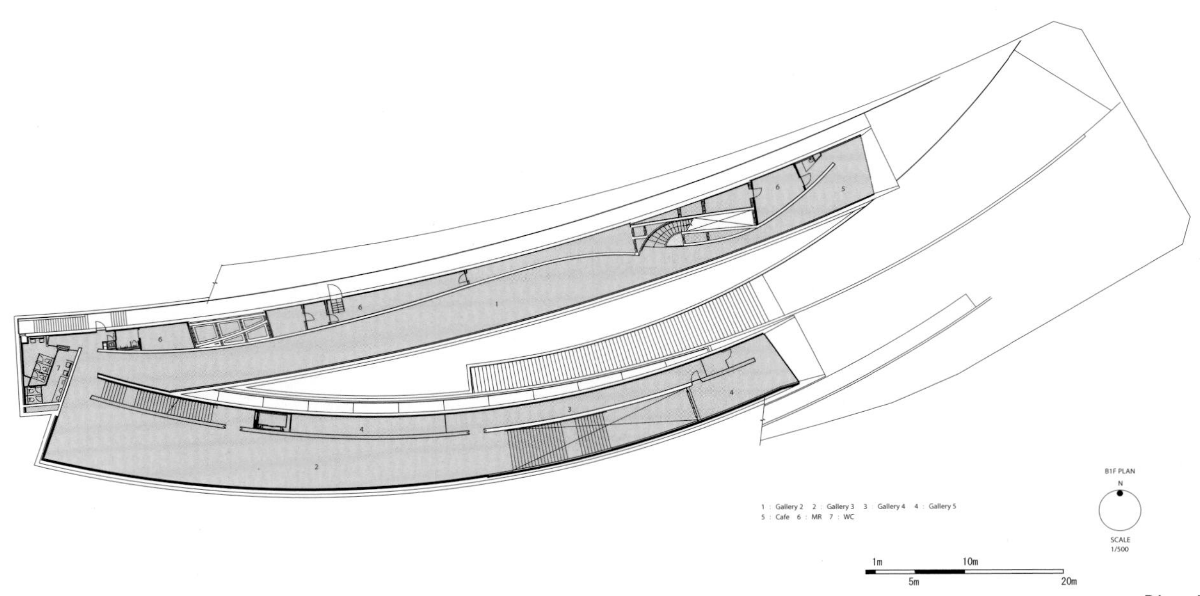

Plan Level -1 地下一层平面

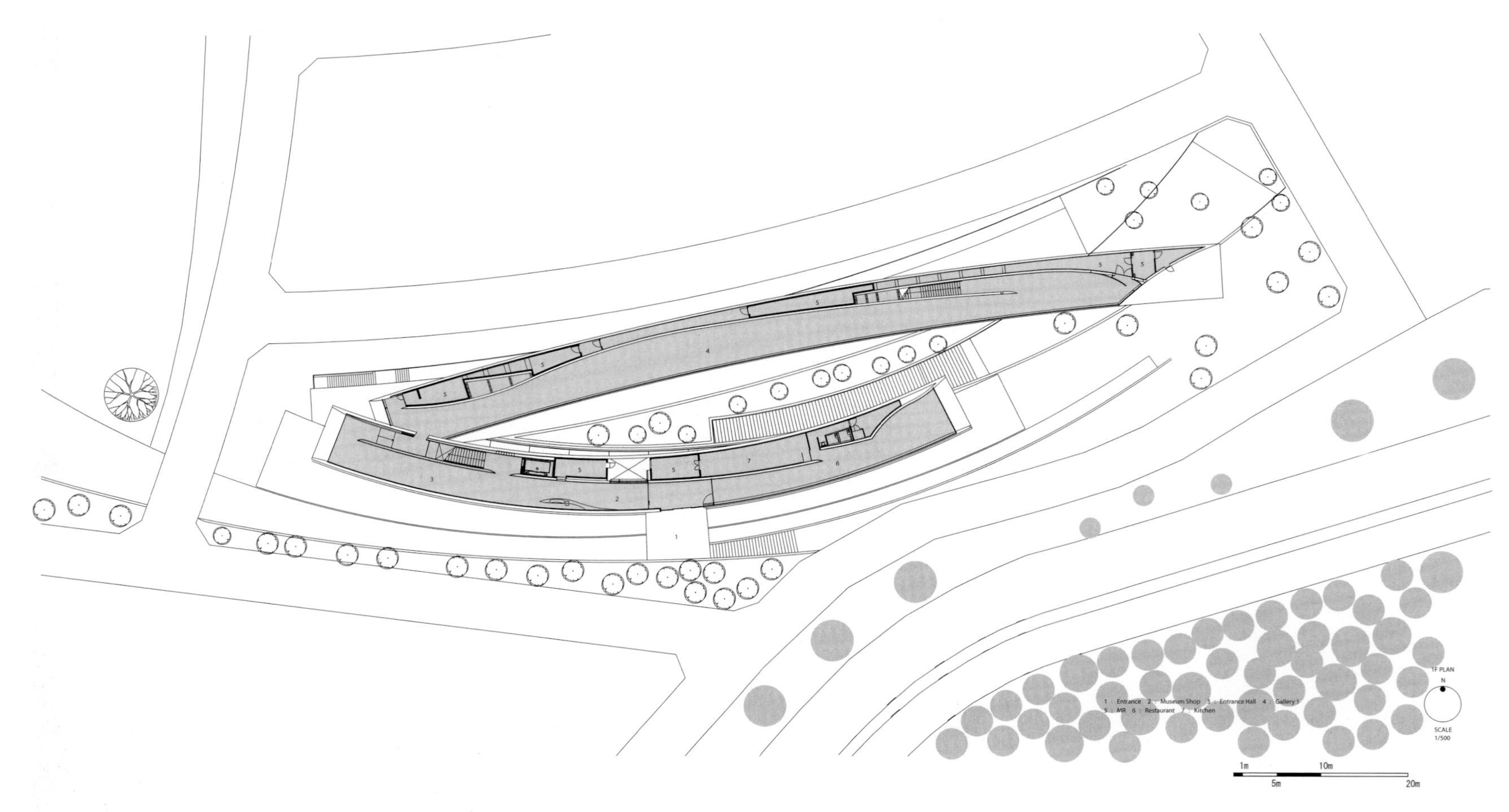

Plan Level 1 一层平面

HOKIMUSEUM

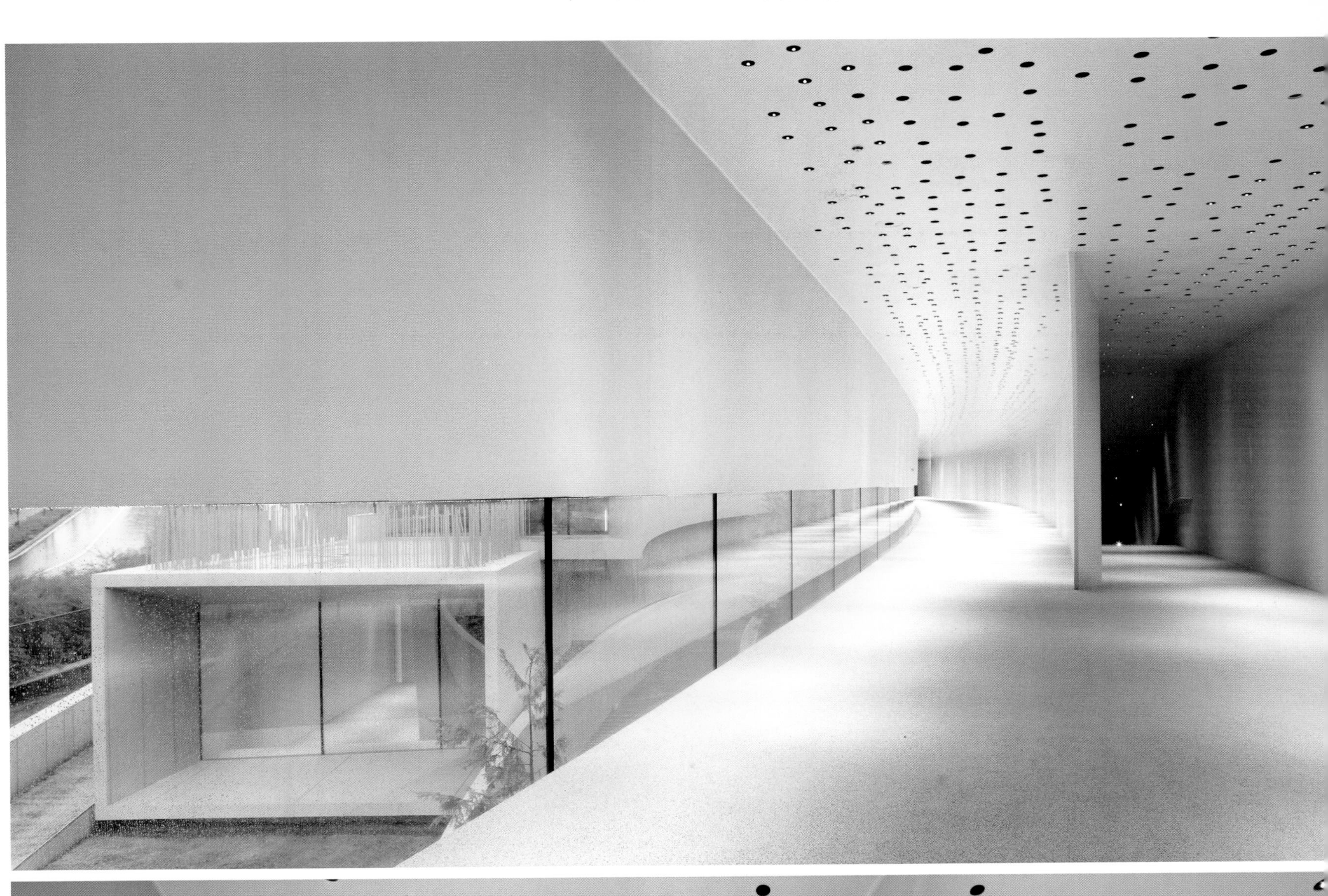

Musée National Des Beaux-arts Du Québec

魁北克国立美术馆

OMA　大都会建筑事务所

Location: Québec City, Canada
Program: $12{,}000\ m^2$ museum expanision composed of three stacked galleries of decreasing size: contemporary exhibitions ($1{,}500\ m^2$), permanent contemporary collection ($950\ m^2$), design and Inuit Galleries ($550\ m^2$)
Status: Competition 2010, 1st place. To be Compled in the: fall of 2013
Image Courtesy: OMA

区位：加拿大，魁北克
功能：$12\,000\ m^2$博物馆扩建部分，
包括三个缩小了的当代展览馆（$1\,500\ m^2$），
永久性当代收藏馆（$950\ m^2$），设计及因纽特画廊（$550\ m^2$）
状态：2010年竞赛第一名，预计2013年秋竣工
图片提供：大都会建筑事务所

The new building for the Musée National Des Beaux-arts Du Québec – the museum's fourth building in an increasingly complicated site, interconnected yet disparate – is a subtly ambitious, even stealthy, addition to the city. Rather than creating an iconic imposition, it forms new links between the park and the city, and brings new coherence to the MNBAQ.

The intricate and sensitive context of the new building generated the central questions underpinning the design: How to extend Parc des Champs-de-Bataille while inviting the city in? How to respect and preserve St-.Dominique church while creating a persuasive presence on Grande-allée? How to clarify the museum's organization while simultaneously adding to its scale?

Our solution was to stack the required new galleries in three volumes of decreasing size – temporary exhibitions (50 m x 50 m), the permanent modern and contemporary collections (45 m x 35 m) and design / Inuit exhibits (42.5 m x 25 m) – to create a cascade ascending from the park towards the city. The building aims to weave together with the city, the park and the museum; it is simultaneously an extension of all three. While they step down in section, the gallery boxes step out in plan, framing the existing courtyard of the church cloister and orienting the building towards the park. The park spills into the museum (through skylights and carefully curated windows) and the museum into the park (though the extension of exhibitions to the terraces).

The stacking creates a 14 m#high Grand Hall, sheltered under a dramatic 20 m cantilever. The Grand Hall serves as an interface to the Grande-allée, an urban plaza for the museum's public functions, and a series of gateways into the galleries, courtyard and auditorium.

Complementing the quiet reflection of the gallery spaces, a chain of programs—foyers, lounges, shops, bridges, gardens—along the museum's edge offer a hybrid of activities, art and public promenades. Along the way, orchestrated views outside reconnect the visitor with the park, the city, and the rest of the museum. The new building connects with the museum's existing buildings by a passageway rising 8.2 m over its 55 m length. Through its sheer length and changes in elevation, the passage creates a surprising mixture of gallery spaces that lead the visitor, as if by chance, to the rest of the museum complex.

Section 剖面

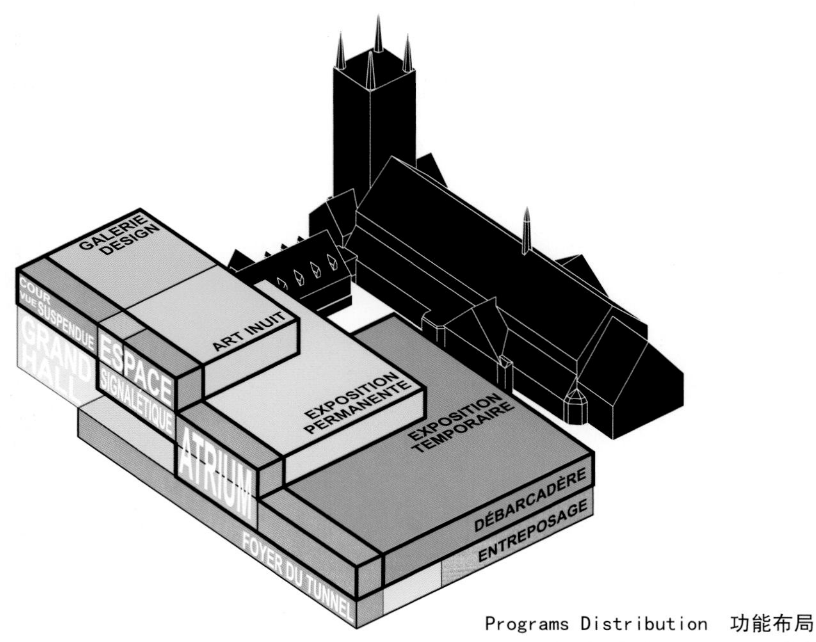

Programs Distribution 功能布局

Courtyard 庭院

魁北克国立美术馆的新楼位于一个更加复杂的基地内的博物馆的第四栋大楼，与原有建筑相互联系却又截然不同。 它怀着雄心，悄然甚至鬼祟地进入了这座城市。与创建一个新的地标不同，它的出现建立了公园与城市间的连接，并将新的一致性注入魁北克艺术博物馆。

新建筑周围错综敏感的背景为设计提出了支撑性的核心问题：如何在引入城市的同时延伸魁北克战争公园呢？如何为大花径创造一个有说服力的外表的同时保存和尊重圣多米尼克教堂呢？如何在清晰化博物馆的组织的同时不断的扩大它的规模呢？

我们的解决方案是将设计要求的美术馆层叠在三个不断递减的体量中——临时展览空间（50米×50米）、近代和当代的永久性收藏（45米×35米）和设计作品及因纽特人的展览空间（42.5米×25米）——创造一个从公园到城市层叠式的上升。这个建筑的目的是将城市、公园和博物馆编织在 一起，它同时延伸这三个展览空间。

在剖面中沿台阶下行的同时，美术馆同时在平面上进行了出挑，勾勒出现存教堂庭院的外围，并将建筑朝向公园。公园景观通过天际线和细致的开窗方式渗透进博物馆内， 而博物馆也通过展览空间向露台的延伸溢入公园。

体块的堆砌创造了一个14米高的大会堂，它遮蔽在一个20米高的夸张的悬臂下。大会堂的作用是连接大花径，作为承载博物馆公共活动的城市广场，为进入美术馆、庭院和观众厅提供一系列的路径。

作为美术馆安静空间的补充，沿着博物馆边缘的一系列功能——休息室、娱乐室、商店、小桥、花园——提供了一个活动、艺术和公共散步广场的混合体。沿路走下去，精心构建的外部风景将游览者与公园、城市和博物馆的其他部分重新结合起来。

新建筑和博物馆现有建筑由一个离地8.2米远、长55米的通道连接在一起。通过它的纯长度和在立面上的变化，该通道提供了一个令人惊讶的混合展览空间，偶然性地将游客引入博物馆综合体的其他部分。

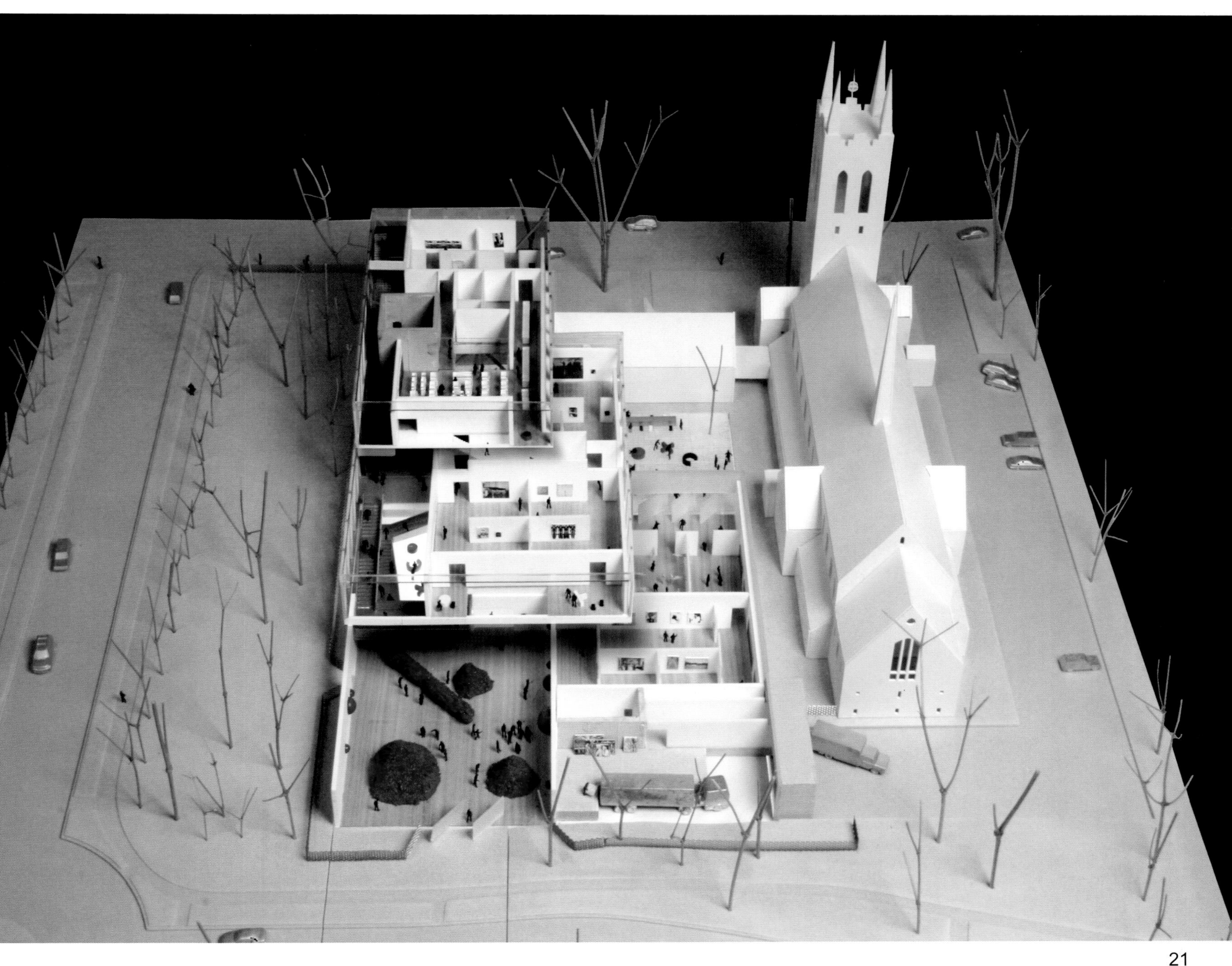

Archaeological Museum of 'Praça Nova do Castelo de São Jorge'

Praça Nova do Castelo de São Jorge 考古遗址博物馆

Carrilho da Graça Arquitectos Carrilho da Graça建筑事务所

Location: Lisbon, Portugal
Typology: Museum
Area: 3,500 m^2
Image Courtesy: FG+SG architecture photography

区位：葡萄牙，里斯本
类型：博物馆
面积：3 500 m^2
图片提供：FG+SG 建筑摄影

An extensive archaeological excavation of this site, begun in 1996, uncovered remnants of its successive periods of inhabitation – Iron Age settlement, Mediaeval Muslim occupation and a Fifteenth Century Palace – and the most significant artifacts removed, protected and now exhibited at the Castle's Museum, leaving the exposed archeological site open to an intervention of protection and musealization.

This intervention addressed the themes of protection, revelation and readability of the palimpsest that any such excavation represents, with a pragmatical approach aimed at clarifying the palindromic quality of interpretation that the exposed structures suggested in their spacial distribution.

Thus, the first action was the clear delimitation of the site with a precise incision, comparable to that of a surgical intervention on a living body. A membrane of corten steel was inserted to contain the higher perimetrical surface, allowing both access and a panoramic view of the site, the materiality of these walls slowly evolving and changing over time as a living material. The same precision of cut was used in the inserted elements that allow the visitor to comfortably wander through the site – the limestone steps, landings and seatings – setting them apart from the roughness of the excavated walls.

Stepping down to the site, to its simultaneously first material level and last period of occupation – the remnant pavement of the Fifteenth Century Palace of the Bishop of Lisbon – a hovering structure protects the existing mosaics, its underside covered in a black mirror that allows the visitor to see reflected the vertical perspective of the pavements that the eye level of their placement denies.

Further down the site and its timeline, the necessary canopy for the protection of the Eleventh Century Muslim domestic structures and its frescoes was taken as an opportunity to reproduce, through conjectural interpretation, its spacial experience as a series of independent rooms arranged around a patio that introduced light and ventilation into an otherwise exteriorly isolated dwelling. Professedly abstract and scenographic, the white walls that stage the domestic spatiality of the two excavated dwellings float above the visible foundations of the original walls, touching the ground on the mere six points where the evidence of the primeval limits is absent, while its translucent covering of polycarbonate and wood filters the sunlight.

Underlying the whole site, the evidence of the Iron Age settlement is exposed and protected through a self-contained volume that, in a spiraled movement, extends from the perimetrical corten walls to embrace the depth necessary to its revelation. Massive and dramatic, the volume is pierced with horizontal slits that invite the curiosity for the observation of its interior, leading the visitor around the excavated pit to the point where the view is unobstructed and both the physical and time distance of the exhibited structures is made obvious.

The palimpsest of the site History is thus decoded and the possibility of its palindromic time-space reading is made clear: not only through the informational signage at the disposal of the visitor, but also, and significantly, through the experience construed by its material protection and musealization.

1 . porta de acesso principal
2 . átrio
3 . cozinha
4 . compartimento de funcionalidade indefinida
5 . pátio
6 . salão
7 . alcova
8 . despensa
9 . latrina
10 . compartimentos de apoio ás casas
11 . rua

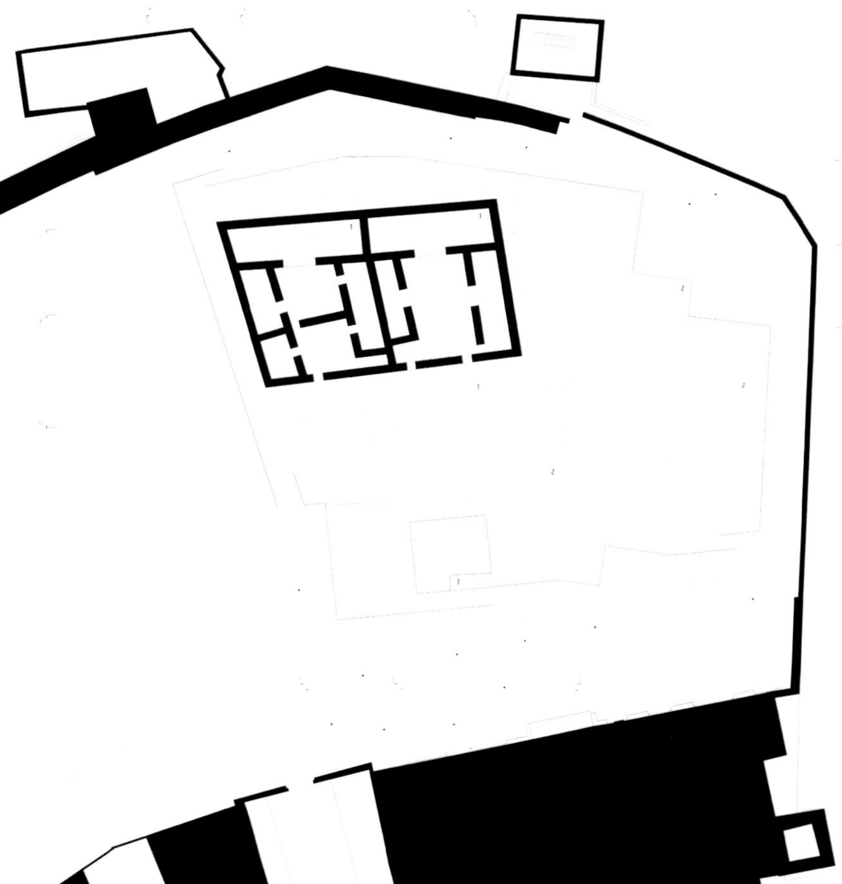

Site Plan 总图

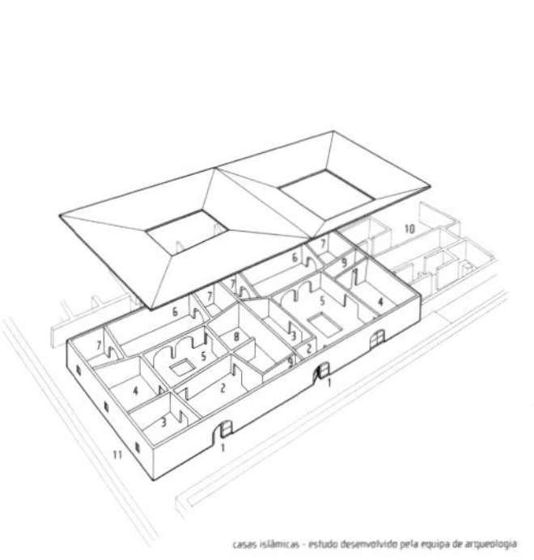

casas islâmicas - estudo desenvolvido pela equipa de arqueologia

1996年开始的广泛的考古挖掘出土了这一地区一段连续的历史时期的生活残余，从石器时代，到穆斯林统治的中世纪，再到十五世纪，其中最有历史价值的手工艺品现存放于城堡博物馆用于保护和展览，而剩下的遗址则希望被露天保护并在原址建立博物馆。

这样的介入是为了强调主题是保护遗迹展露其最真实的最直观的一面，就像一本历史的重新本。遗址代表的一种实用主义的取向，通过裸露遗迹的空间分布明确其中的特点。

因此，工程面临的第一项任务就是以精确的切割指明遗址的界限，就如同外科手术在人体上明确介入区域一样。嵌入的耐候钢膜用来包裹较大周长的表面，使得墙面的材料犹如富有生命似的，随着时间的推移逐渐发生变化，并在遗迹的入口和全景中得到展示。同样精密的切割还用于嵌入的供游客使用的部分。这些部分能够让旅客舒适地漫步于遗迹之间，包括石灰岩台阶、楼梯平台以及座位，将他们与发掘出来的粗糙墙面区别开来。

步入遗址，到达了遗址的第一层同时也是遗迹最后的使用时期：十五世纪里斯本大主教的宫殿遗迹。一个悬架的结构保护了现存的马赛克墙砖，下侧装配着一面黑色的镜子使得参观者能够通过镜面的反射看到视线所不能及的步行通道下方的垂直视角。

再往下便是一个用来保护十一世纪穆斯林统治时期内部结构以及壁画必需的敞篷。这被视为通过推测的解读而进行的重新创作的机会。遗迹中围绕的一个平台排列这一系列独立的房间。通过独特的空间设计，能够为这些与外界相对隔绝的房间引入光线以及通风。那些具有抽象感，及透视效果的白色墙面展示了挖掘出来的两所民居的空间性。这白色的墙面是悬浮于可见的原始墙面的地基之上的，与地面只有六个接触点，而地面上没有原始界限的痕迹，同时半透明的聚碳酸酯及木材遮盖确保了阳光的渗入。

位于整个遗址最下方的是石器时代古人类遗址。这部分的遗址是一个自控保护的区域。螺旋型台阶延伸至边沿的耐候钢墙面，这样的深度才能充分的显露出其历史的真实。这个庞大而富于戏剧性的体量被一条横向的裂缝贯穿，这样勾起了人们一探其内部结构的好奇心，指引着那些围绕遗址周围转圈的游客们进入到内部，在那里视线将毫无阻拦。整个遗址物理分布以及时间分布的格局将一览无余。

这个遗址历史进程的重写实现了，而时空往复的解读则更为清晰。不仅是通过参观者能够看到的信息型的指示标记，而且很有深意地通过遗址内各种出于保护或者使其博物馆化而使用的材料创造出重临现场的真实解析。

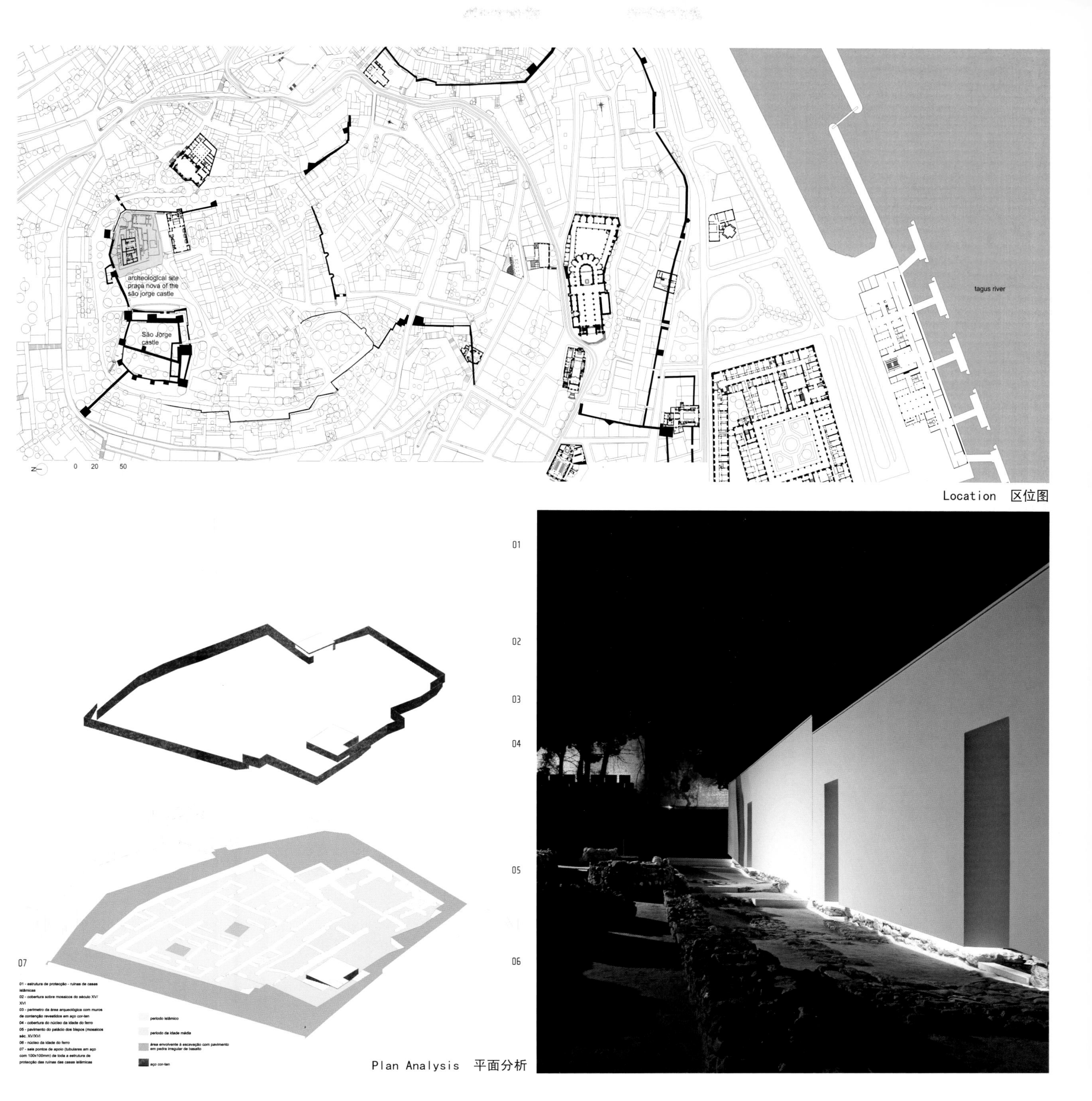

Location 区位图

Plan Analysis 平面分析

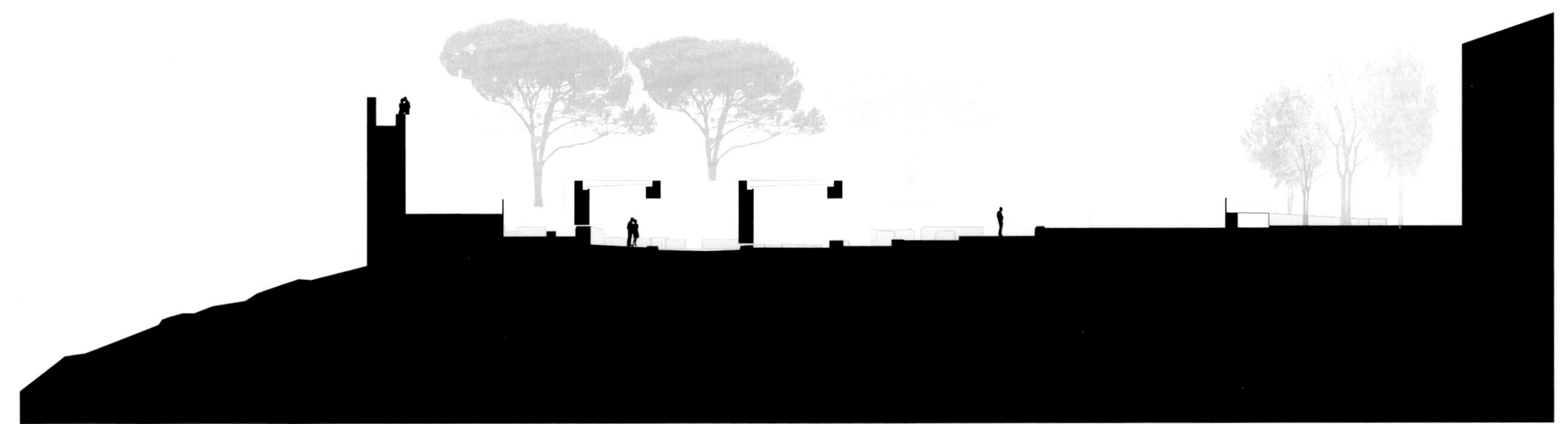

Section 剖面

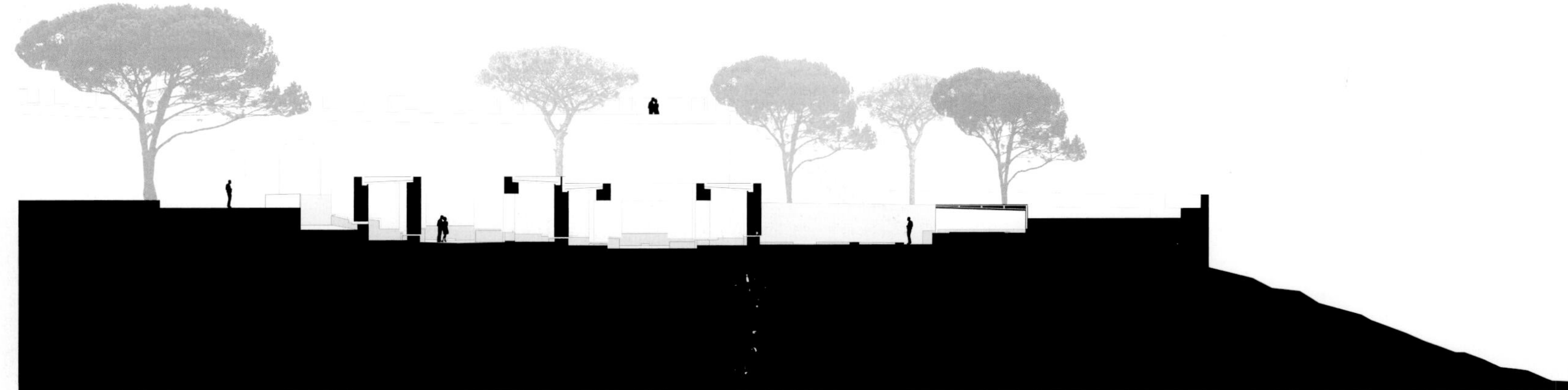

Section 剖面

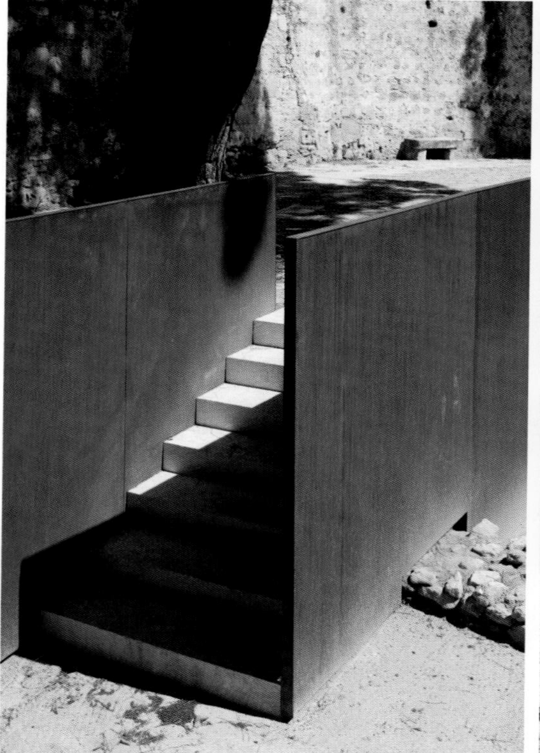

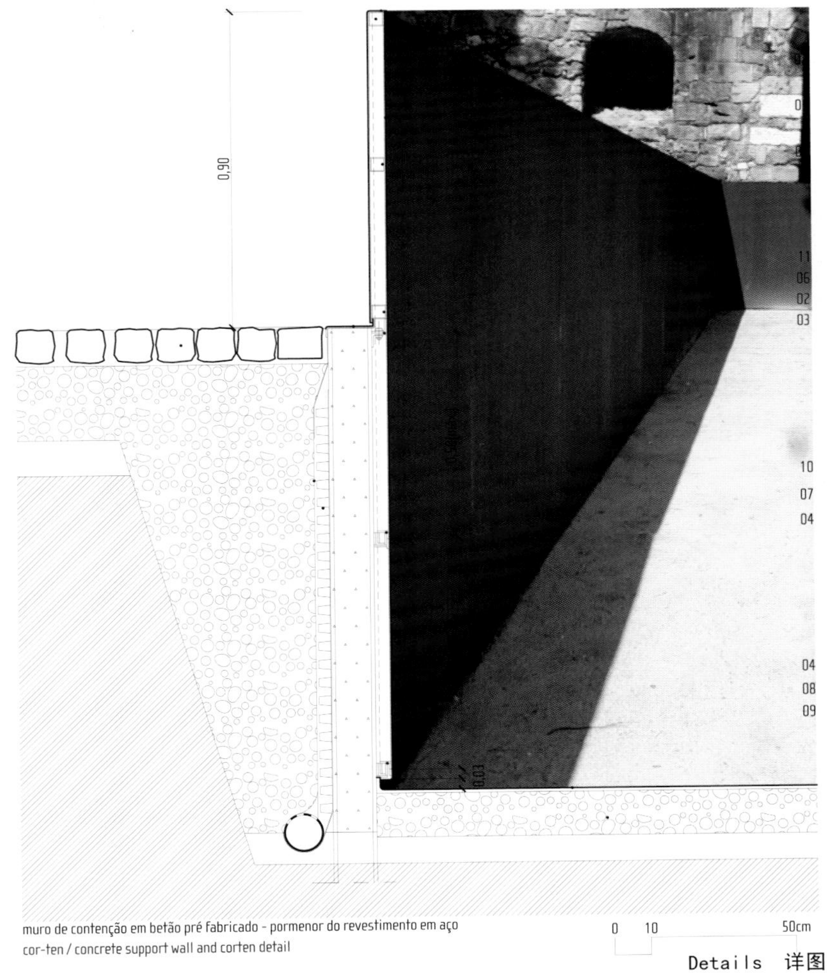

muro de contenção em betão pré fabricado - pormenor do revestimento em aço cor-ten / concrete support wall and corten detail

China Science & Technology Museum

中国科技博物馆

RTKL Associates Inc.　RTKL国际有限公司

Location: Beijing, China
Typology: Museum
Size: 102,193 m^2
Image Courtesy: Fu Xing

区位：中国，北京
类型：博物馆
规模：102 193 m^2
图片提供：傅兴

Located on the north end of Beijing's Olympic Park, the China Science and Technology Museum is dedicated to improving China's foundation for science and technology education and to inspiring a new generation of enthusiasts. RTKL's design scheme for the 102,000 m^2 museum creates an architectural dialogue that combines the scientific thinking of the ancient Chinese with the advances of modern technology.
Drawing inspiration from childhood puzzle games, in which each piece of the puzzle exists as a fractional yet integral component of an overall picture, the building's architecture hints at the concepts of interconnectivity, integrity and ultimate oneness. These metaphorical references manifest themselves in the form of a giant geometric cube, composed of a series of interlocked building blocks that fit together like pieces of a puzzle. The museum features state-of-the-art exhibits, conference facilities, classrooms and training areas, as well as retail, food venues and 3D and 4D simulation theaters.

位于北京奥林匹克公园北端的中国科技博物馆，为改善中国科技教育的基础，激励新一代热情高涨的研究者，贡献着自己的力量。RTKL建筑设计咨询有限公司为这个占地102 000平方米的博物馆所设计的建筑方案，创造了一个建筑对话，这一对话将古代中国人的科学思想与现代技术的先进性相结合。

孩提时代的拼图游戏是该设计方案的灵感来源，每一小片拼图都作为整个图像的一个部分而非整体存在着，该建筑的建筑线索即在于内在联结、整体性及最终一体化的理念。这些隐喻的参考物以巨大几何立方体的形式诠释了自身，这个立方体是由一系列内在联锁的建筑体构成的，这些建筑体就好比拼图游戏中的一块块相吻合的小块拼图。该博物馆以一流的会议设施、教师和培训区、零售餐饮场所及3D和4D仿真剧院为特色。

中國科學技術館

中國科學技術館

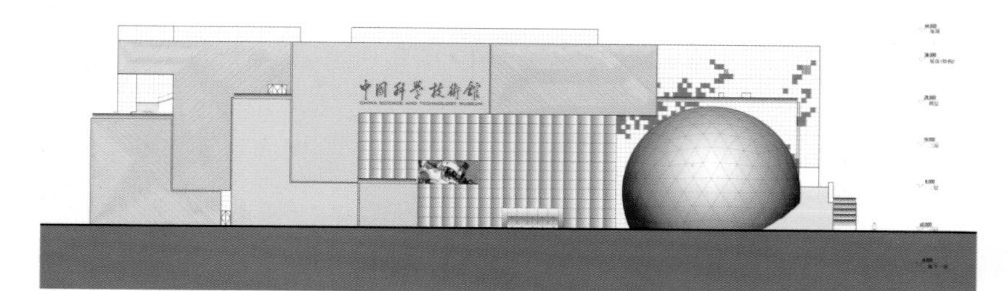

West Elevation 西立面

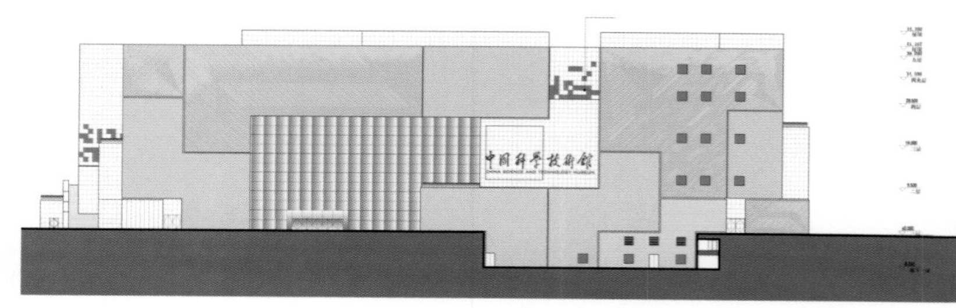

East Elevation 东立面

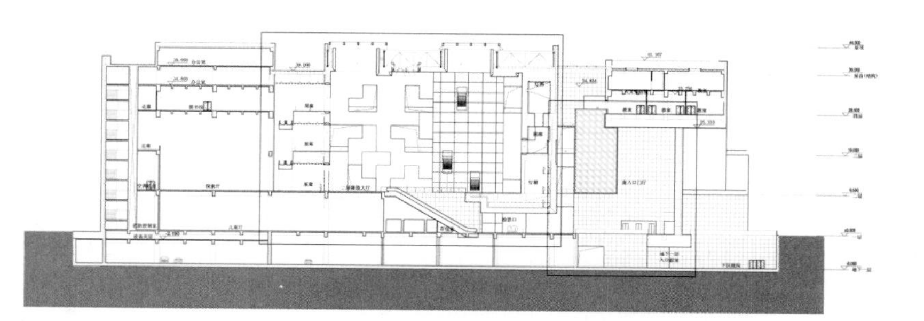

Section 剖面

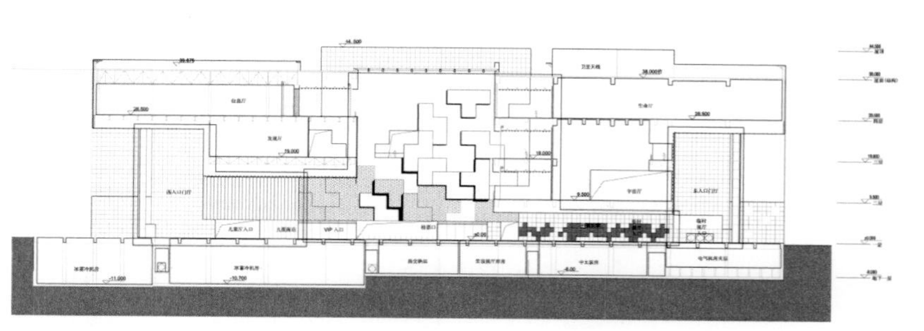

Section 剖面

Mech. Rm.
空调机房
Open to Sunken Court
下沉庭院上空
Reception Area
接待区
B.O.H
后勤
B.O.H
后勤
门卫
Loading Dock
货车位
Fire Control Rm.
消防控制室
Stor. Rm.
中转库
Mech. Rm.
空调机房
DN
下
Children's Hall
儿童厅
Temp. Exhibition Hall
临时展厅
Open to Sunken Court
下沉庭院上空
Children's Store
儿童商店
Mech. Rm.
空调机房
Temp. Exhibition Lobby
临时展厅前厅
First Floor Lobby
一层集散厅
Security
保安
VIP
Nurse Station
医务
Mech. Rm.
空调机房
Cloak
存包
First Floor Lobby
一层集散厅
Reception
接待
Control Rm.
灯光音响控制室
Main Stage
舞台
Auditorium Foyer
报告厅前厅
Stor. Rm.
备用间
Mech. Rm.
空调机房
Ticketing
售票
Dome Cinema Lobby
影院前厅
VIP
Museum Store
商品服务区
Display Case
橱窗
Display Case
橱窗
Storage
库房
Open to Sunken Court
下沉庭院上空
DN
下
Open to Sunken Court
下沉庭院上空
DN
下

Plans 平面

4 D 影院
4D Theater

The Los Angeles Museum of the Holocaust

洛杉矶大屠杀博物馆

Belzberg Architects　　Belzberg建筑事务所

Location: Los Angeles, USA
Typology: Museum
Status: Completed in 2010
Image Courtesy: Belzberg Architects, Benny Chan, Iwan Baan

区位：美国，洛杉矶
类型：博物馆
状态：2010年建成
图片提供：Belzberg建筑事务所，Benny Chan，Iwan Baan

The new building for the Los Angeles Museum of the Holocaust (LAMH) is located within a public park, adjacent to the existing Los Angeles Holocaust Memorial. Paramount to the design strategy is the integration of the building into the surrounding open park landscape. The museum is submerged into the ground allowing the park's landscape to continue over the roof of the structure. Existing park pathways are used as connective elements to integrate the pedestrian flow of the park with the new circulation for museum visitors. The pathways are morphed onto the building and appropriated as surface patterning. The patterning continues above the museum's galleries, further connecting the park's landscape and pedestrian paths. By maintaining the material pallet of the park and extending it onto the museum, the hues and textures of concrete and vegetation blend with the existing material palette of Pan Pacific Park. These simple moves create a distinctive façade for the museum while maintaining the parks topography and landscape. The museum emerges from the landscape as a single, curving concrete wall that splits and carves into the ground to form the entry. Designed and constructed with sustainable systems and materials, the LAMH building is on track to receive a LEED Gold Certification from the US Green Building Council.

新建的洛杉矶大屠杀博物馆（LAMH）坐落在一个公共花园内，毗邻业已存在的洛杉矶大屠杀纪念碑。设计策略的关键在于它将建筑物本身同周围开阔的公园景观融为一体。博物馆是地下建筑，这种设计使得公园景观在博物馆的屋顶之上得以延续。现有的公园小径作为连接设置将公园内的人流与参观新建博物馆的人流合为一股。小径演变成了建筑物之上的设计元素，并被纳入表面图案结构的组成部分。该图案结构延伸至博物馆画廊的上方，进一步将公园景观同游人小径连接在了一起。通过维护公园的材料板并将其延伸至博物馆上部，混凝土及植被的色调及纹理与泛太平洋公园现有的材料板相混合。这些简简单单的改动在为博物馆创造了一个与众不同的外观的同时，也保留了公园地形及景观。博物馆从公园景观之中显露出来，自成一体，它是由一个单独的、弯曲的混凝土墙面向地面裂开并雕刻成为入口。配置了可持续系统，运用了可持续材料的洛杉矶大屠杀博物馆（LAMH）已经走上了接受美国绿色建筑委员会所颁发的美国民间绿色建筑认证奖金奖的红地毯。

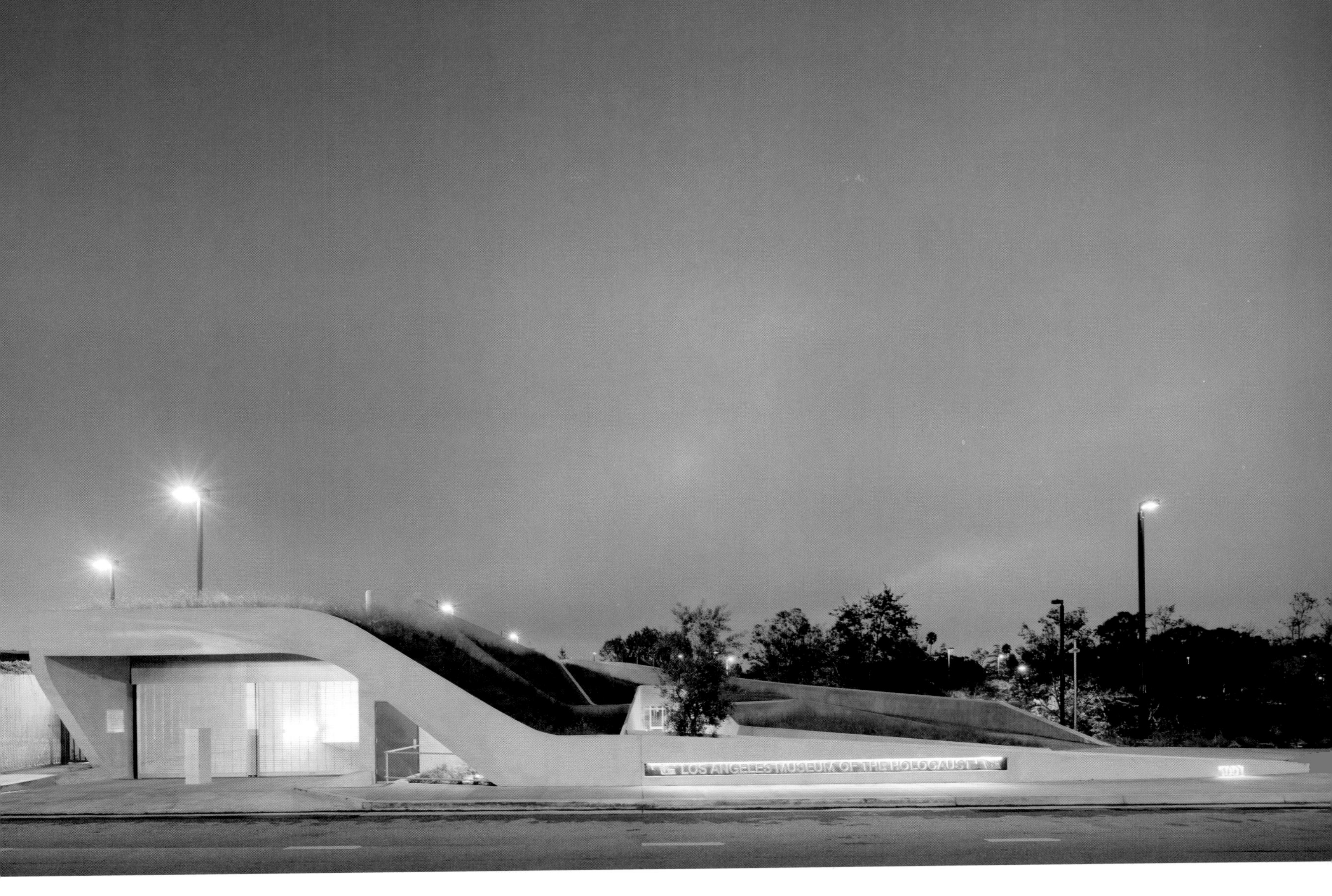

Site Plan

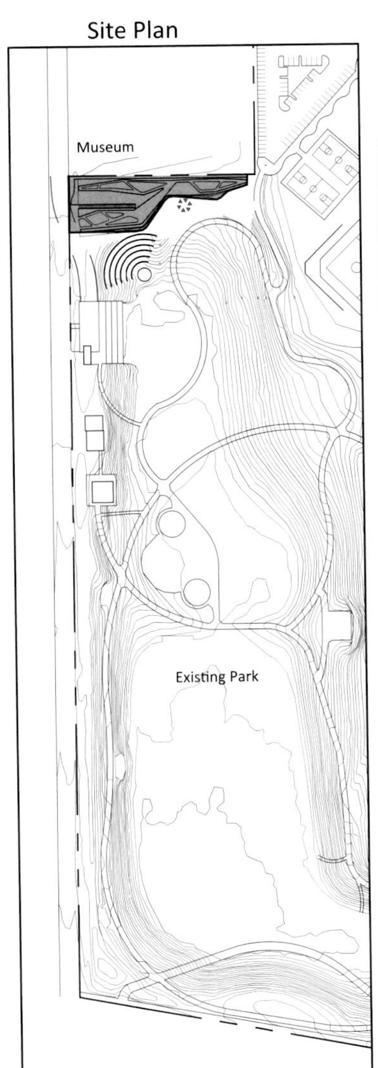

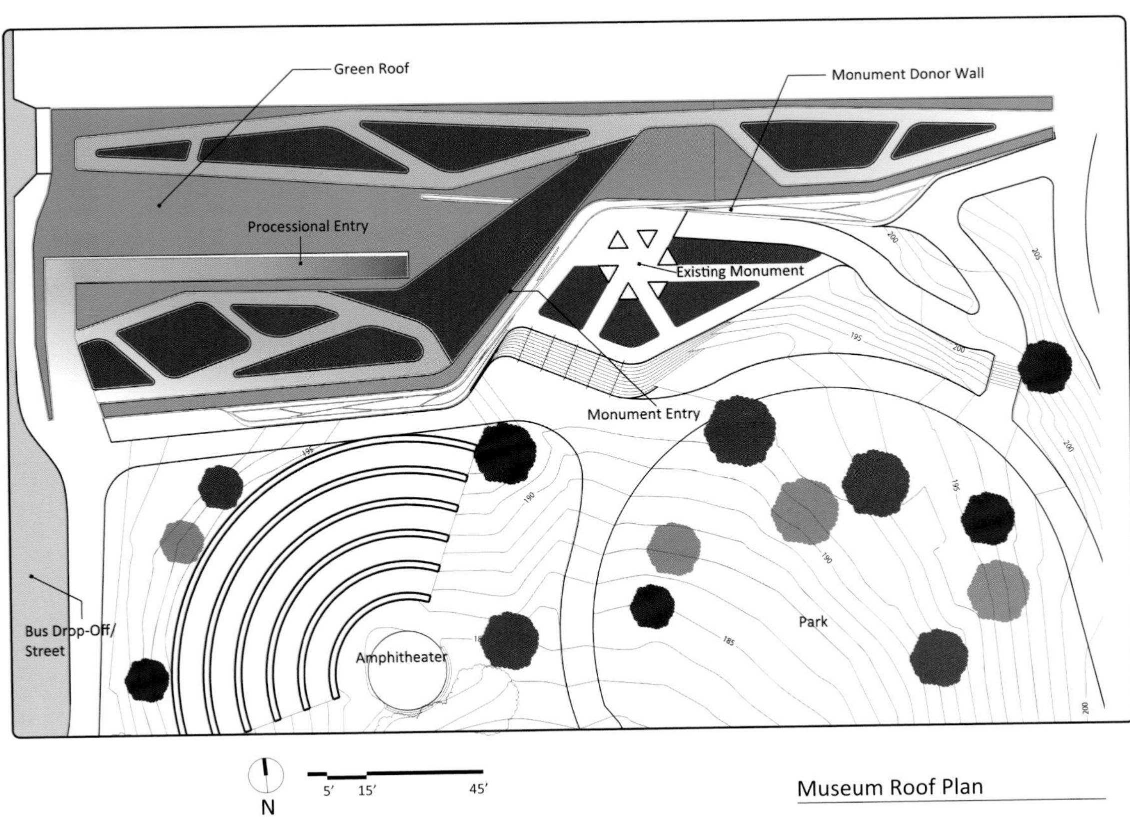

Museum Roof Plan

THIS WALL OF REMEMBRANCE HAS BEEN
DEDICATED BY
THE SOL KEST AND JONA GOLDRICH FAMILIES
IN MEMORY OF THEIR
PARENTS, RELATIVES AND FRIENDS
WHO PERISHED IN THE HOLOCAUST.
THE KEST FAMILY
THE GOLDREICH FAMILY
THE CREATION OF THIS MONUMENT
IS A TRIBUTE TO THE PERSEVERANCE AND
PERSONAL COMMITMENT OF:
GOVERMENTAL OFFICIALS
EDMUND D. EDELMAN
SUPERVISOR, 3RD DISTRICT
COUNTY OF LOS ANGELES
DAVID ROBERTI
CALIFORNIA STATE SENATOR
23RD DISTRICT
ZEV YAROSLAVSKY
COUNCILMAN, 5TH DISTRICT
COUNTY OF LOS ANGELES
OTHER ELECTED OFFICIALS
FEDERAL GOVERNMENT
HENRY WAXMAN - CONGRESSMAN, 24TH DISTRICT
STATE OF CALIFORNIA
PETE WILSON - GOVERNOR
BURT MARGOLIN - ASSEMBLYMAN, 45TH DISTRICT
COUNTY OF LOS ANGELES
GLORIA MOLINA - SUPERVISOR, 1ST DISTRICT
KENNETH HAHN - SUPERVISOR, 2ND DISTRICT
DEANE DANA - SUPERVISOR, 4TH DISTRICT
MICHAEL ANTONOVICH - SUPERVISOR, 5TH DISTRICT
CITY OF LOS ANGELES
TOM BRADLEY - MAYOR
JOHN FERRARO - COUNCILMAN, 4TH DISTRICT
COMMITTEE MEMBERS

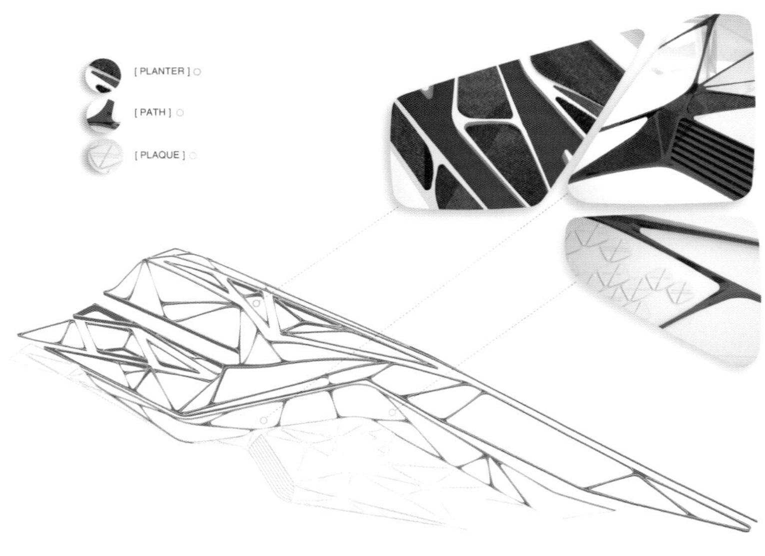

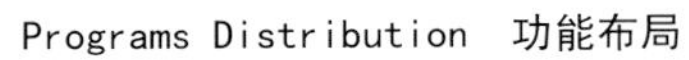
Programs Distribution 功能布局

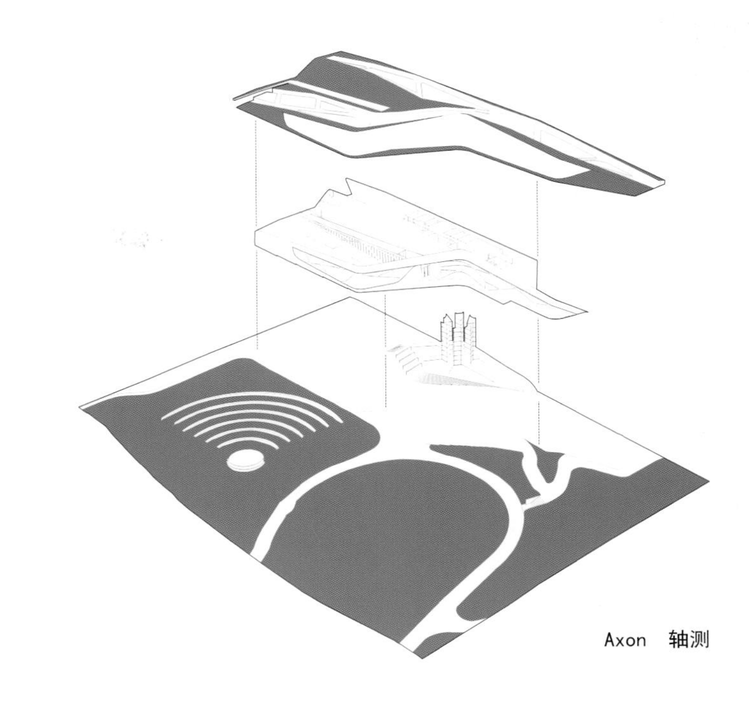

Axon 轴测

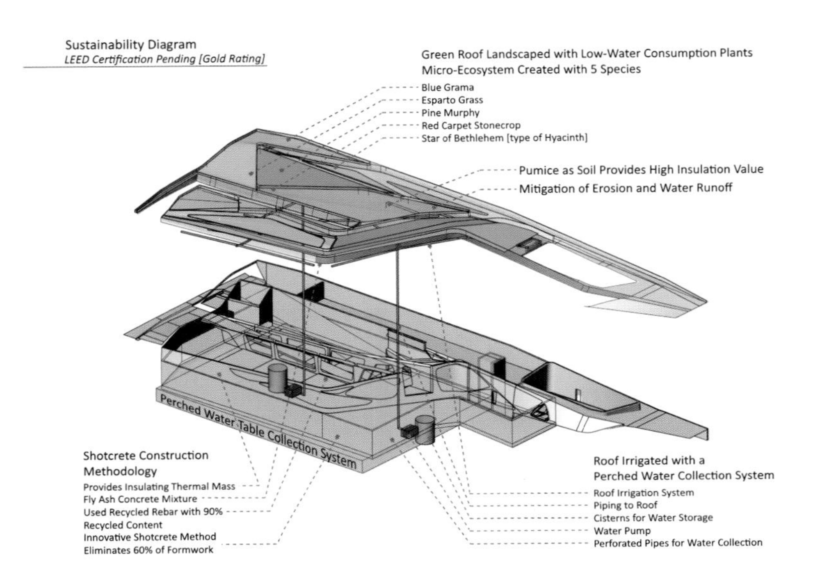

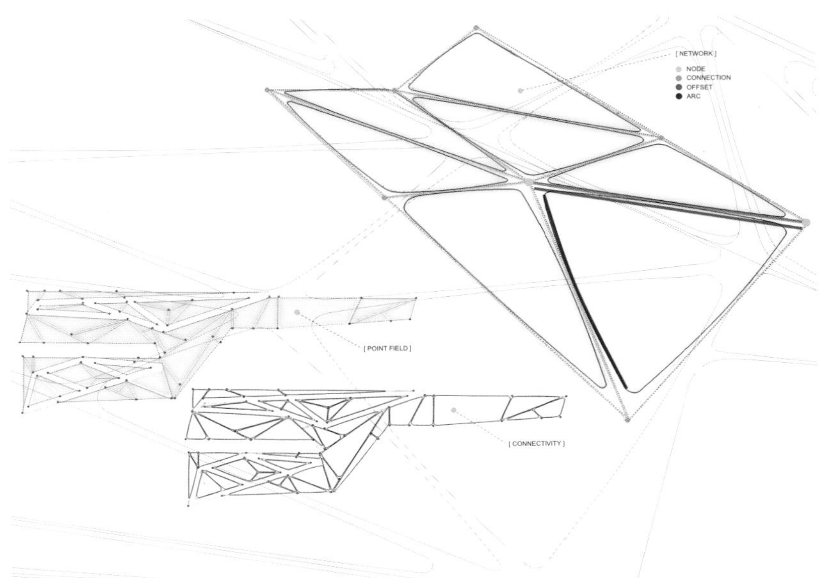

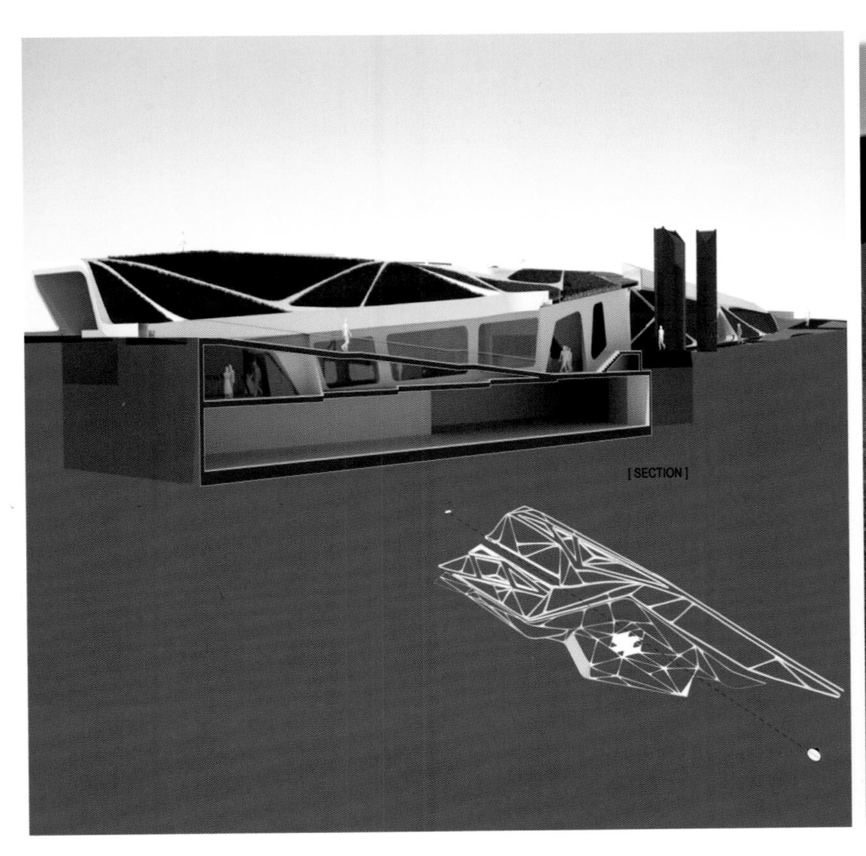

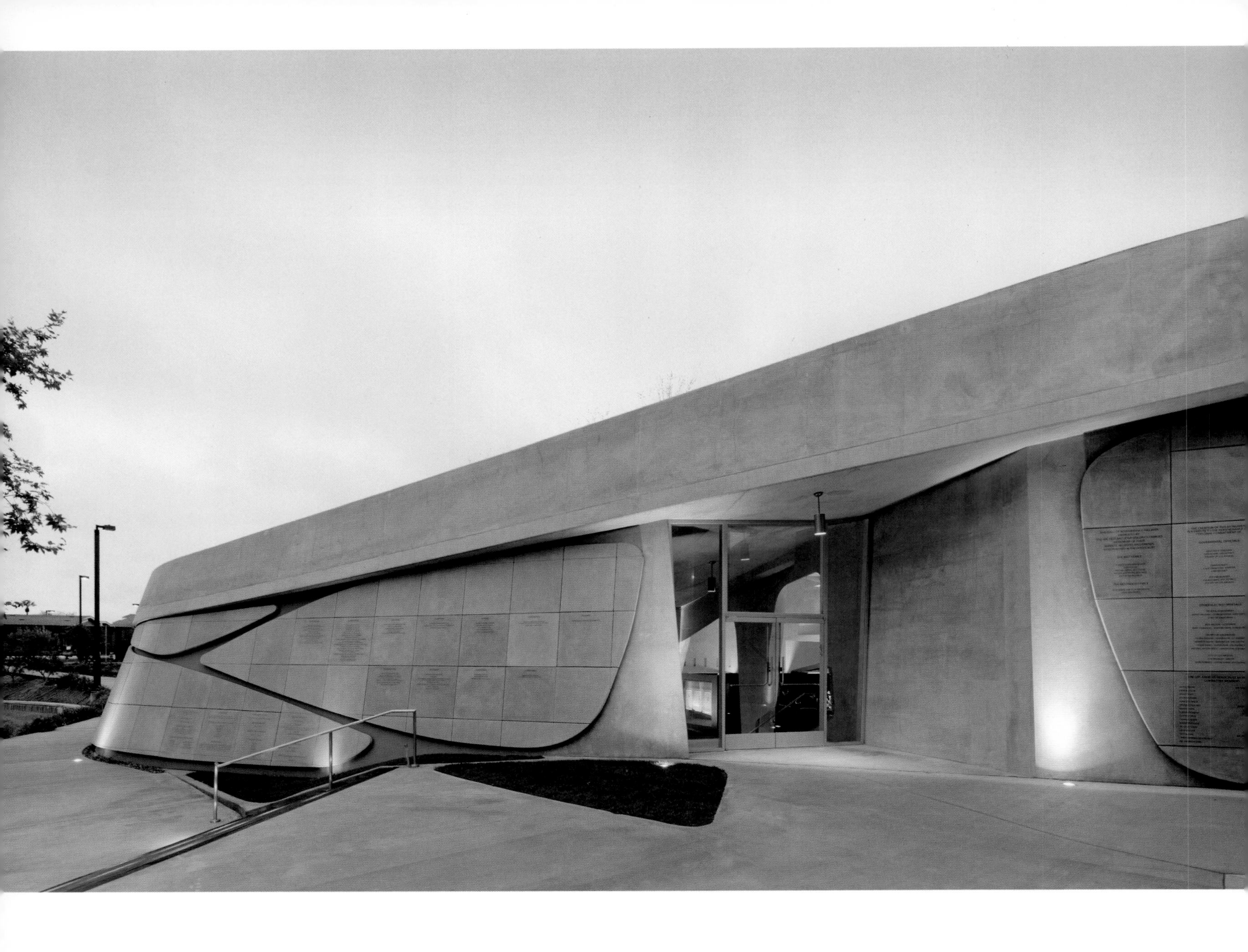

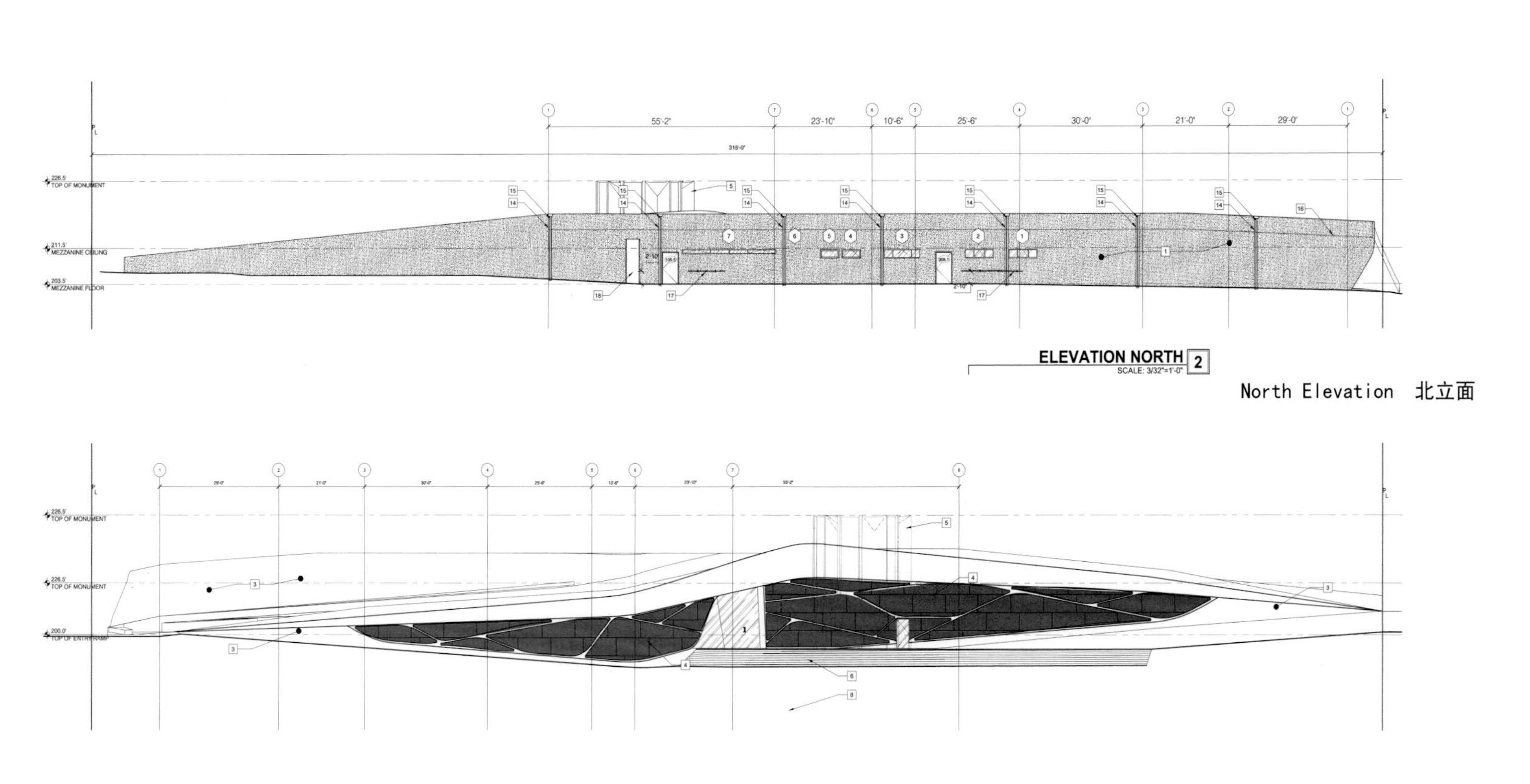

North Elevation 北立面

South Elevation 南立面

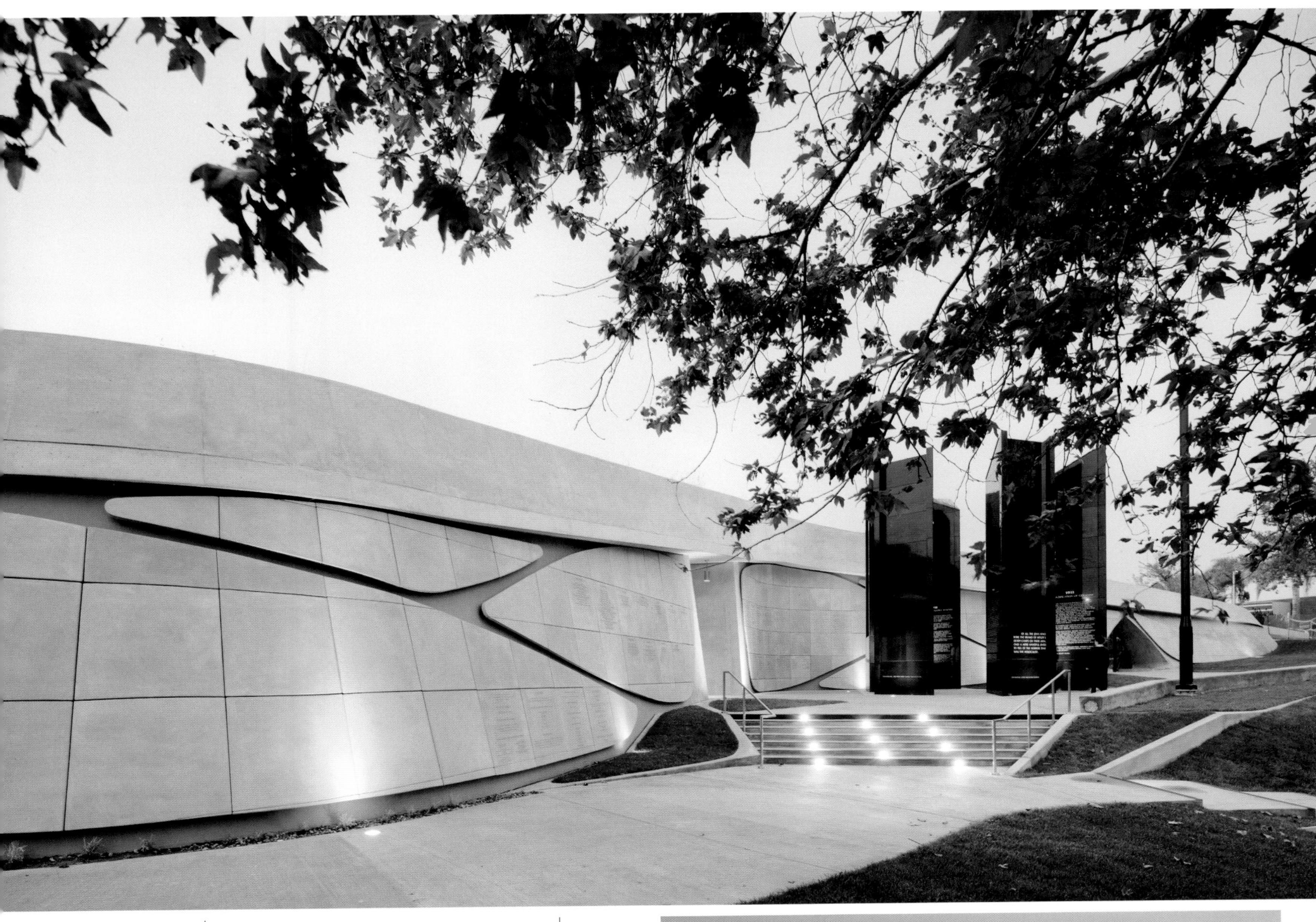

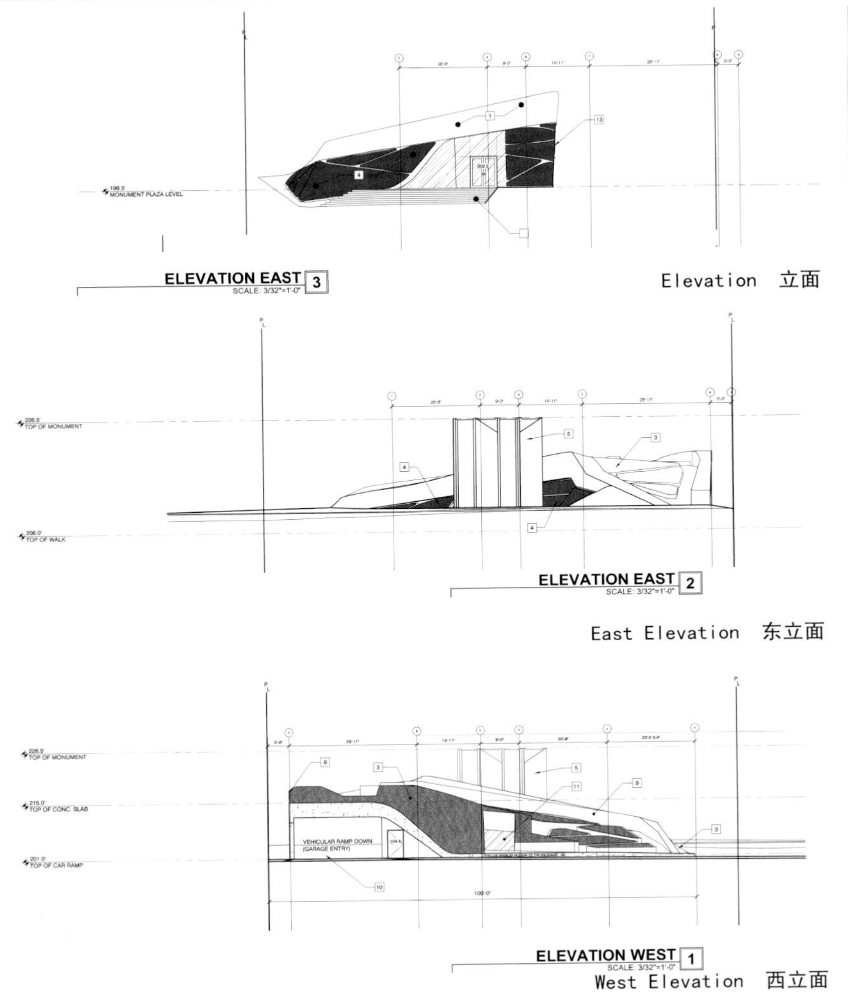

Elevation 立面

East Elevation 东立面

West Elevation 西立面

LOS ANGELES MUSEUM OF THE HOLOCAUST

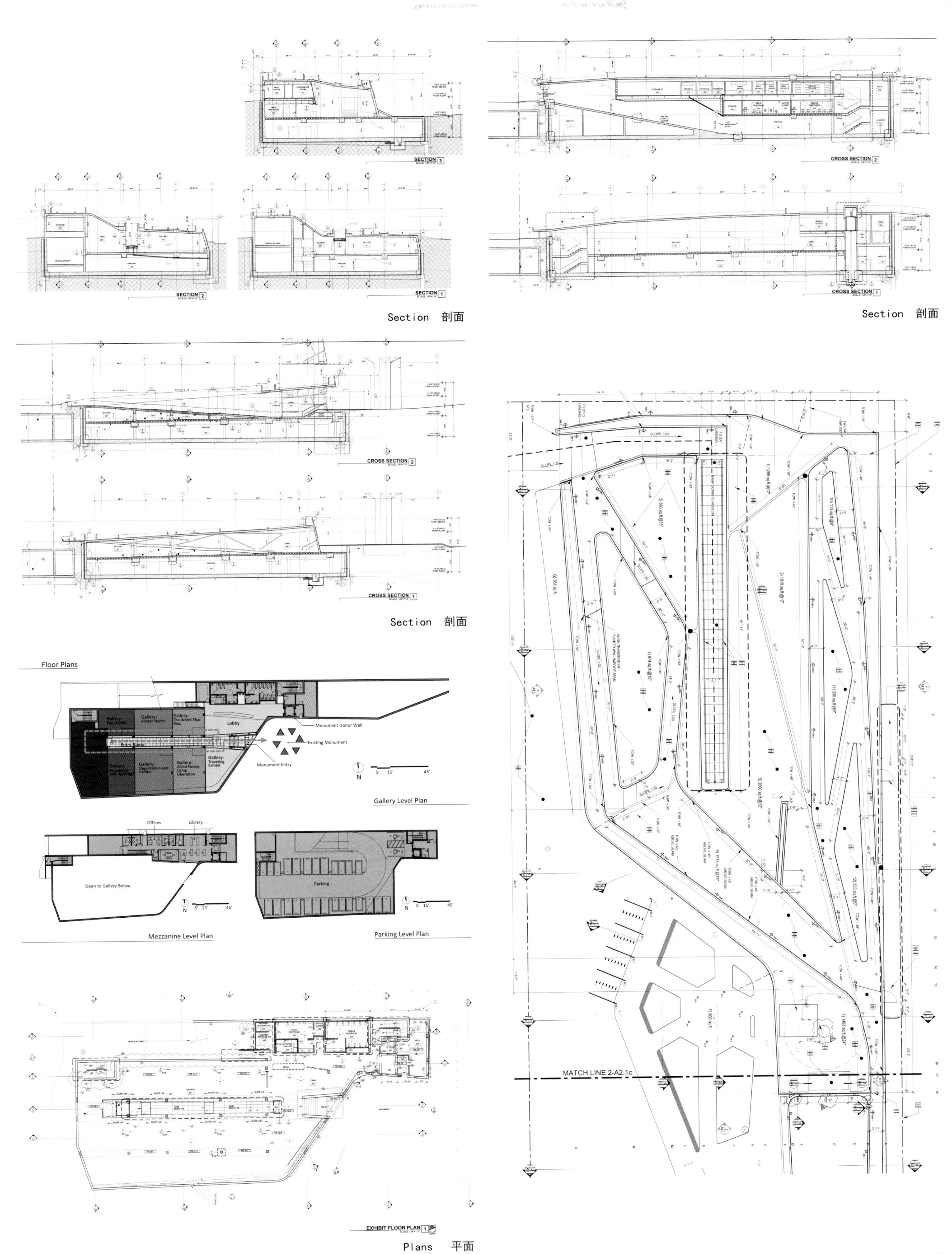

Section 剖面

Section 剖面

Section 剖面

Plans 平面

Serlachius Museum Gösta

GÖsta Serlachius 博物馆

HEIKKINEN-KOMONEN ARCHITECTS　　HEIKKINEN-KOMONEN建筑事务所

Location: Gösta, Finnland
Building Area: 4,700m²
Gross Floor Area: 3,600m²
Status: Competition, 2nd prize
Image Courtesy: HEIKKINEN-KOMONEN ARCHITECTS

区位：芬兰，GÖsta
占地面积：4700m²
总建筑面积：3600m²
状态：竞赛第二名
图片提供：HEIKKINEN–KOMONEN 建筑事务所

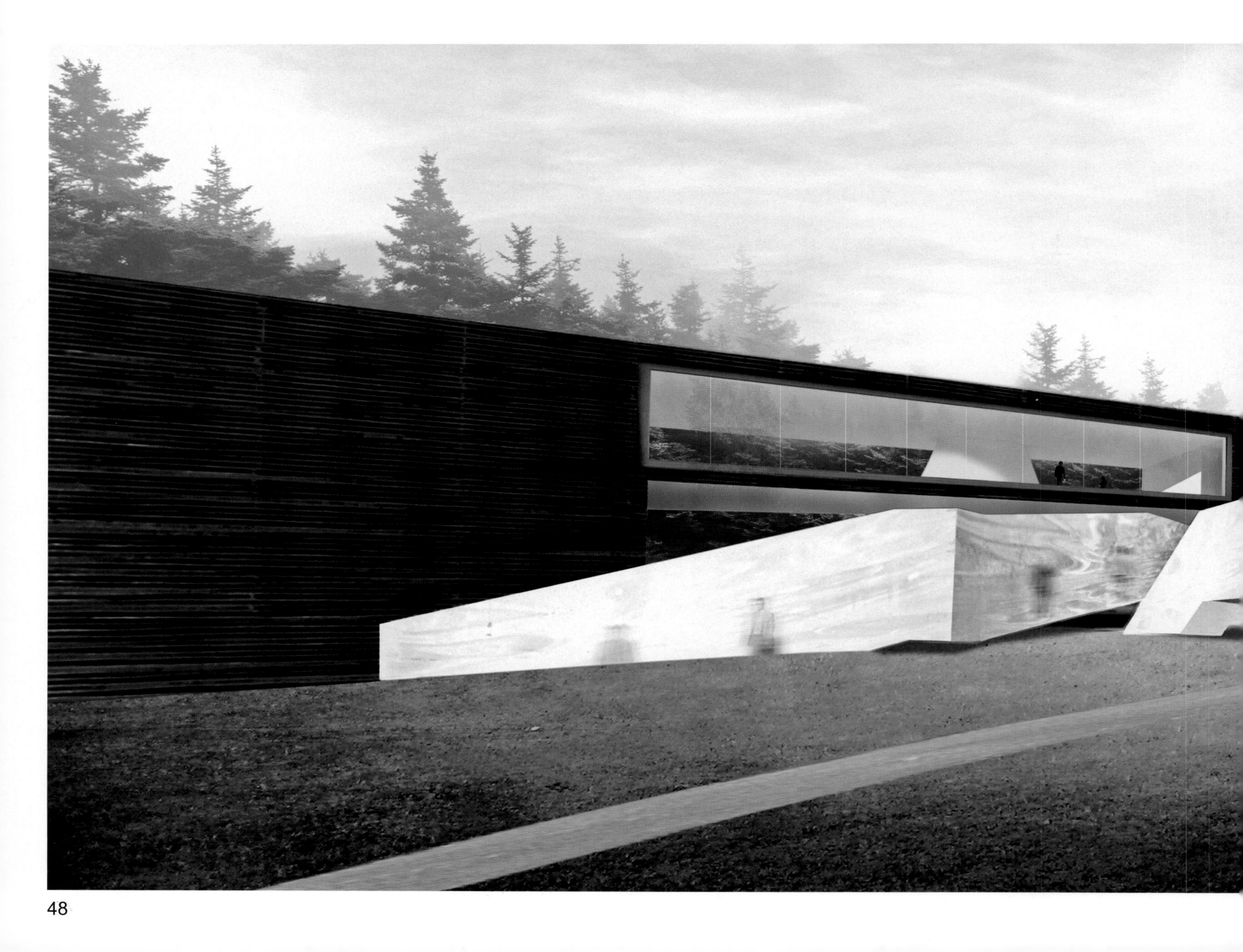

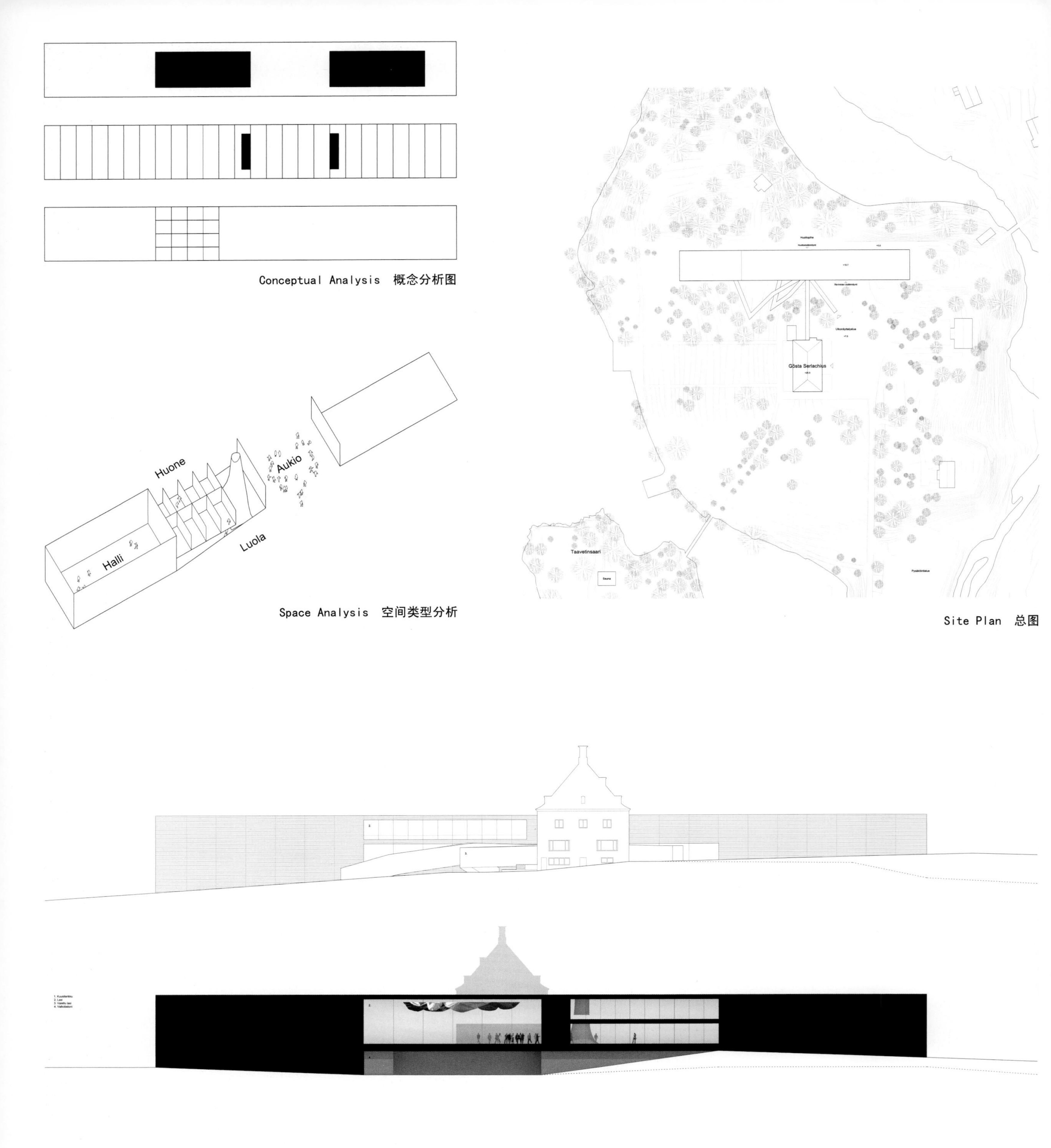

Conceptual Analysis 概念分析图

Space Analysis 空间类型分析

Site Plan 总图

Elevation and Section 立面与剖面

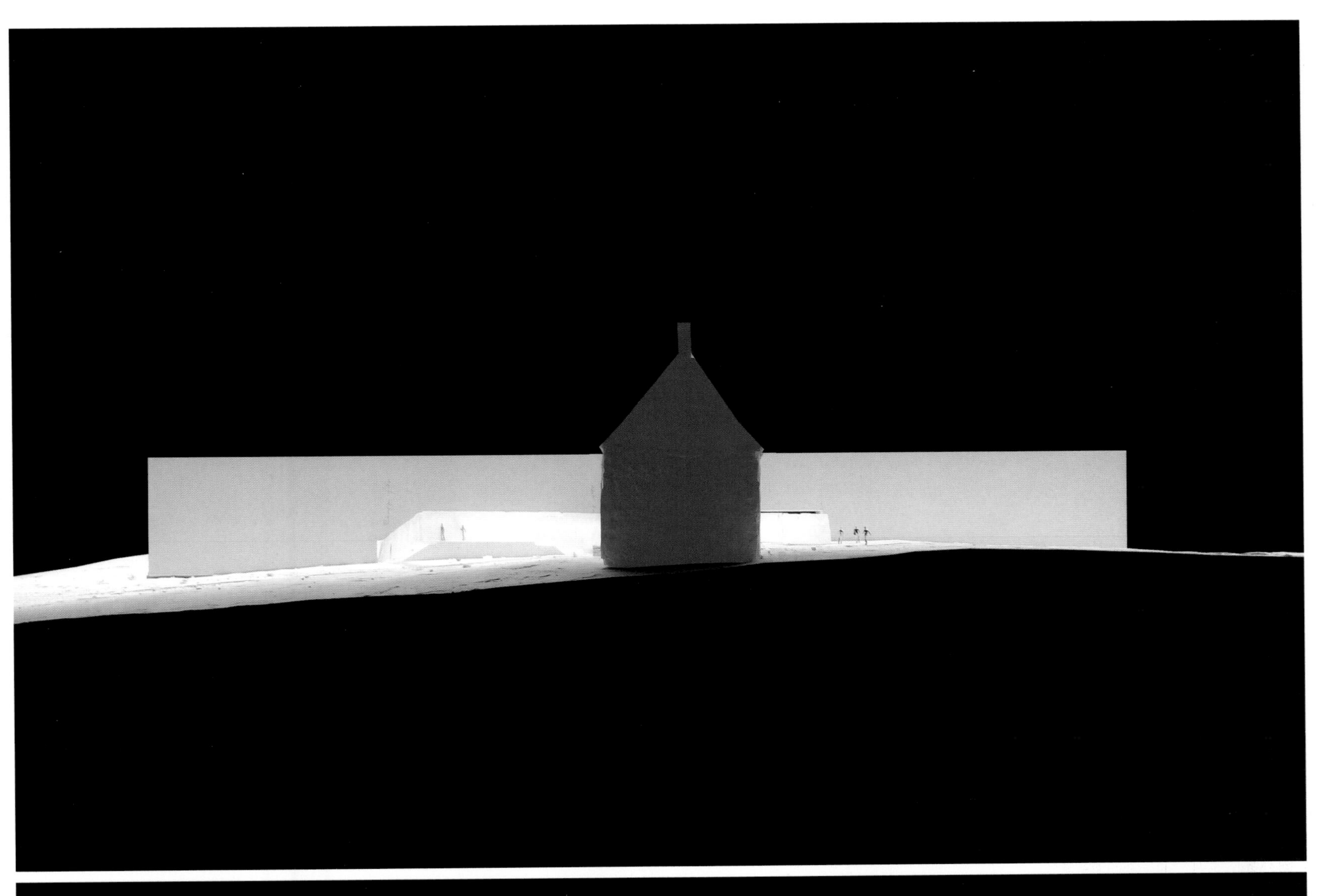

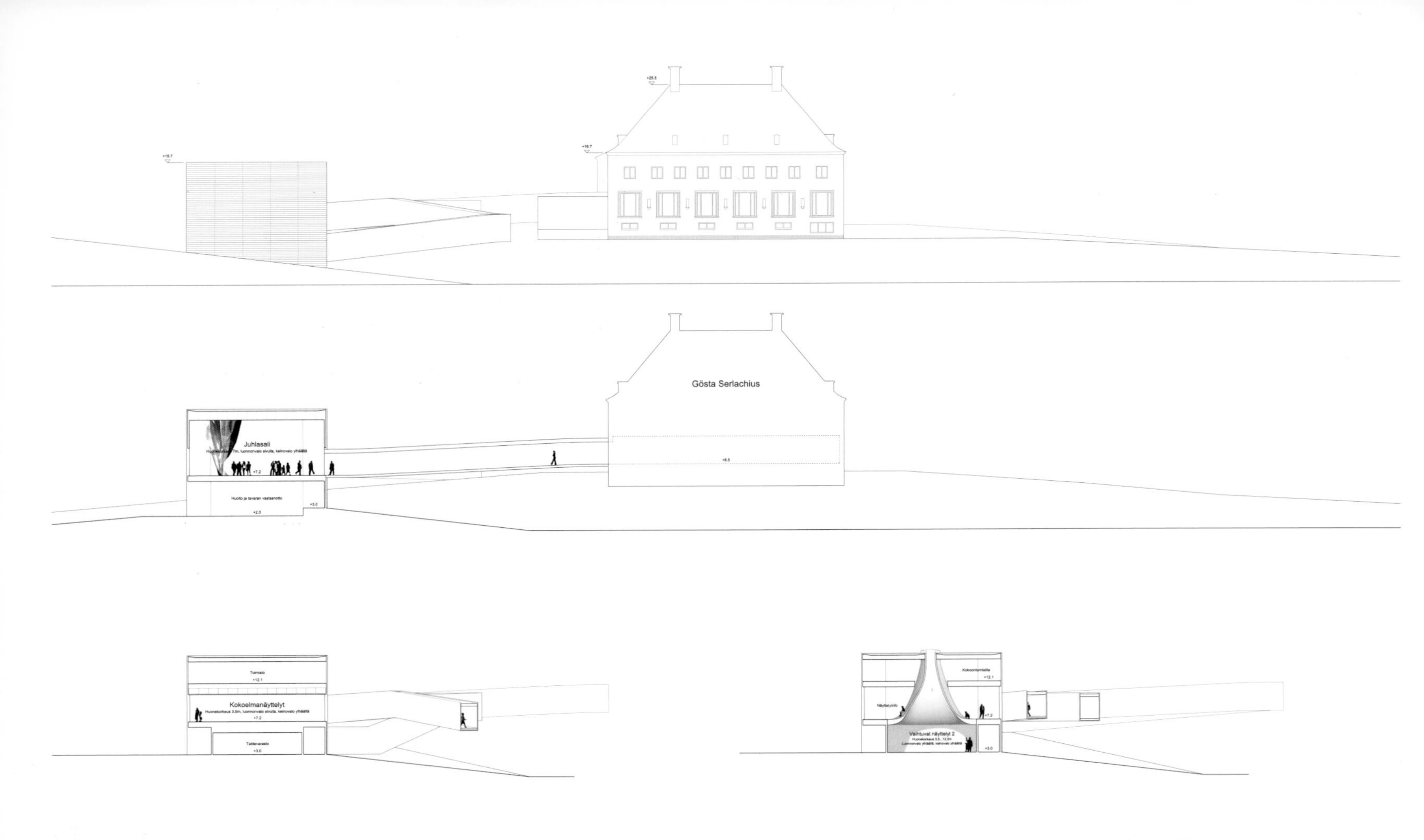
Gösta Serlachius
Juhlasali
Kokoelmanäyttelyt

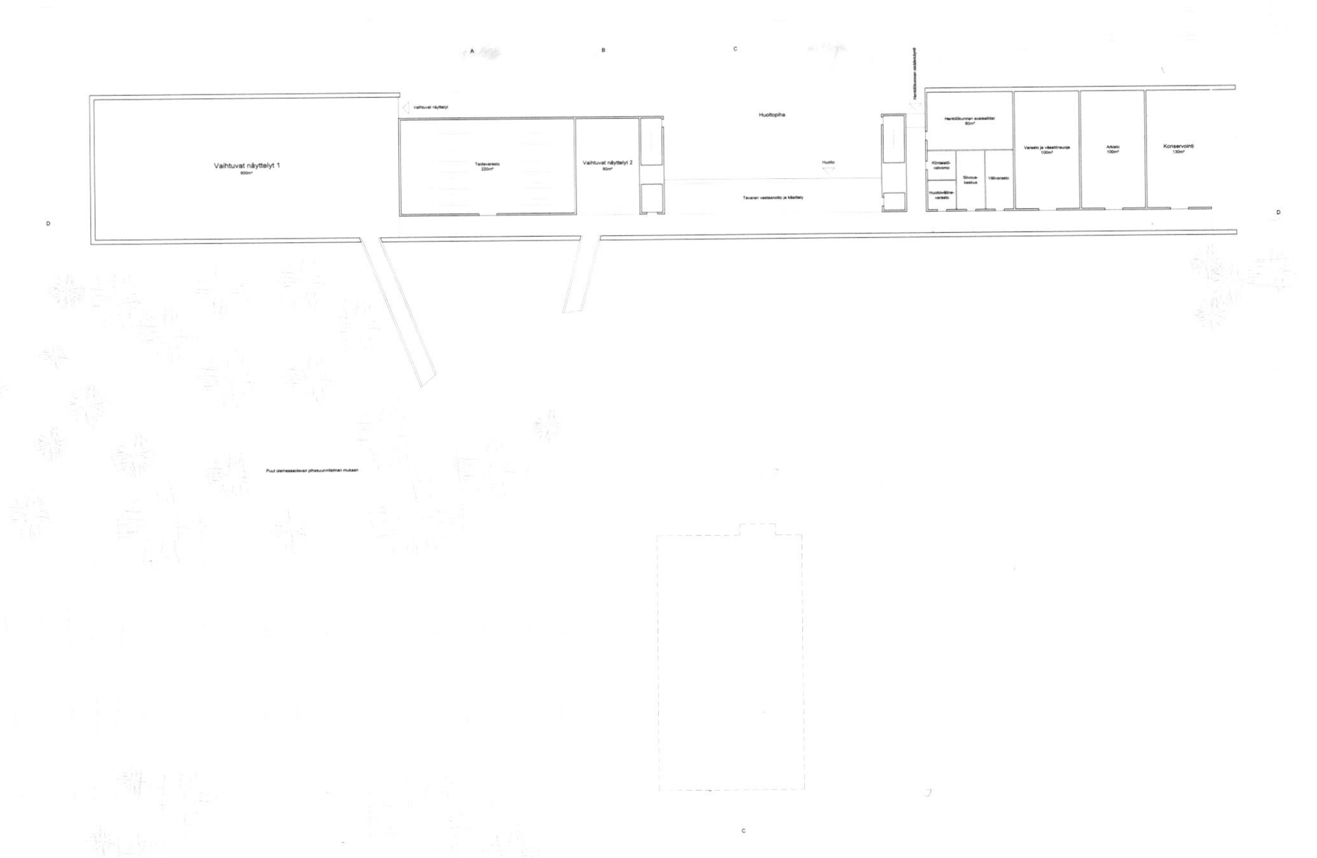

Plan Level 01 一层平面

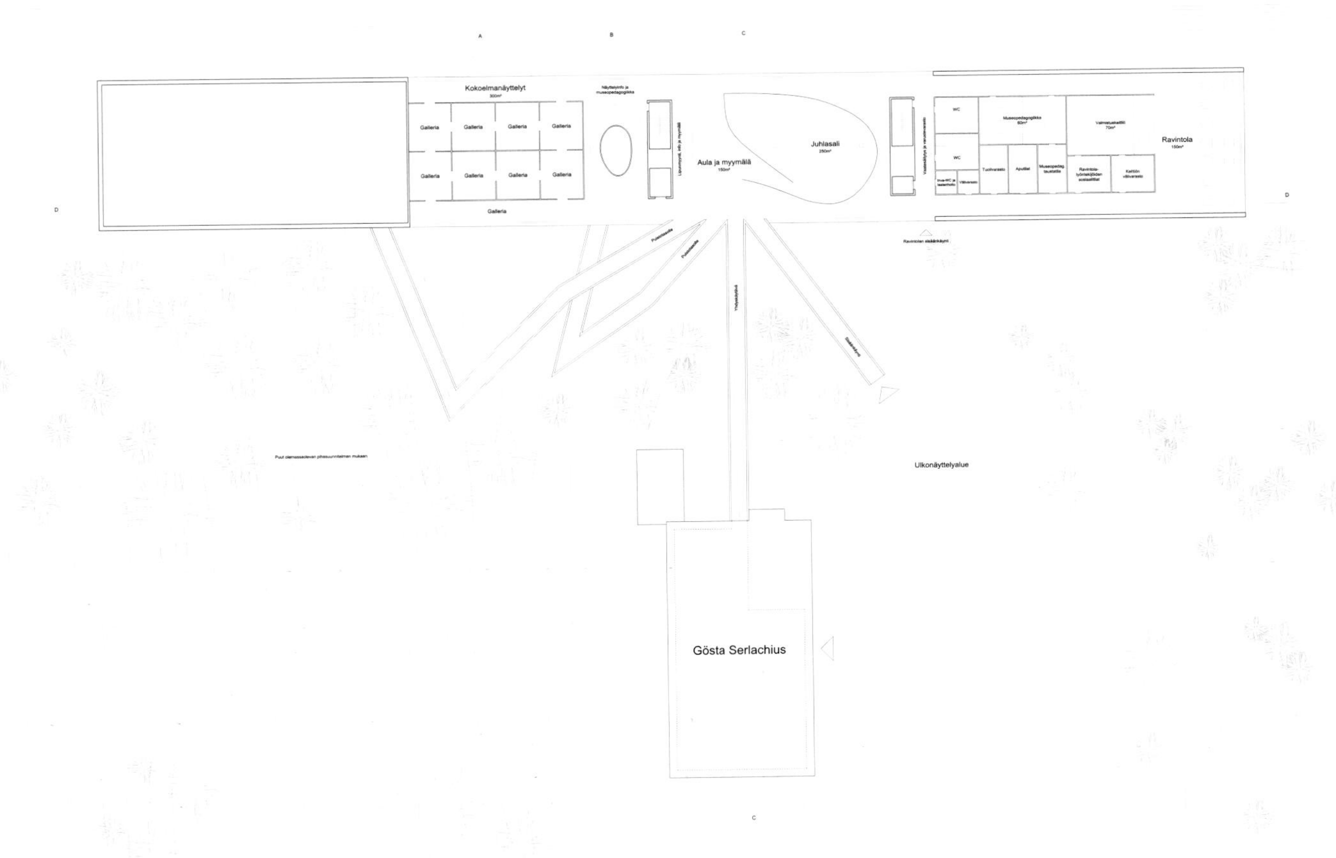

Plan Level 02 二层平面

Ancient Animal Fossil Museum of Hezheng, Gansu Province

甘肃和政古动物化石博物馆

ISOZAKI + HuQian Partners　　矶崎新+胡倩工作室

Location: Hezheng, Gansu Province
Typology: Museum
Site Area: 130,000 m^2
Building Area: 12,000 m^2
Gross Floor Area: 29,655 m^2
Status:Under construction
Image Courtesy: ISOZAKI + HuQian Partners

区位：中国，甘肃，和政
类型：博物馆
用地面积：130000 m^2
建筑面积：12000 m^2
总建筑面积：29655 m^2
状态：施工中
图片提供：矶崎新+胡倩工作室

Geographical Conditions

Located in Hezheng, Gansu Province in the northwest of The People's Republic of China, this plot has already finished its first and second phase of the construction of the Ancient Animal Fossil Museum and started its operation. The third phase of construction project is "the window of the world of ancient animal fossil".

The museum stores up a big collection of orictocoenosis of ancient mammals as well as the world's oldest fossil specimens which are buried under the enumerable lands which are home to abundant treasure trove, especially those buried in soil layer with a history of 30 million to 2 million years. According to the design condition, the constructive object is to build a unique construction which can bring these fossils under public attention and be itself as precious as the fossils are.

Building Lot

This plot was, at the first place, predetermined to be the hinterland of the existing museum due to the proposal of changing this area into the hills on the opposite bank of the retarding reservoir, connecting the first and second phase of the existing museum, the wide cultural expo garden in front of the plot and the retarding reservoir into an integral whole, thus forming a cultural facility area. The proposal has received recognition. The new plan of land utilization includes a 25-meter-high cliff. It's no different with arranging constructions down in the cliff. One can clearly see the view from the streets and arterial highway.

Such area will become an important facility group of education and culture in Hezheng County, and meanwhile will function as a visual gate for visitors coming from other provinces into the city streets.

Three exhibition department

The exhibition in the museum consists of three departments: scientific show, educational show and artistic show, all of which intend to offer audience a profound visiting experience. The space of each and every department implements the designs which are in line with their characteristics, thus becomes a multifarious exhibition space.

Academic Exhibition

Academic exhibition department gives a full play to the level difference of lands. Coming down through a spiral exhibition area, the visitors would feel as if they were deep down into the inaccessible stratum, pursuing of an even ancient geology age. The exhibiting configuration is in accordance with the academic classification of the various fauna fossil museum pieces. Visitors can get a deeper understanding of the

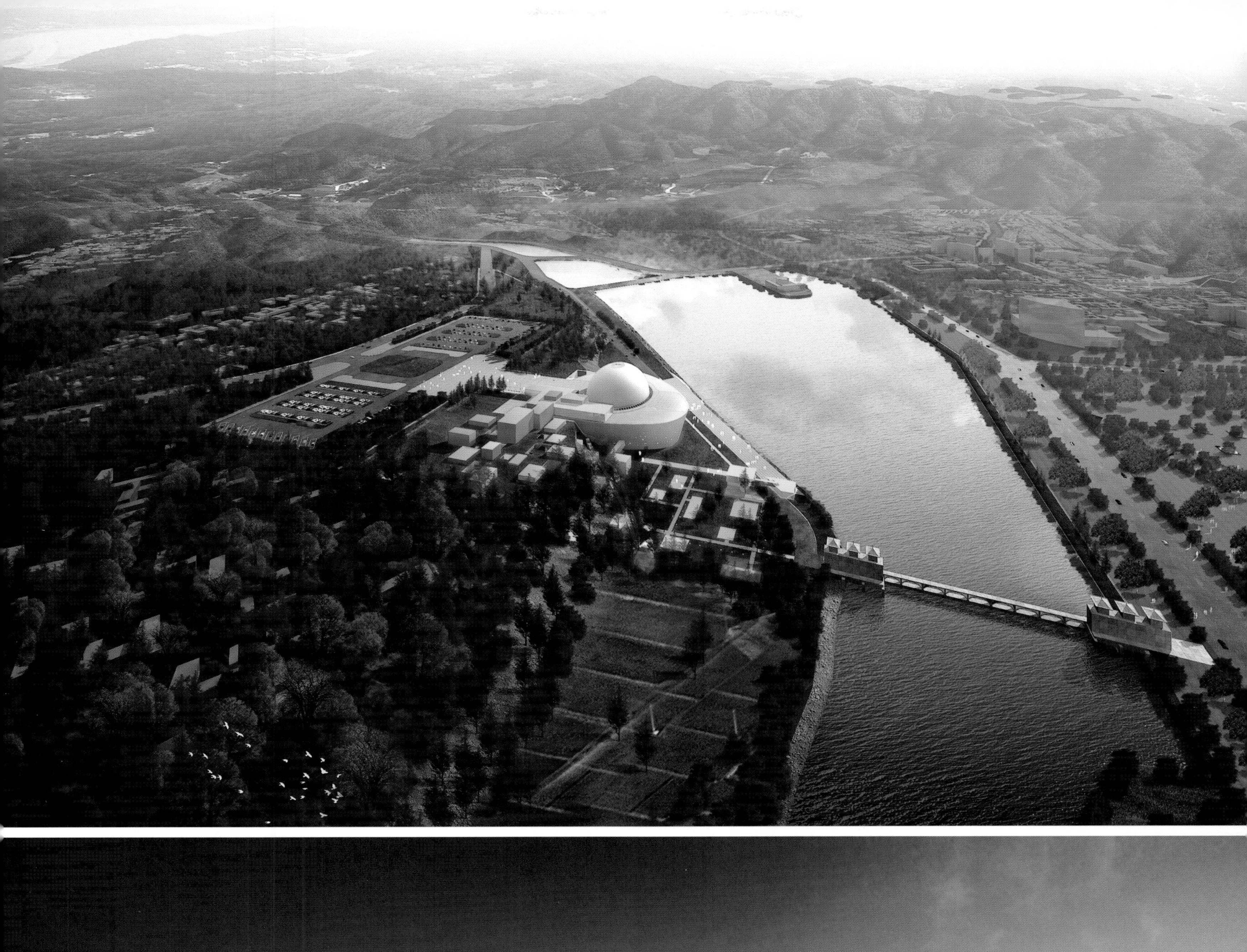

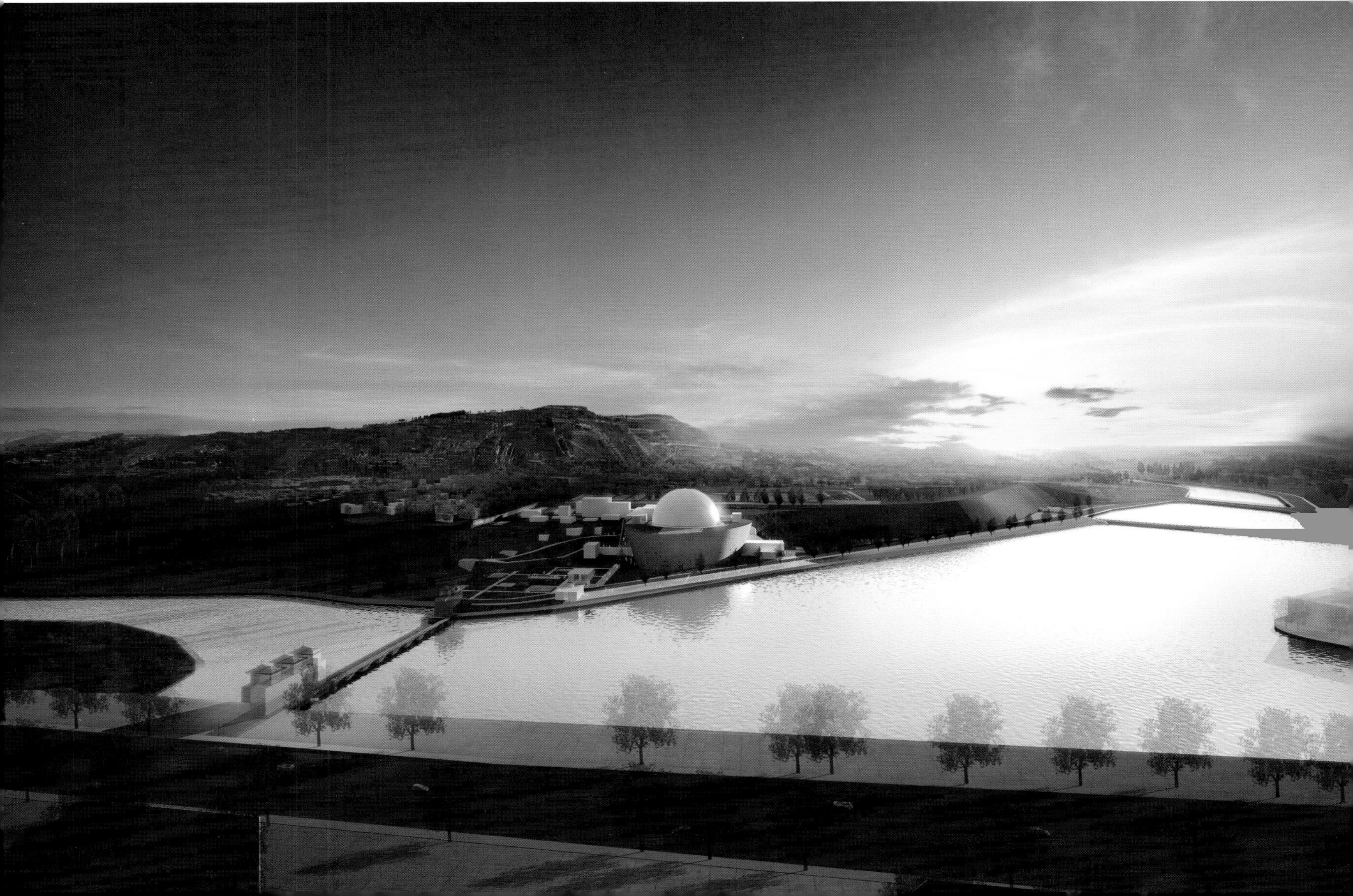

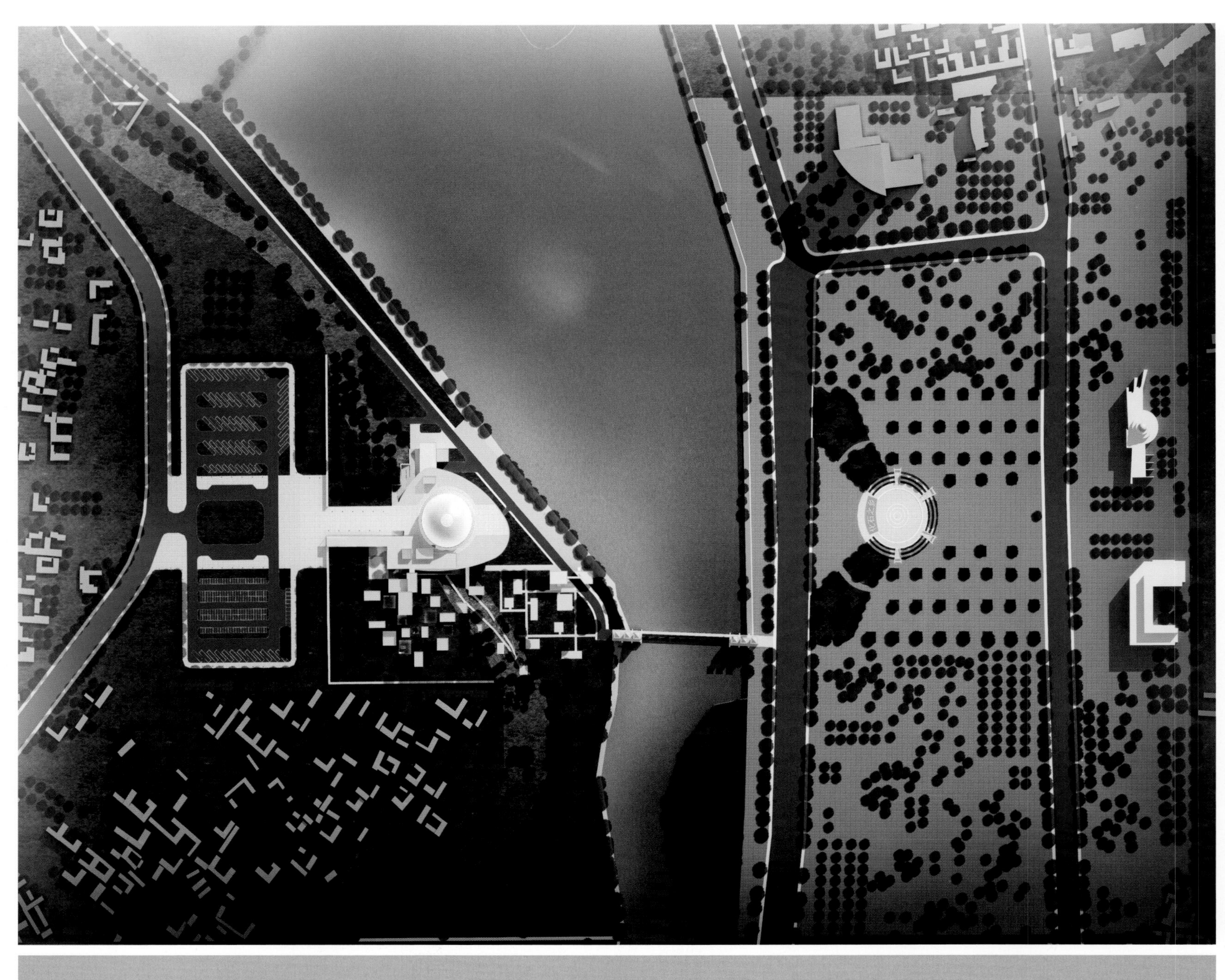

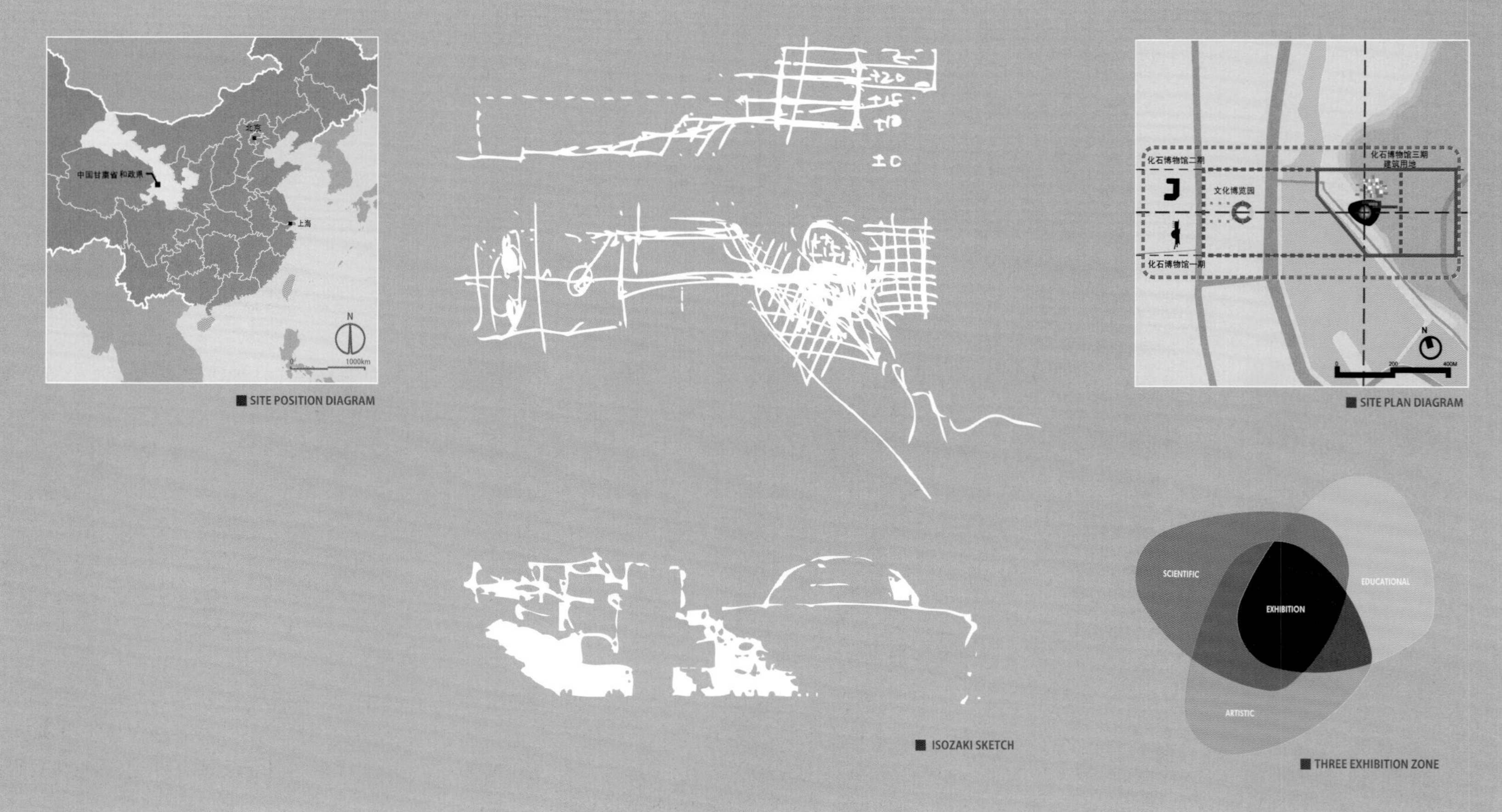

■ SITE POSITION DIAGRAM

■ ISOZAKI SKETCH

■ SITE PLAN DIAGRAM

■ THREE EXHIBITION ZONE

academic themes about the fossils, ground layer, geology timing as well as biological evolution. "Knowledge" is considered to be the theme of the exhibition. Plenty of the primary bone and recovery bone specimens of ancient animal orictocoenosis excavated in the suburb of Hezheng will be exhibited as many as possible. In addition, the museum will build a visiting route only for those who tour around the "best among the six worlds" showpieces, or various visiting routes around the bridge-shaped passage which cycles the middle-layer academic exhibition area.

Educational Exhibition

The educational exhibition department will exhibit showpieces of touching experience. Children and the ordinary visitors will gain all the feelings through the real touch. Besides, the museum also attempts to make use of all sorts of electronic machines' man-machine conversation to show the exhibition space through their experimental display technique. Furthermore, an education data space about the museum pieces is also established in the deepest part under the "best among the six worlds" spherical hall which is a collection of various exhibition rooms. Such education data space can also offer educational and academic intelligence about the treasures in here.

Artistic Exhibtion

The artistic exhibition department has already planned to invite 10 modern artists from home and abroad to apply the fossils collected in this museum as subjects and themes, giving free exhibitions of these fossils in certain areas of neutral exhibition rooms (which are called "the water cube" exhibition space). These artists who found a new way of regarding subjects and the world have focused their attention onto the fossil treasure unearthed in Hezheng County, they created an original exhibition space which any other museums in the whole world can never be compared to.

地理条件

所在地位于中华人民共和国西北部的甘肃省和政县。该地块已经建成古动物化石博物馆的第一期、第二期的建筑并运营，第三期的建筑计划为“古动物化石世界之窗”。

该博物馆收藏有从世界上屈指可数的拥有丰富埋藏物的土地中，特别是约3 000万至200万年前的土层中出土的大量的古哺乳动物化石群，以及世界上最古老的化石标本。设计条件上，计划建成展示这些化石并与此珍贵价值相符的具有很高独特性的建筑。

建筑用地

当初，该地块预定作为已建成博物馆的腹地，因有提议将该地变更为滞洪水库对岸的丘陵，将已建成的博物馆一期、二期，地块前面的宽阔的文化博览园至滞洪水库都连成一体，组成一个文化设施区，该提议已获得认可。新规划的用地包括有25米左右高的山崖，犹如在山崖下布置建筑物，从市街及干线道路可以清楚地看到。

由此构成的该区域，将成为和政县重要的教育及文化设施群，并起到从外省进入和政县市内街道时视觉大门的作用。

三个展示部门

本博物馆的展示，由学术展示（Sientific）、教育展示（Educational）、艺术展示（Artistic）的三个部门构成，意图给参观者带来深奥的参观体验。各部门的空间，实施符合各个展示性格的特有的设计，成为具有多样性的展示空间。

学术展示

学术展示部门，最大限度地利用地块的水平高低差，参观者通过螺旋状下降的展示区，犹如亲身进入更深的地层来追溯更远古的地质学年代。展示构成与各个动物群的化石珍藏品的学术分类一致，对化石和地层、地质学时间及生物进化的本博物馆的学术主题，参观者可以获得更深的理解。该部门考虑用“知识”作为展示主题，将和政县近郊出土的大量的古动物化石群的原生骨骼标本及复原骨骼标本等学术上贵重的展示品，尽可能多地展示出来。另外，根据参观者的要求，开设有仅参观本博物馆收藏的“六个世界第一”的展示品，或者环绕地下中间层学术展示区的桥形通道一周等的多种参观路线。

教育展示

教育展示部门，展示被称为“触摸体验”的展示品，通过实际触摸，孩子们及一般的参观者可获得五官感受，此外，还考虑尝试利用各种电子机器的人机对话的实验性展示手法展示空间。另外，在收容“六个世界第一”各展览室的球形大厅的最下部，也设有关于本博物馆珍藏品的教育资料空间，可提供关于本博物馆珍藏品的教育及学术上的情报。

艺术展示

艺术展示部门，已计划由10位国内外著名的现代艺术家，采用本博物馆收藏的化石为题材及主题，在各自的一定大小的中立展览室(在现代美术馆建筑上称作“水立方”的展示空间)中，自由地摆放展示。发现对物与世界新看法的专业艺术家们，将光线聚焦在和政县出土的化石珍藏品上，在这个博物馆打造出全世界博物馆都无可比拟的独创的展示空间。

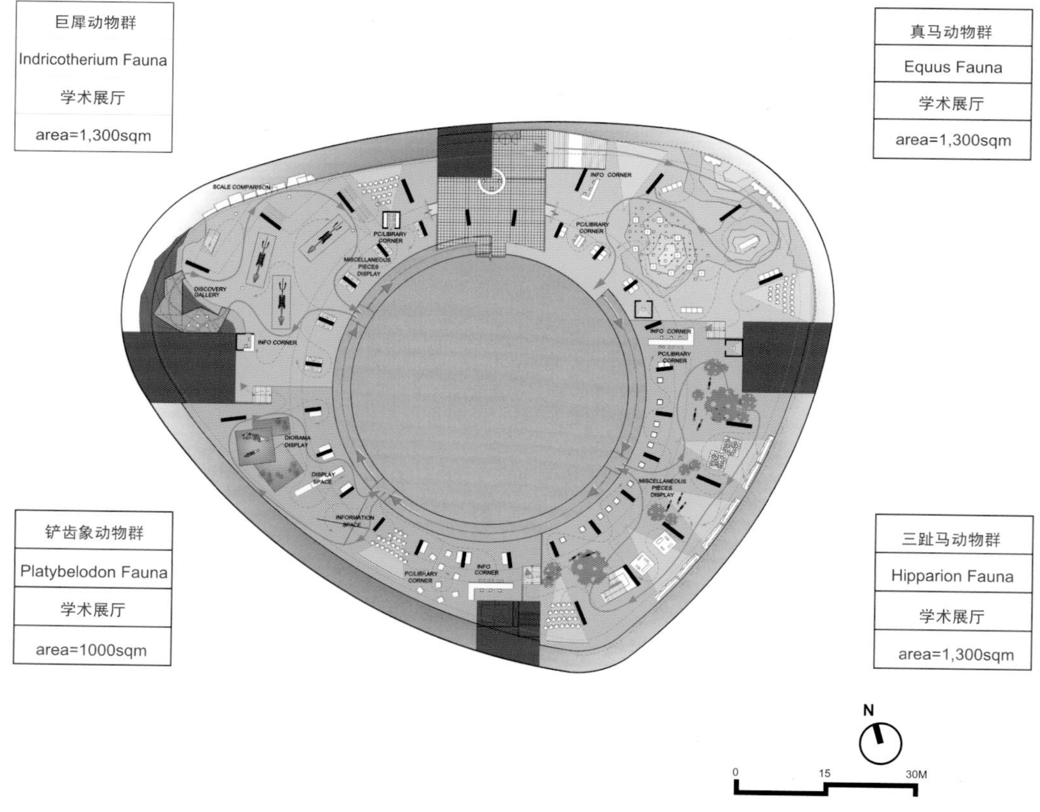

N

0 15 30M

■ **SCIENTIFIC EXHIBITION DIAGRAM**

■ **DRAWING scale 1/1500**

Scientific Exhibition Diagram 科学展示空间分析

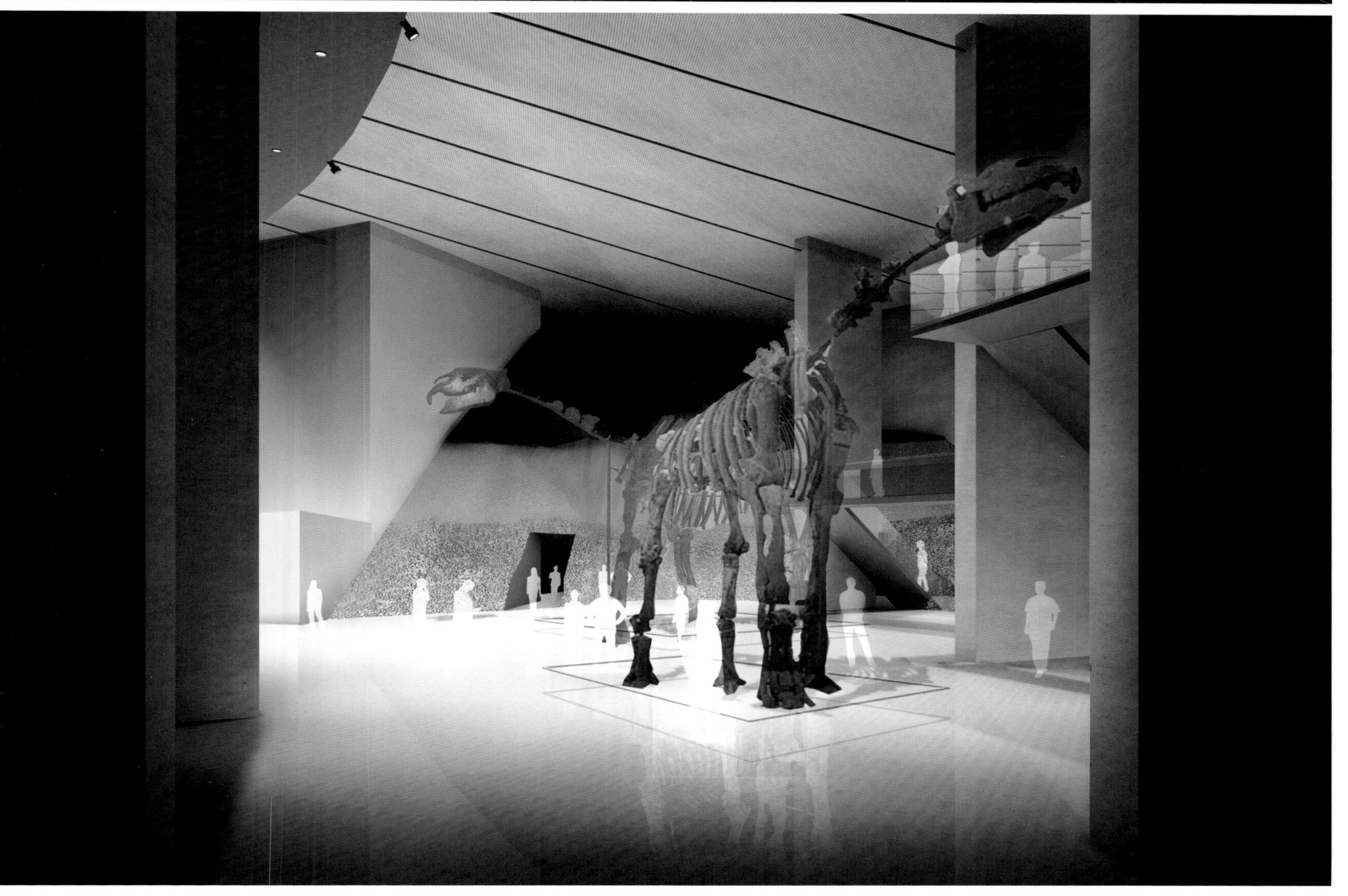

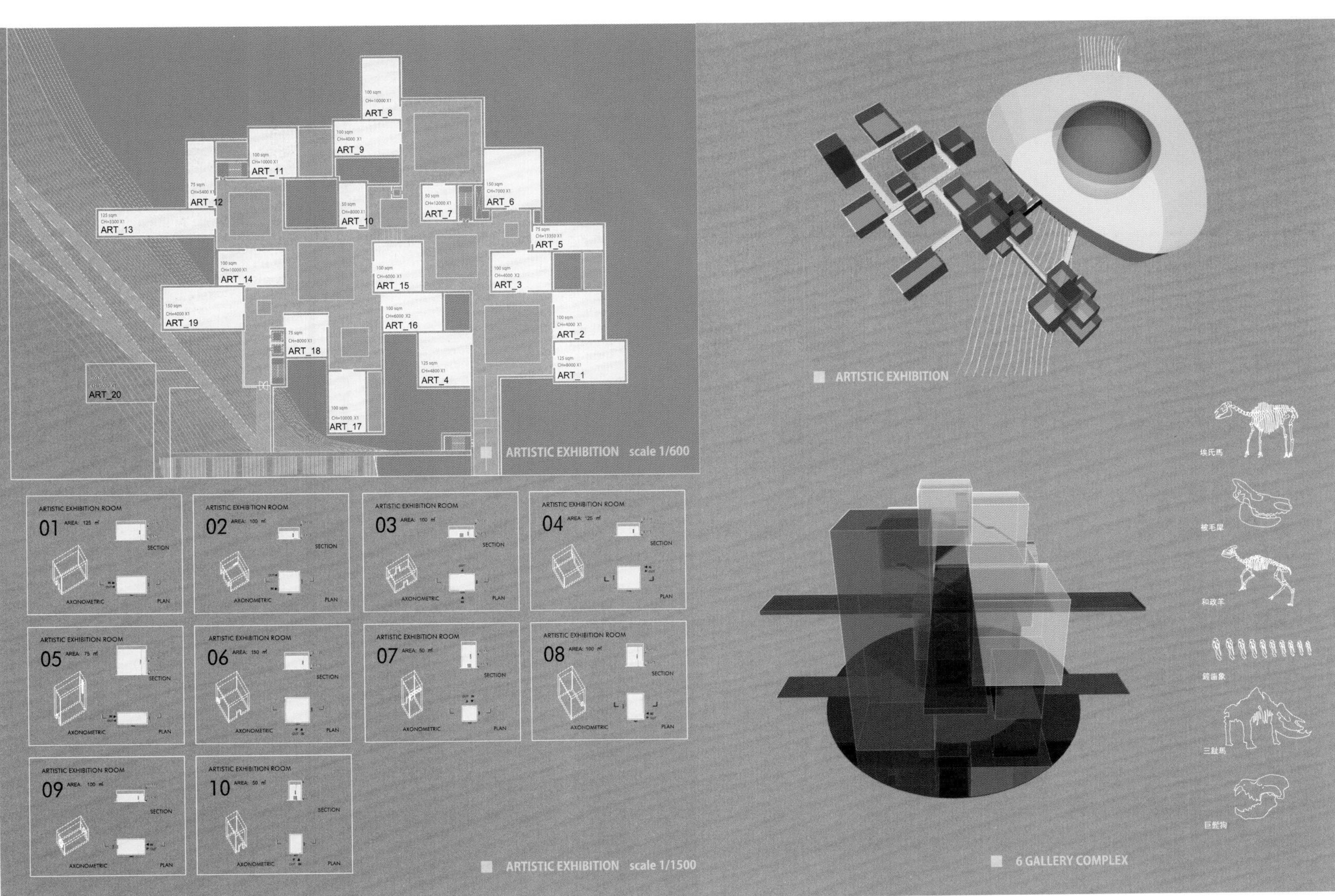
ARTISTIC EXHIBITION scale 1/600
ARTISTIC EXHIBITION
ARTISTIC EXHIBITION scale 1/1500
6 GALLERY COMPLEX

Bechtler Museum

贝希特勒博物馆

Studio arch. Mario Botta　　Mario Botta建筑工作室

Location: Charlotte (NC), USA
Typology: Museum
Site area: 1,912 m²
Gross surface: 2,490 m²
Status : Completed in 2009
Image Courtesy: Joel Lassiter, Enrico Cano

区位：美国，夏洛特市（NC）
类型：博物馆
建筑面积：1 912 m²
总表面积：2 490 m²
状态：2009年建成
图片提供：Joel Lassiter, Enrico Cano

The museum is in downtown Charlotte, a city that has undergone a rapid urban development in recent years. The new museum will house the works of art of the Bechtlers' collection with important artists such as Tinguely, Niki de Saint Phalle, Picasso, Giacometti, Matisse, Mirò, Degas, Warhol, Le Corbusier, Leger....

The four-storey structure is characterized by the soaring glass atrium that extends through the museum's core and diffuses natural light throughout the building thanks to a system of vaulted skylights. Despite the modest dimensions, its great plastic force is created by the play of solids and voids. It can be thus defined as an architecture-sculpture where the voids, carved inside the primary volume, mark a new urban space sheltered by the fourth floor gallery which juts out from the core of the building, cantilevered and supported by a column rising from the plaza below.

The choice of the materials for the interior spaces and the terra cotta exterior cladding give the museum a rigorous, though elegant, simplicity.

该博物馆位于夏洛特市中心。该市近年来发展速度极快。

新建博物馆将用来展示贝希特勒收藏的如汤格利、尼基·德·圣法尔、毕加索、贾科梅蒂、马蒂斯、米尔、德加、沃霍尔、勒·柯布西耶这类重要人物创作的艺术作品。

一座高耸的玻璃中庭贯穿四层高的博物馆中心，阳光透过拱形天窗，洒向整座建筑。位于四楼的画廊从建筑中心展开，以悬垂的姿势由大厦底部的柱子支撑着。建筑体量适中，虚实对比形成了巨大的张力。在画廊的覆盖下形成了一个新的城市空间，强化了建筑的雕塑感。

室内材料的选择和室外赤褐色的表层使博物馆看起来庄严简洁而不失优雅。

S. TRYON
ST
500
ONLY

CHARLOTTE 2010 M.B.

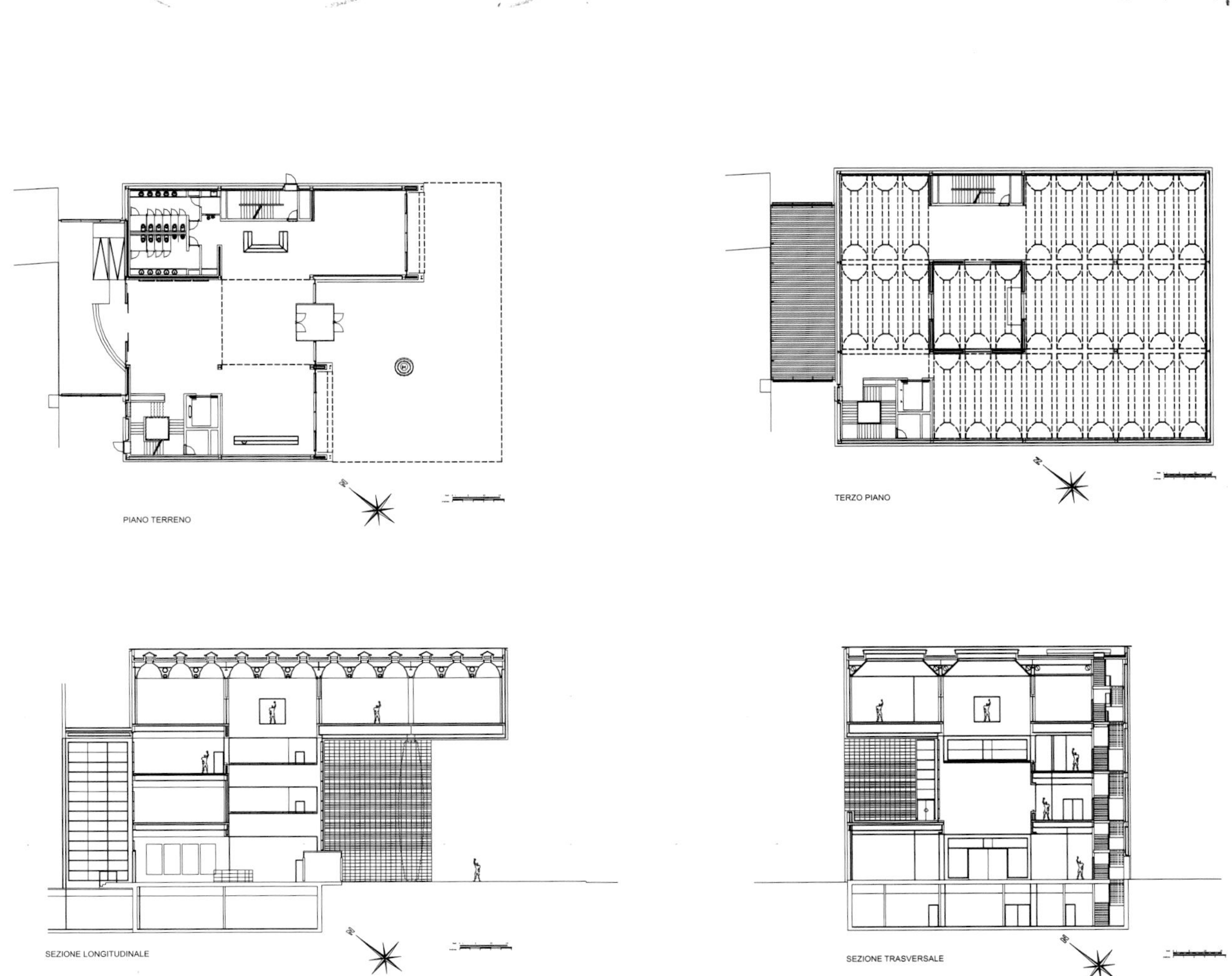
PIANO TERRENO
TERZO PIANO
SEZIONE LONGITUDINALE
SEZIONE TRASVERSALE

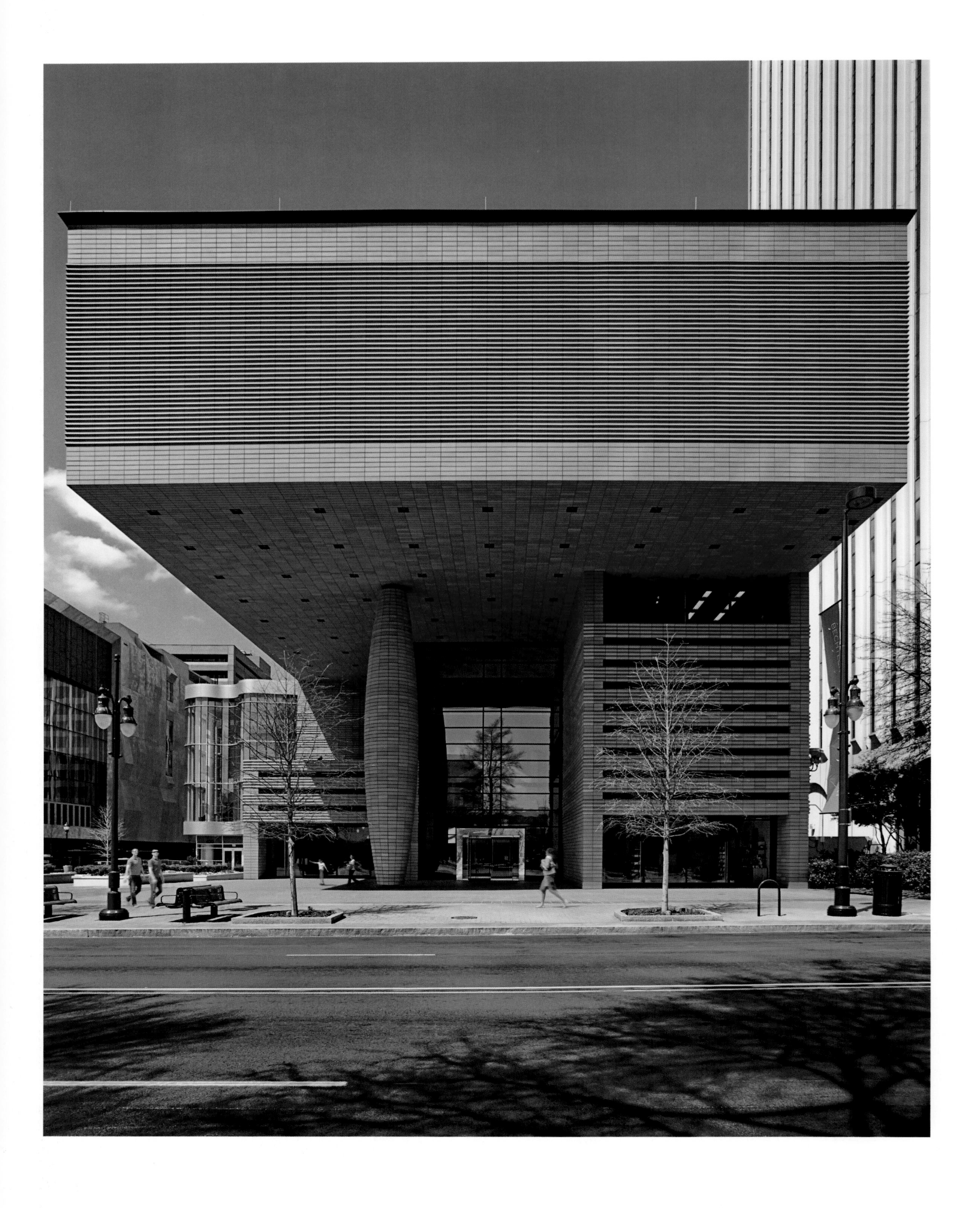

EXIT

Museum of Liverpool

利物浦博物馆

3XN　　3XN建筑事务所

Location: Liverpool, United Kingdom
Typology: Musuem
Size: 13,000 m^2
Status: Completed
Photographers: Pete Carr, Phillip Handforth, Richard White

区位：英国，利物浦
类型：博物馆
规模：13 000 m^2
状态：已完成
摄影师：Pete Carr, Phillip Handforth, Richard White

The new Museum of Liverpool, opening on July 19th will not only tell the story of its importance as one of the world's great ports or about its cultural influence, such as with the Beatles phenomenon. It will also serve as a meeting point for history, the people of Liverpool and visitors from around the globe. Therefore, according to the architect, Kim Herforth Nielsen, the structure functions as much more than just a building or a museum. This is one of the largest and most prestigious projects in 3XN's 25 year history. The Museum's design is a result of a very rigorous process, where it was of utmost priority to listen to the city inhabitants, learn the city's history and understand the potential of the historical site that the Museum now sits upon. The result is a dynamic low-rise structure which enters into a respectful dialogue with the harbor promenade's taller historical buildings. This interaction facilitates a modern and lively urban space. The design is reminiscent of the trading ships which at one time dominated the harbor, while the façade's relief pattern puts forward a new interpretation of the historical architectural detail in the 'Three Graces.' The enormous gabled windows open up towards the city and the harbor, and therefore symbolically draw history into the Museum, while at the same time allow the curious to look in.

这个新建的利物浦博物馆将于2011年7月19日正式对公众开放，它的重要性不仅仅是因为它立于世界上著名的港口或是它的文化影响力，更重要的是它将是一个历史，利物浦当地居民和世界各地游客交会的空间。因此根据主建筑师的说法它已经不是一个简单的建筑或博物馆了。这也是事务所25年中最大最著名的项目之一了。博物馆的设计是经历了很多复杂的过程的，但其中最重要的是聆听城市居民的意见，学习城市的历史和理解历史场地的发展潜力。最后设计生成的是一个动感低层的结构，它与周围海港大道上的高大历史建筑形成和谐对话，这种交流形成一种现代和生动的城市空间。设计是对曾经主导海港的货运船只的一种回忆，表皮的样式呈现出对历史建筑细节的现代理解。巨大的尖顶窗户向着城市和海港打开，象征性地将历史拉进博物馆，同时也能满足室外的好奇的人们。

GREAT WESTERN RAILWAY

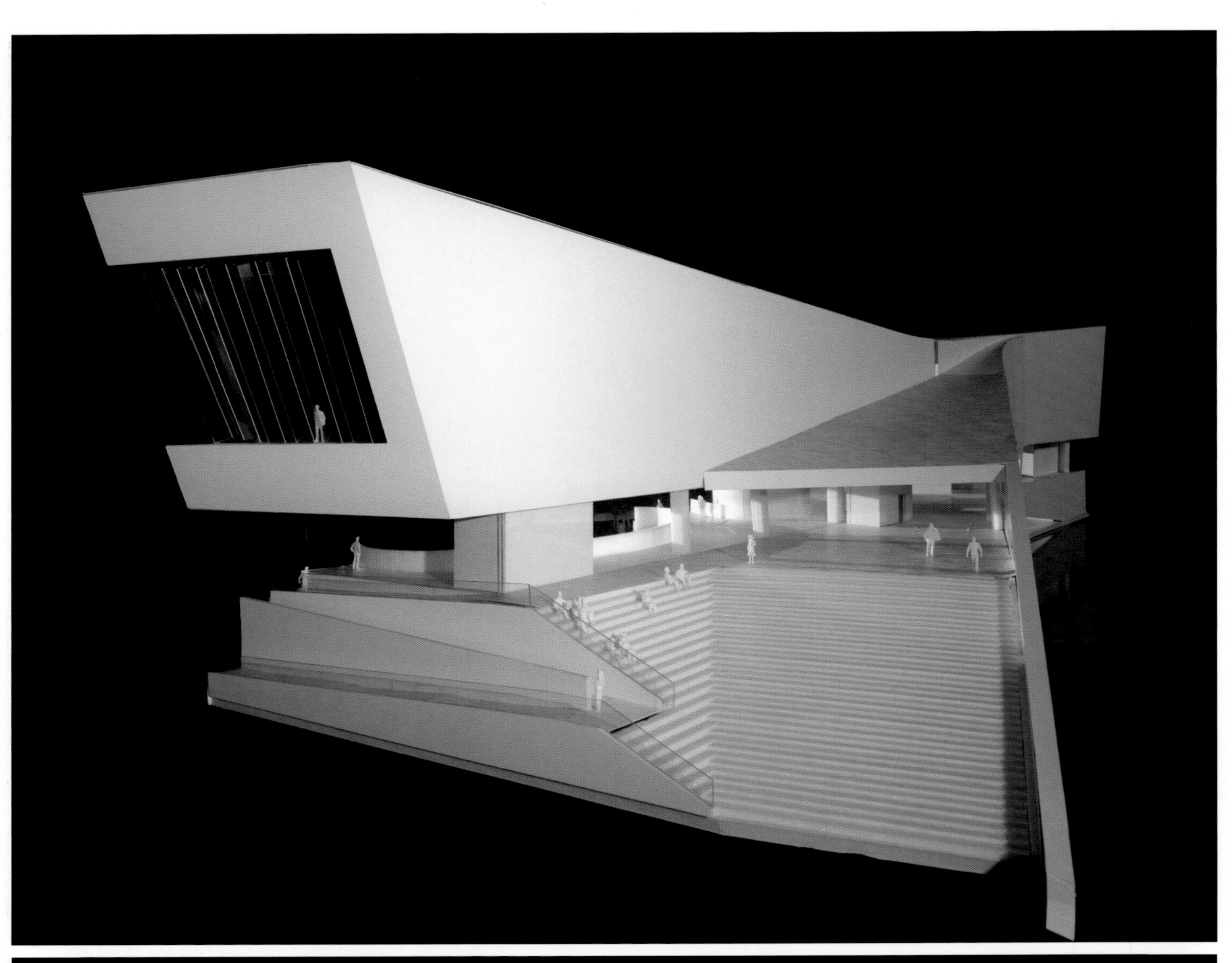

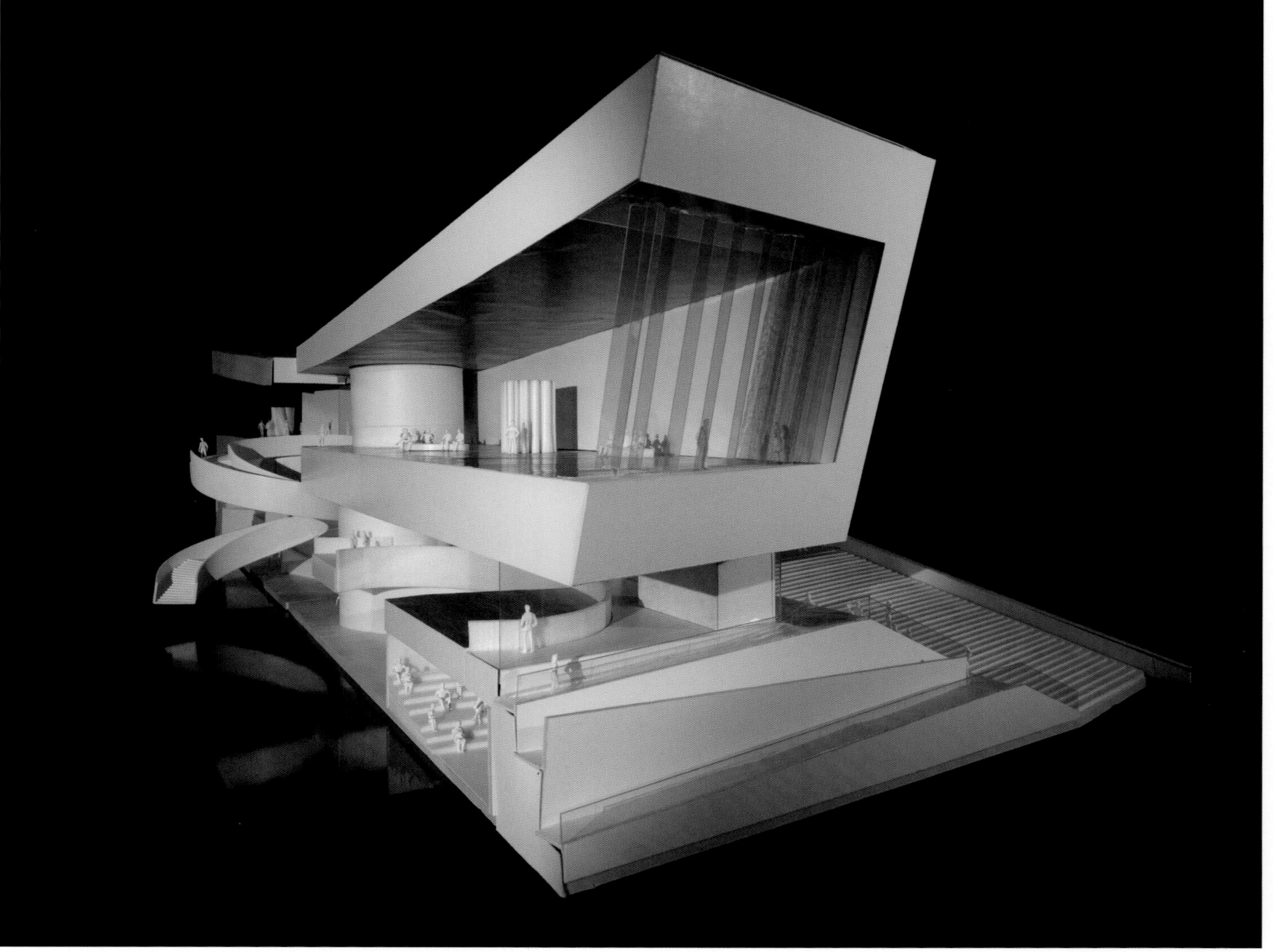

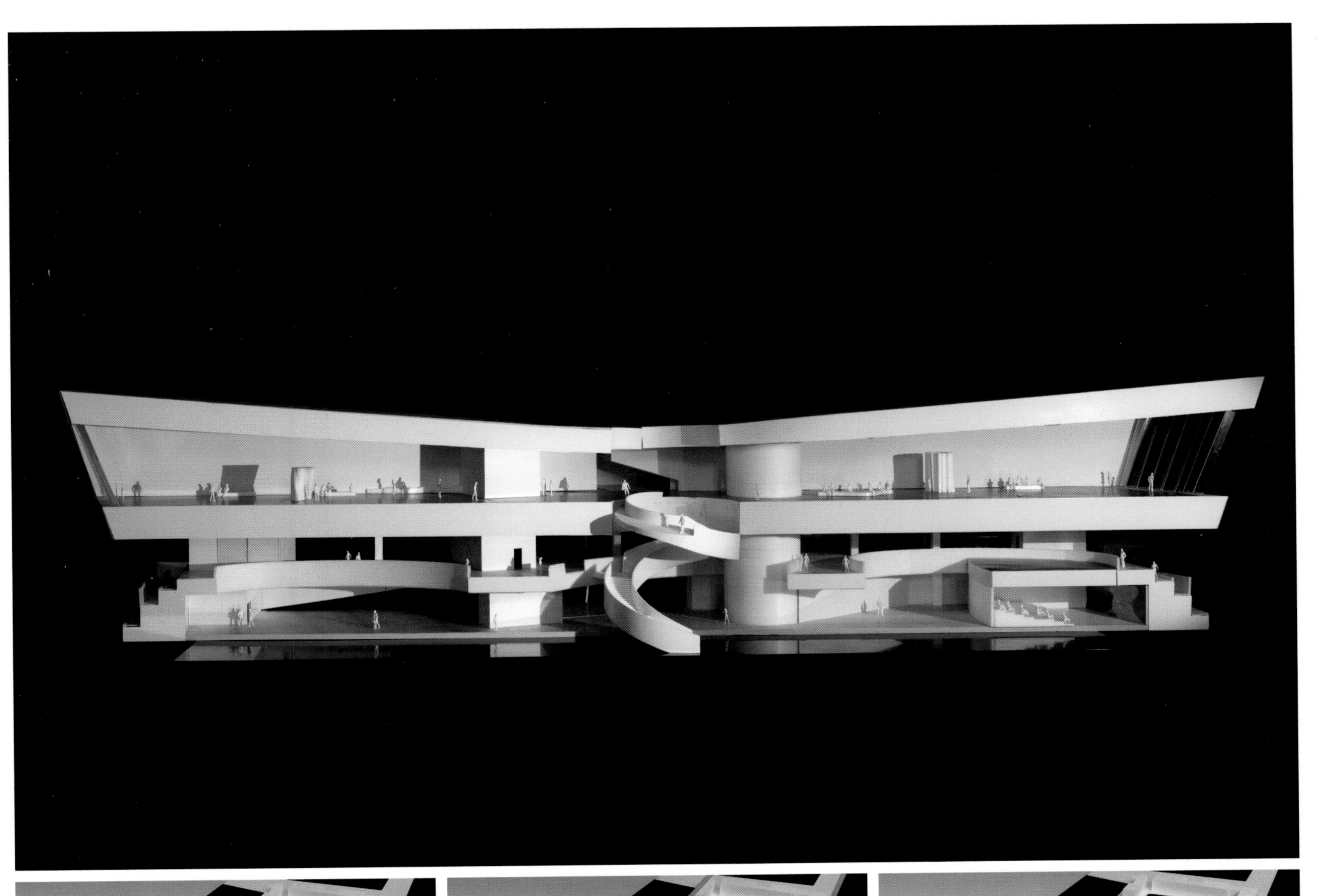

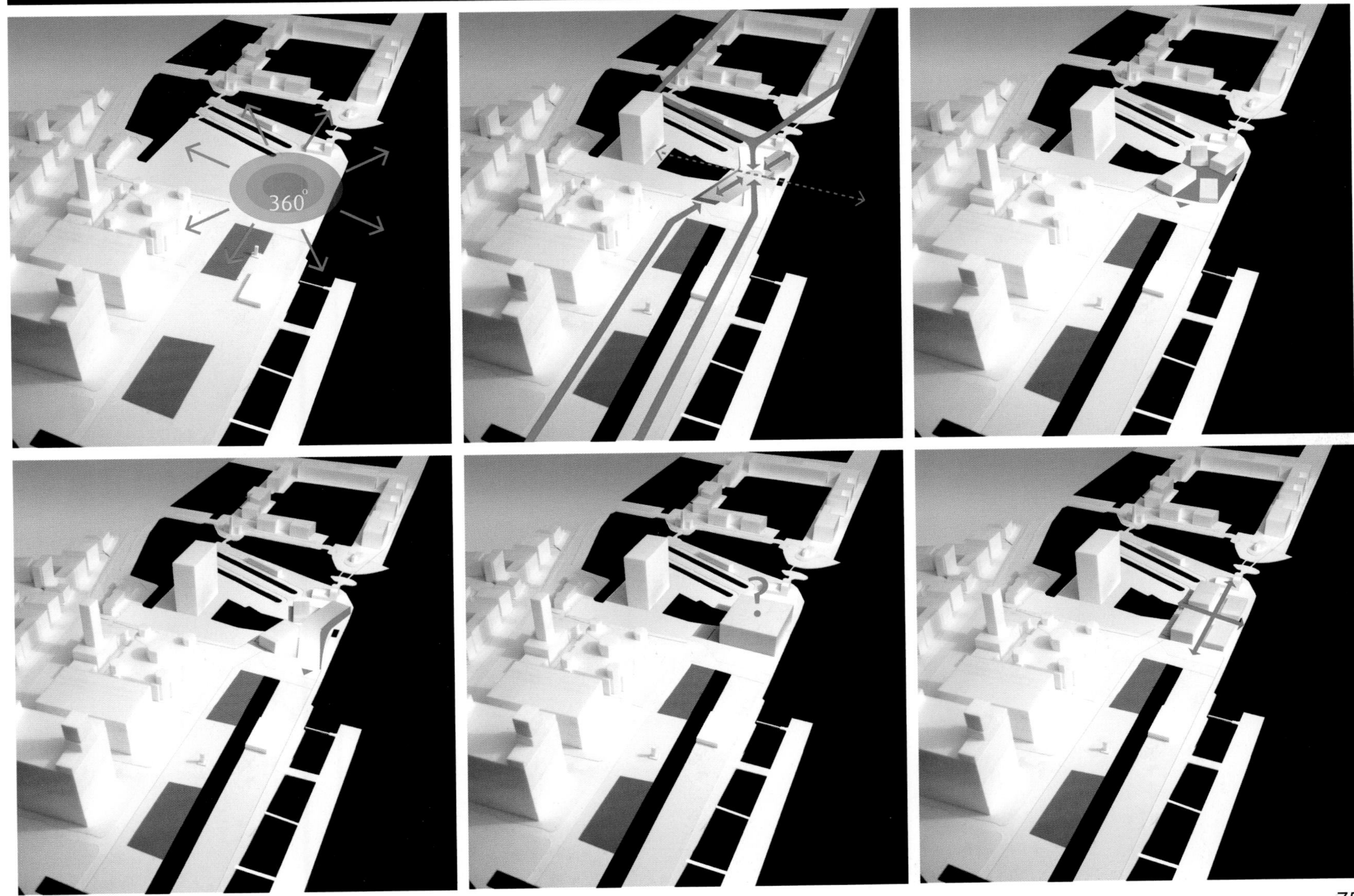
360°

THE
PUMPHOUSE
THE
PUMPHOUSE

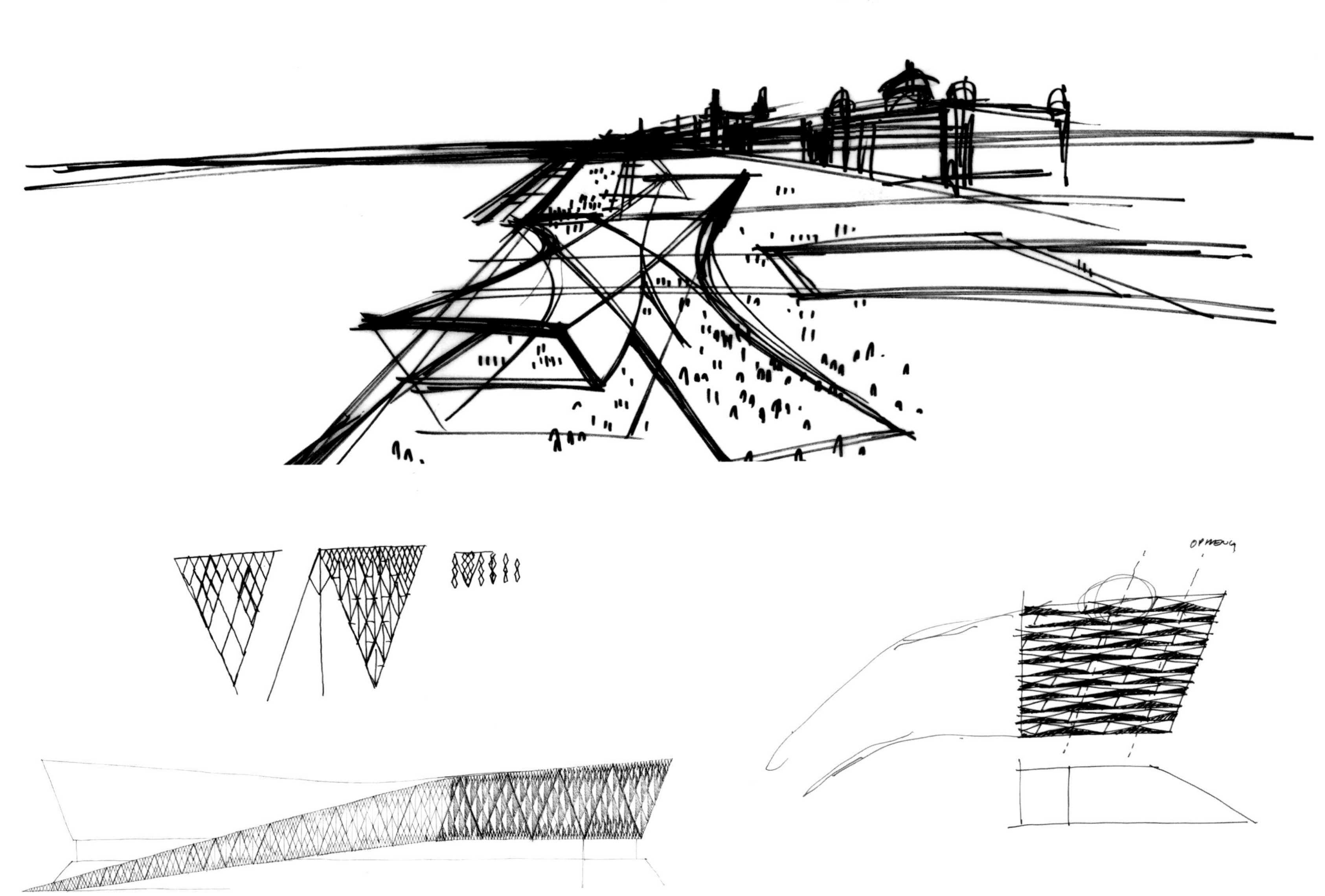

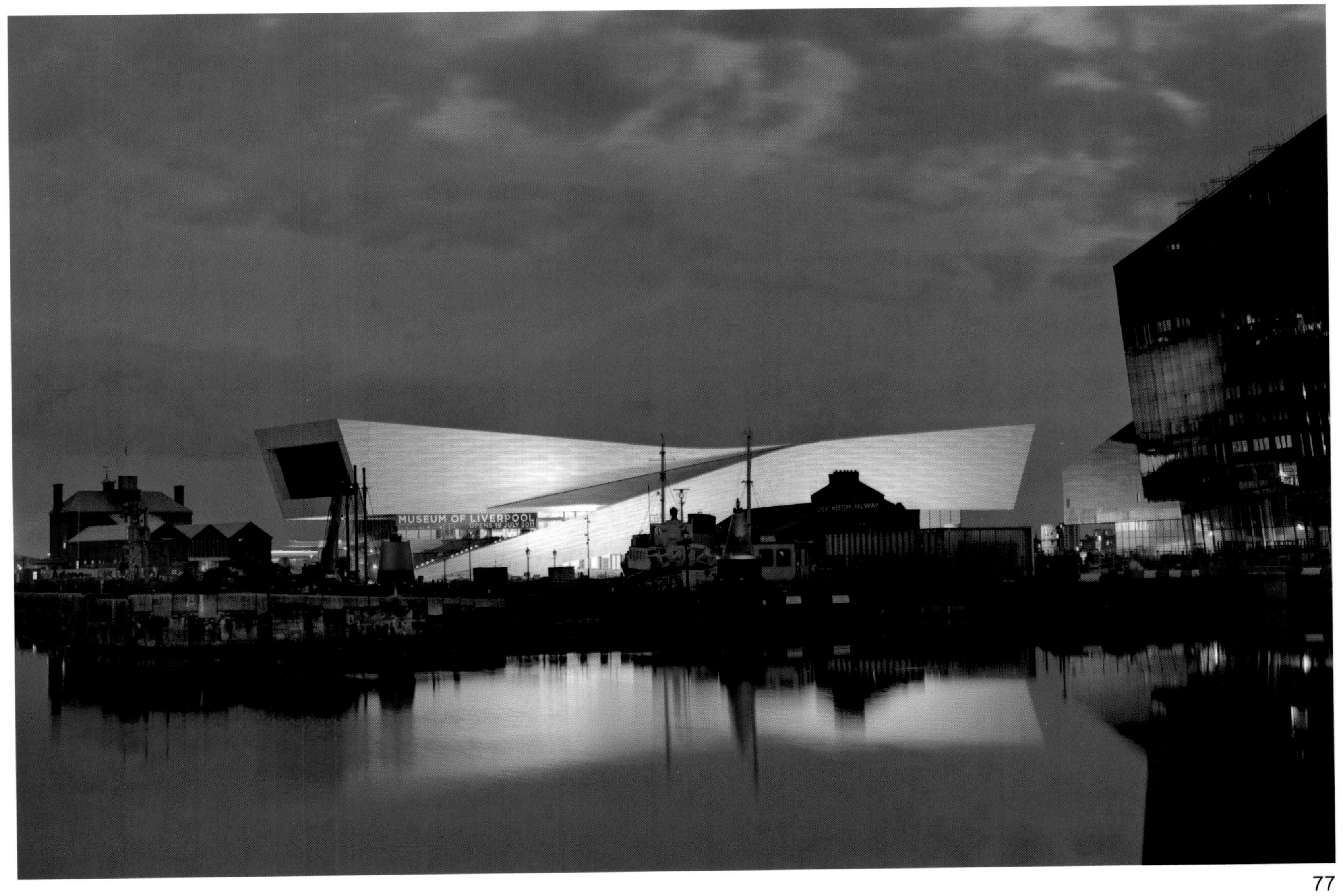
MUSEUM OF LIVERPOOL

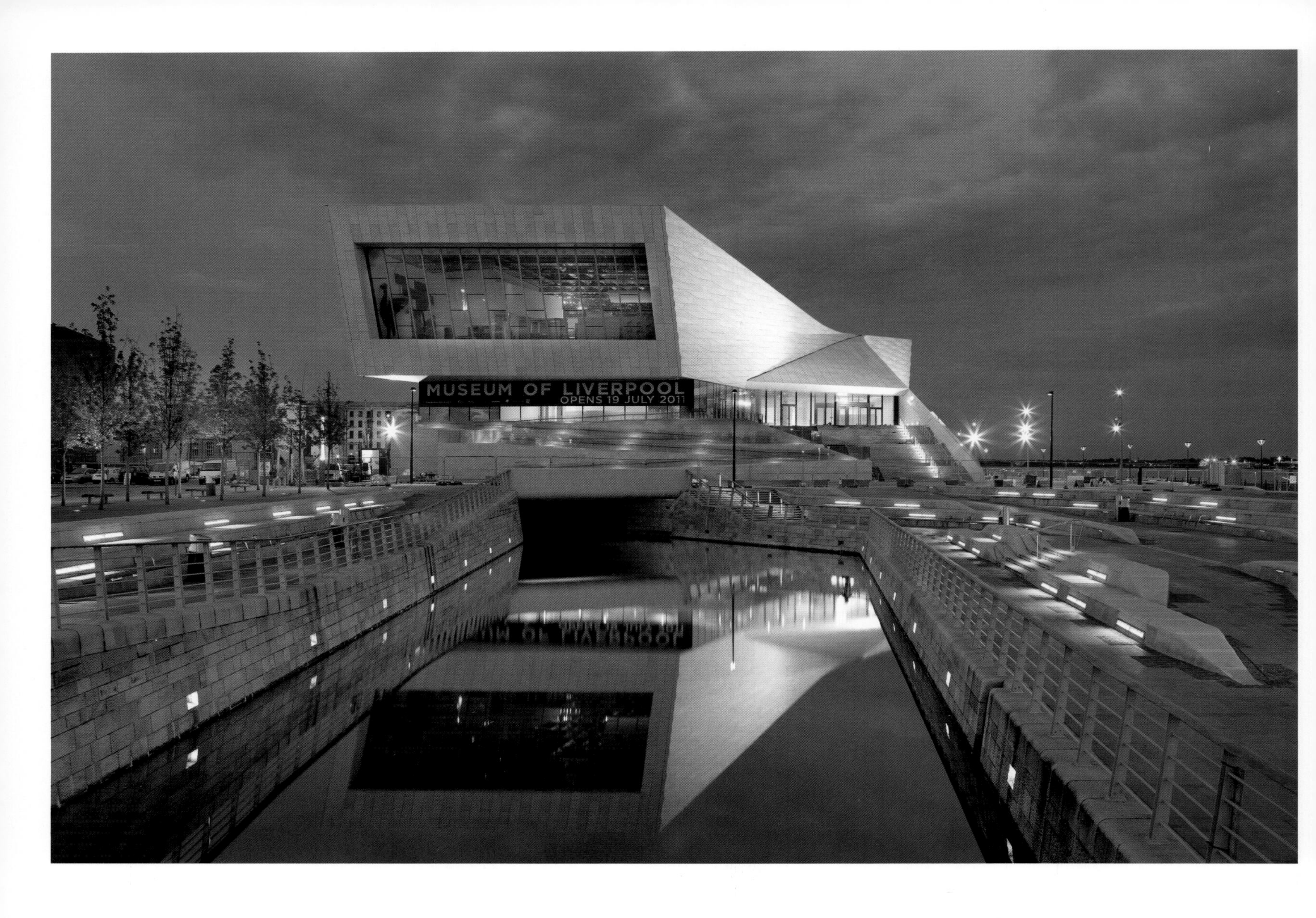

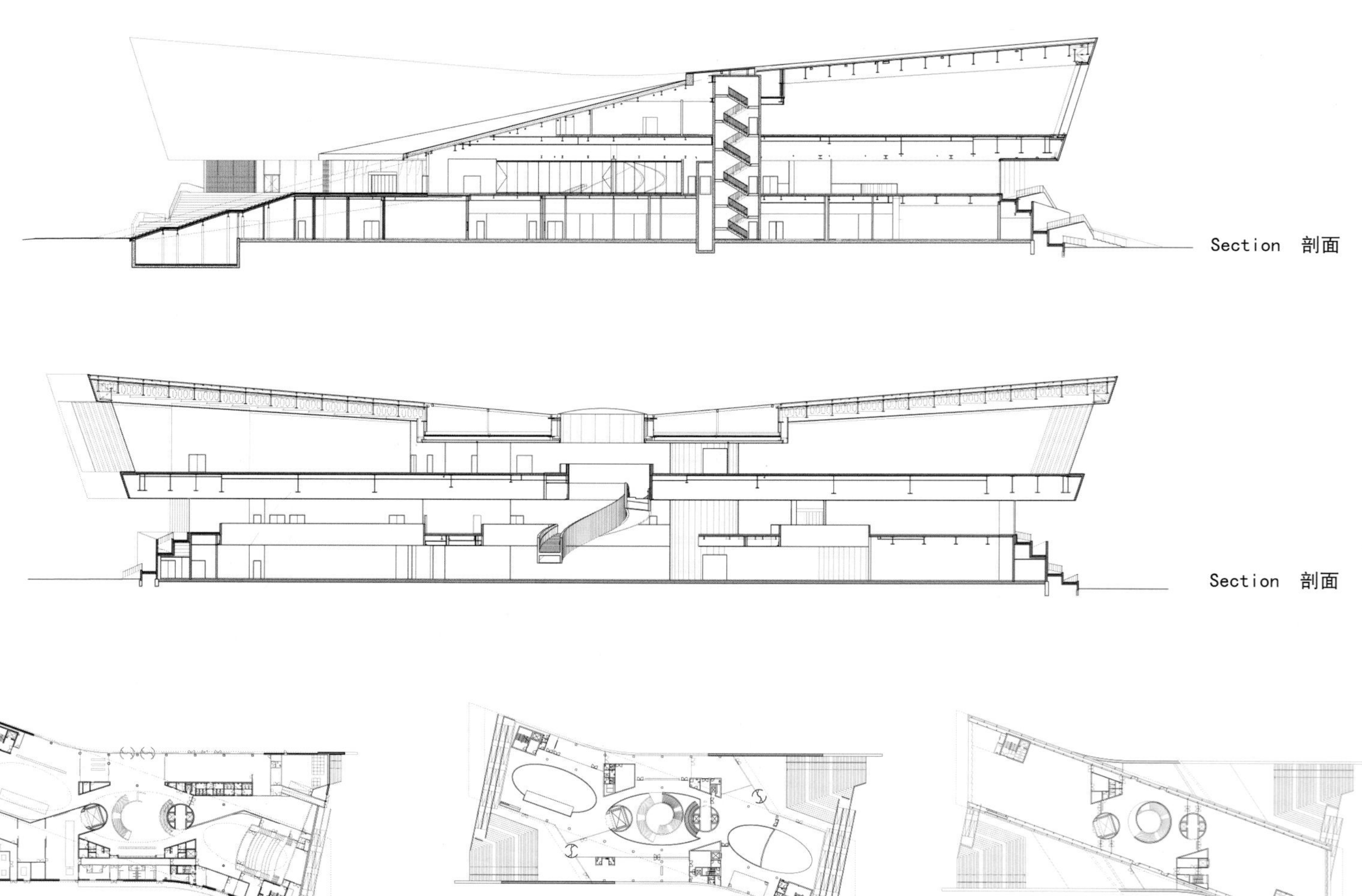

Section 剖面

Section 剖面

Plan Level 01 一层平面

Plan Level 02 二层平面

Plan Level 04 四层平面

MUSEUM OF LIVERPOOL
OPENS 19 JULY 2011

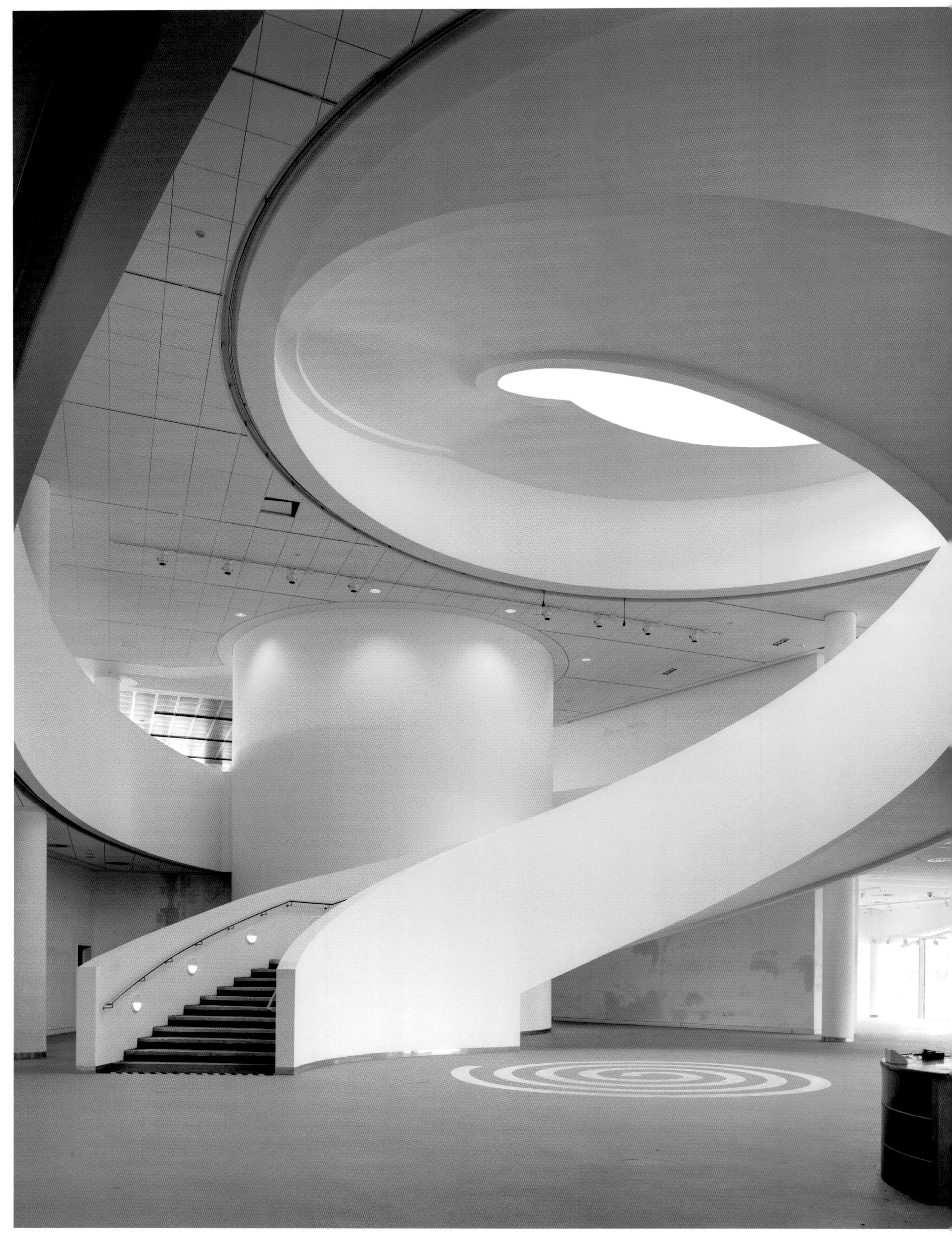

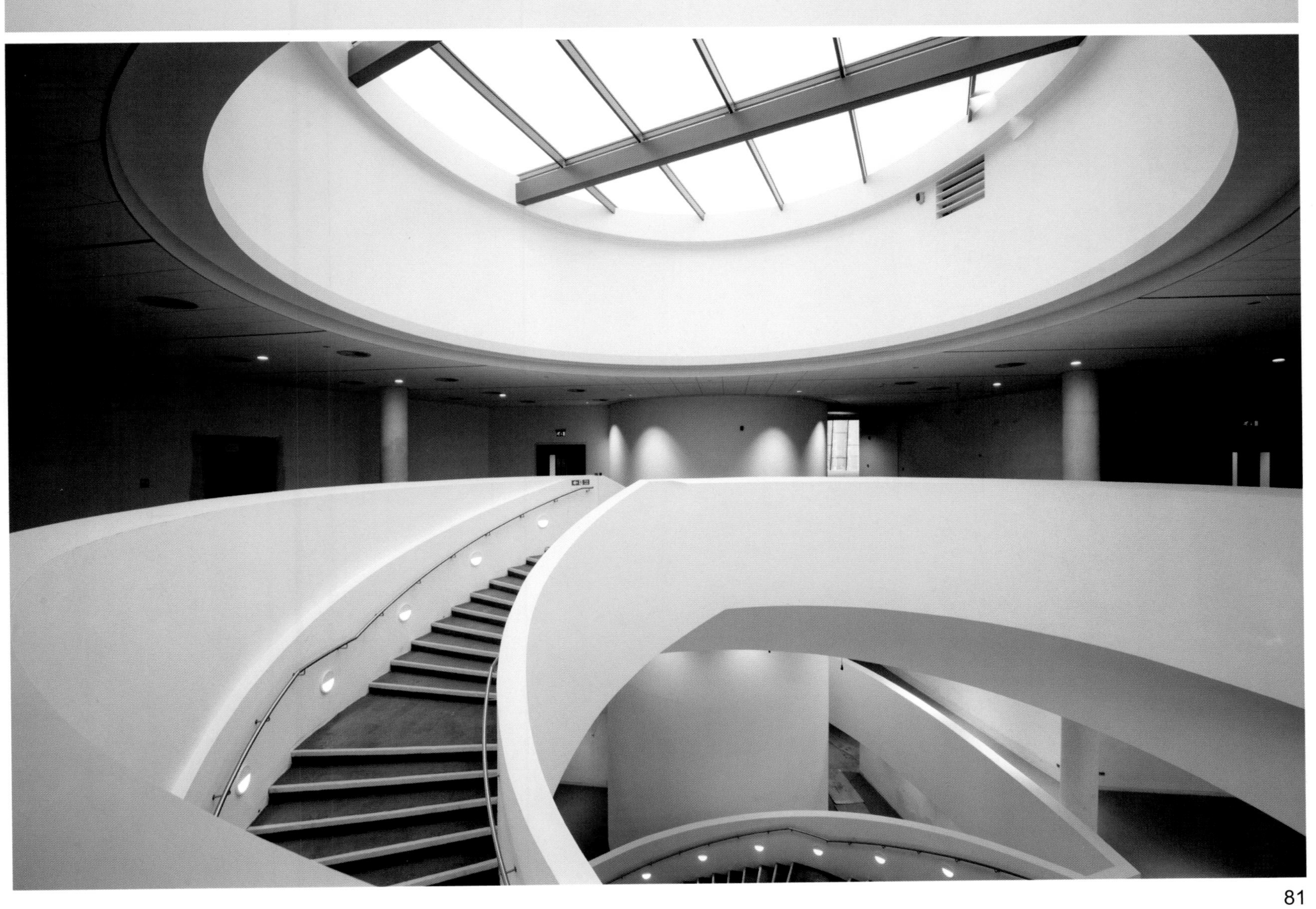

Museum of Art of T. Kantor

T. Kantor艺术博物馆

nsMoon Studio nsMoon工作室

Location: Cracow, Poland
Typology: Museum
Area: 3,852.26 m^2
Status: Planning construction 2012
Image Courtesy: nsMoon Studio

区位：波兰，克拉科
类型：博物馆
面积：3 852.26 m^2
状态：计划2012年施工
图片提供：nsMoon工作室

The Center for the Documentation of the Art of Tadeusz Kantor "CRICOTEKA", the symbol of artistic search for the individual path to creativity, crossing the border between the actor and the audience, between the creator and the recipient, engages everyone in an activity - game - collective play. The municipal public space constitutes both the stage and the audience, it is a venue for constant performance. Its value lies in effacing the borders between the inside and the outside, the space "flows", it is shared. The creators, artists, residents and visitors all take part in the process and activity of creating space which can be shaped with selected objects, structures, means and methods. The center of Kantor's art should promote this message.

According to Kantor's creative visions, the museum building need not be integrated with the surroundings and the existing building does not fulfil its nature through identification with its original function, but through a conscious contrast of form and content they create grounds for conflict and "clash", thus marking their presence in the hitherto indifferent surroundings.

The purpose of creative activity is not to identify the structure as a form but rather the way it interacts with the environments.

Abandoning the idea of developing the entire parcel ground and relieving it for purposes such as performances, exhibitions, shows, happenings in the Vistula riverside "theater" extends the range of influence of the form and its content to the entire urban scale of the city.Transforming the characteristic image of a man carrying a table on his back is a symbol of the "clash" between the matter, the object and a human being who - through his creative effort - invents a new form, endows the new "piece" with meaning and constitutes feedback originating from the combination of the idea of a creator and the choice of "material".The individual zones of the new building do not create a usual functional order but rather delineate the hierarchy of significance of particular programme elements.

塔丢斯·坎特"CRICOTEKA"艺术创造文件材料中心，是个人创造力不断进行艺术性探索的象征，该建筑跨越了演员和观众、创造者和接受者之间的界限，将每个人都吸引到一项活动——游戏——集体表演中去。城市公共空间由舞台和观众构成，这是经常性表演的演出场馆。其价值在于抹去了内外部之间的界限，空间是"流动"的，是共享的。创作者、艺术家、居民和游客都参加到创造空间的过程中去，可以选择不同的对象、结构、手段和方法来创造空间。坎特艺术中心应该推广这个讯息。

根据坎特创意的视角，博物馆的建筑不需要与周围融合，已有的建筑也没有通过原有功能认同来履行其性质，但是通过有意识地对比形式和内容，它们创造了文化冲突与"碰撞"的空间，这让它们至今与周边的环境仍处于漠视的关系中。

创造性活动的目的不是以某种形式来识别结构，而是其与环境的互动方式。

我们放弃了让整个地区发展的构想，为了表演、展览、展示而放弃了一部分创意。维斯瓦河河畔"剧场"表演在形式和内容上的影响力已经达到整个城市地区。转变背桌子的男人这一典型意向，是事件、客体及人类三者相互冲突的象征。这里所说的人类，通过创造性的努力，发明了新形式，赋予新事物以意义，从创造者思想与"材料"选择的结合之中升华出成果。新建筑的个人空间不是采用的常见的功能顺序，而是按重要性划定了各功能元素的层次。

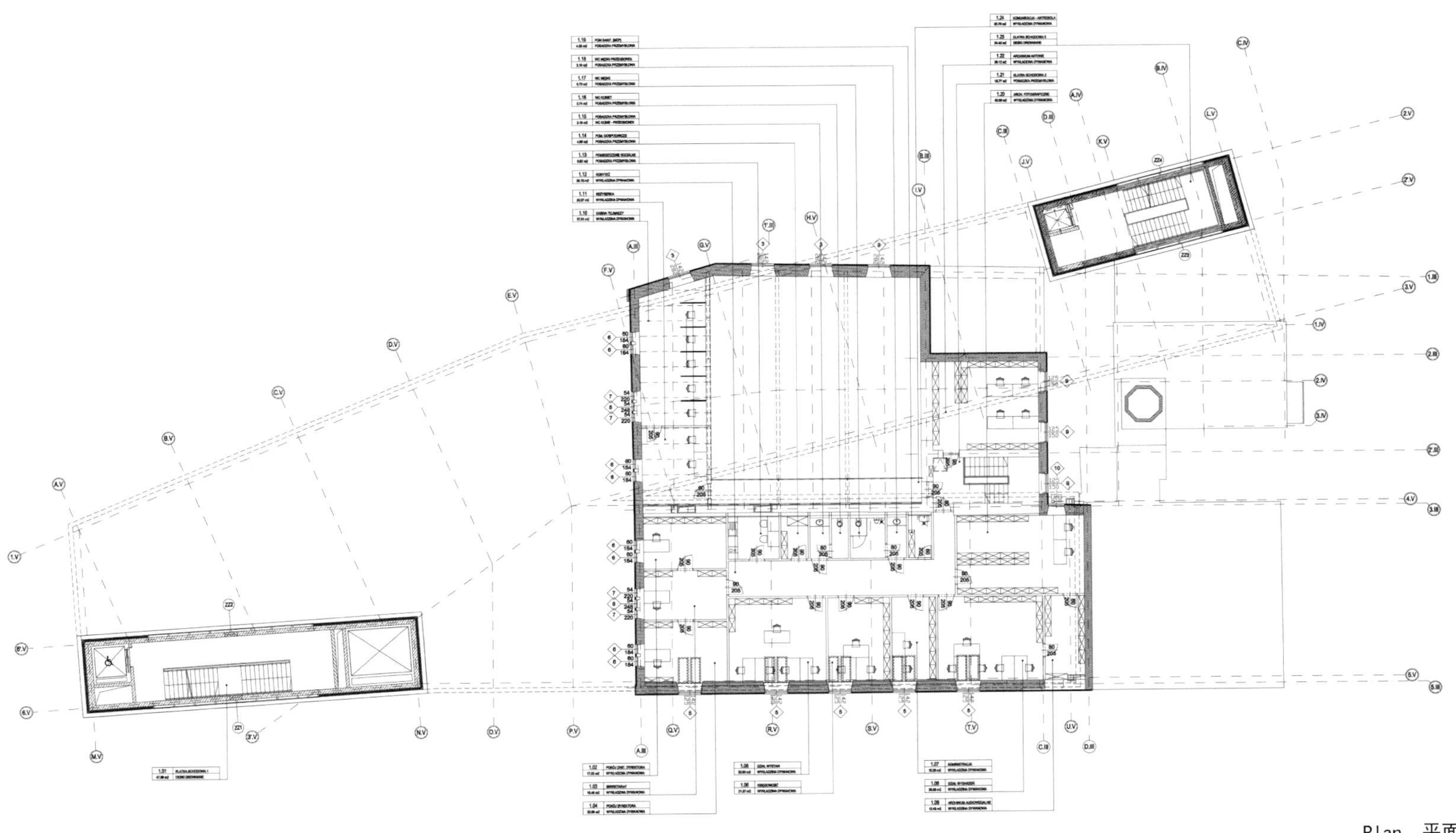

Plan 平面

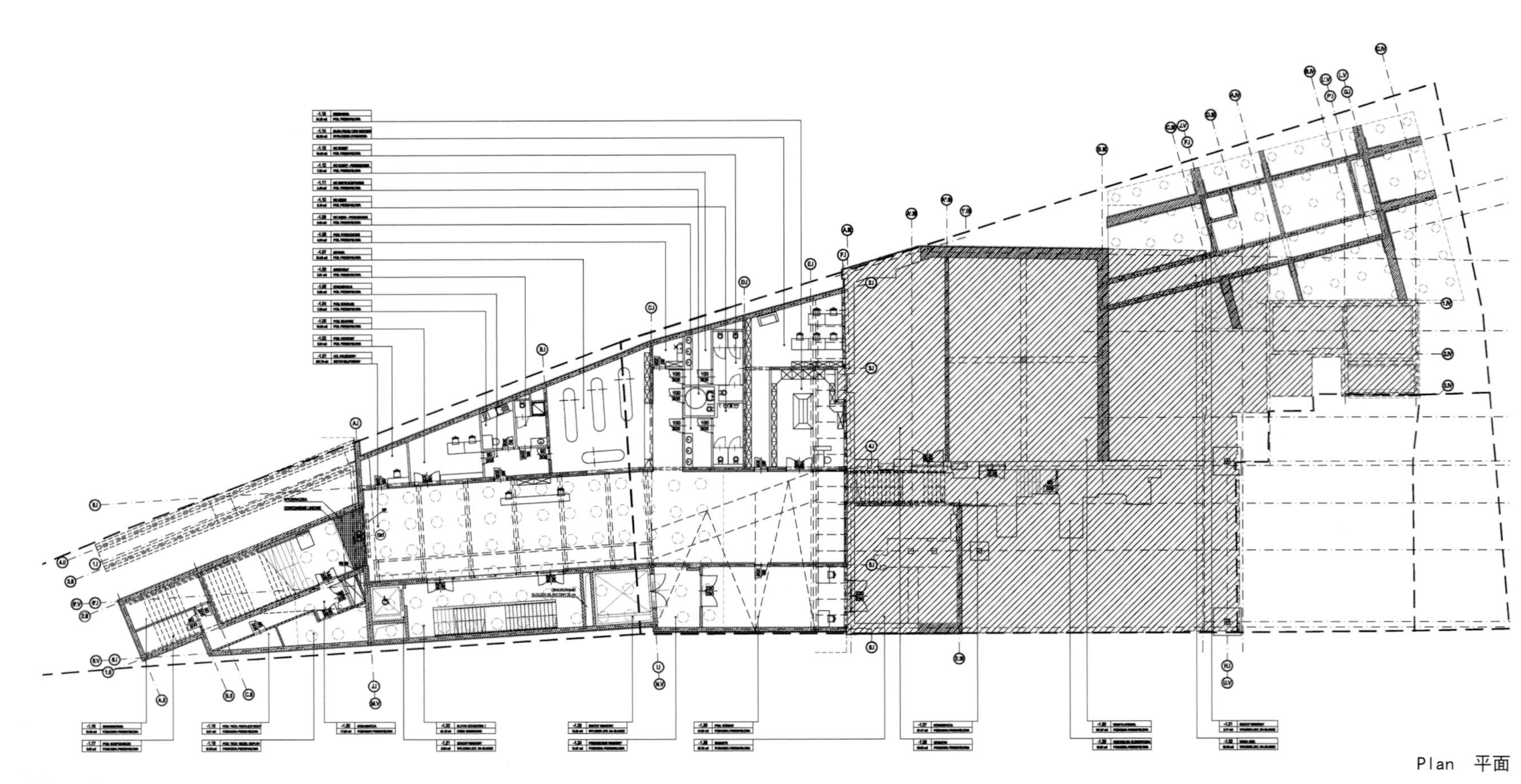

Plan 平面

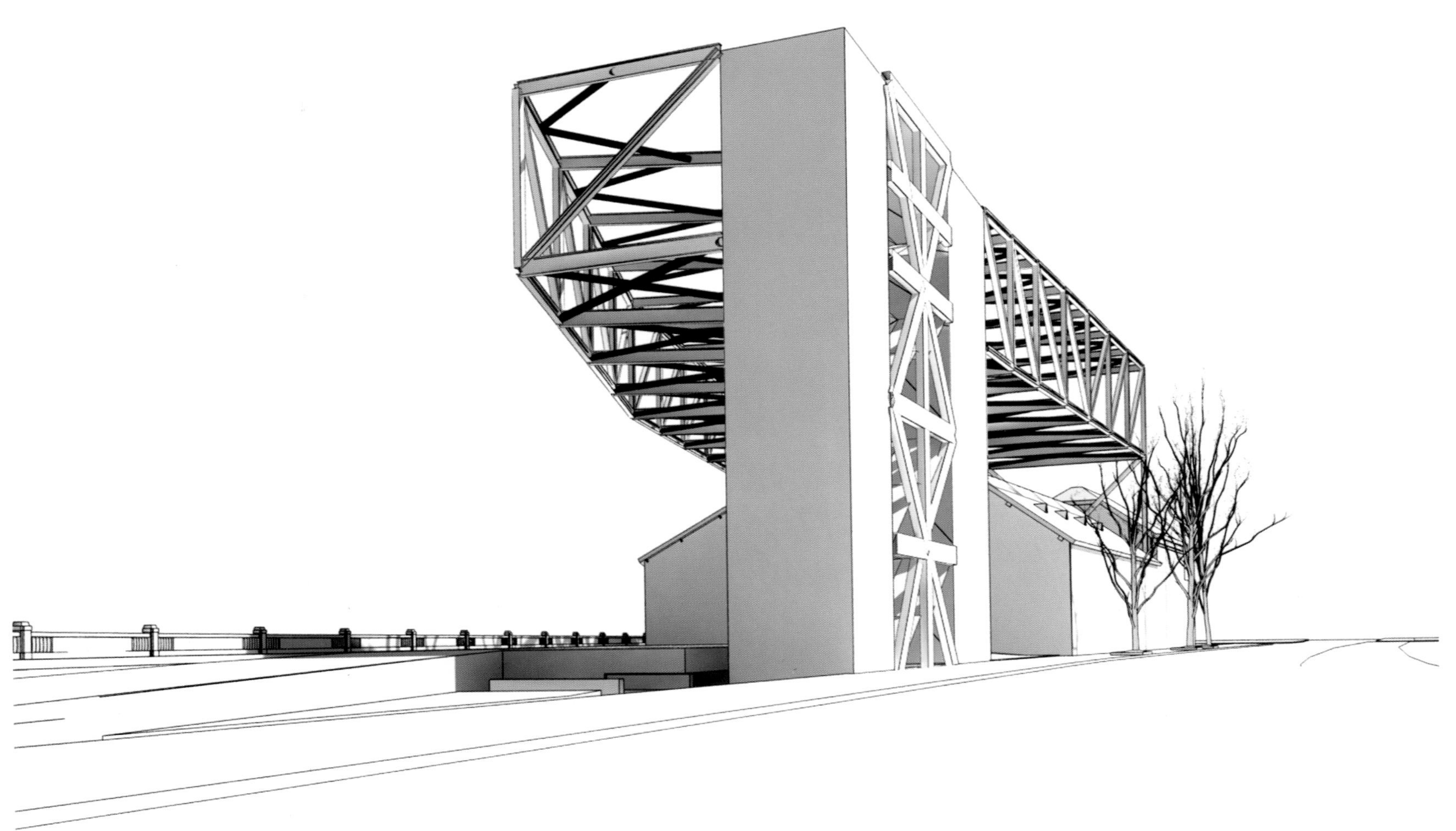

Vistula

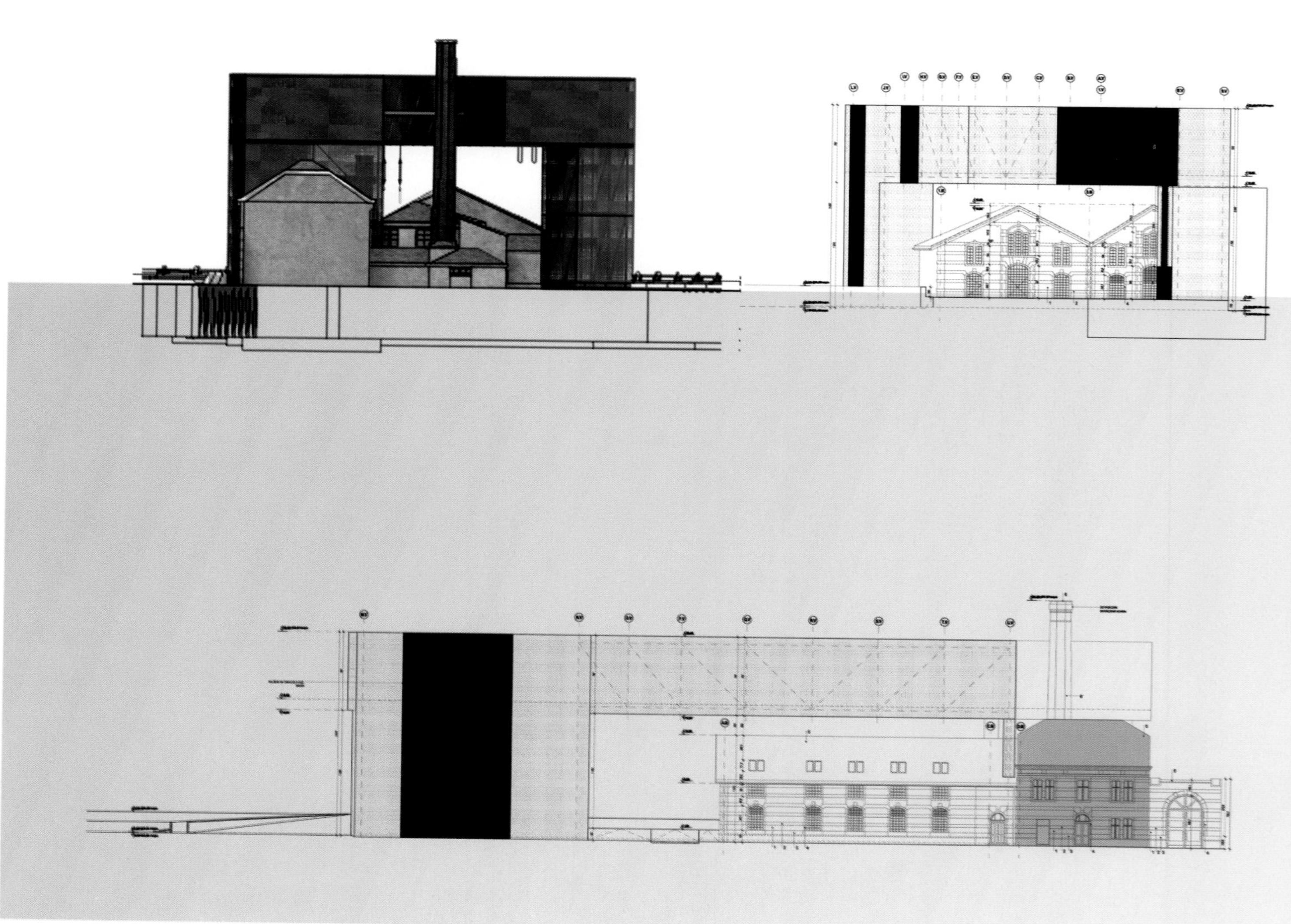

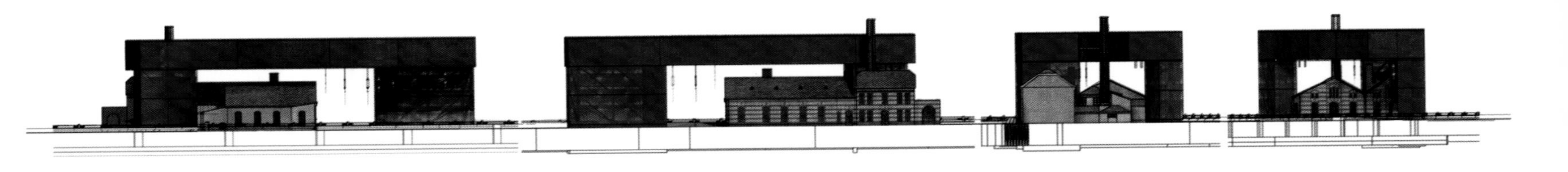

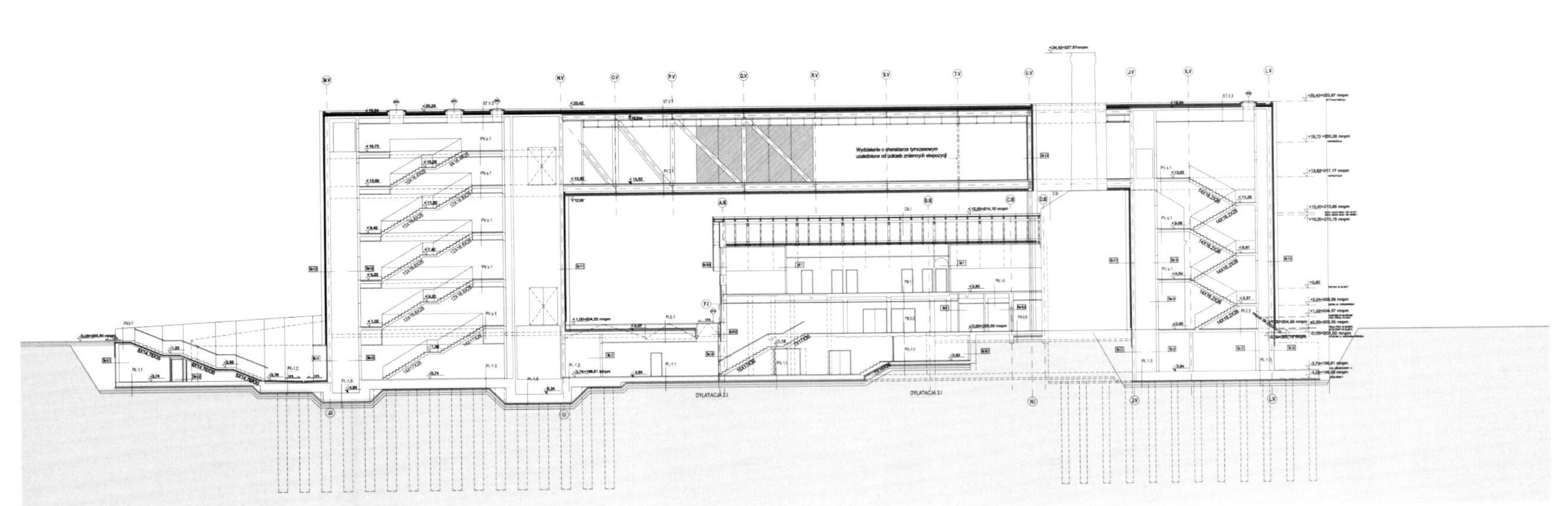

Section 剖面

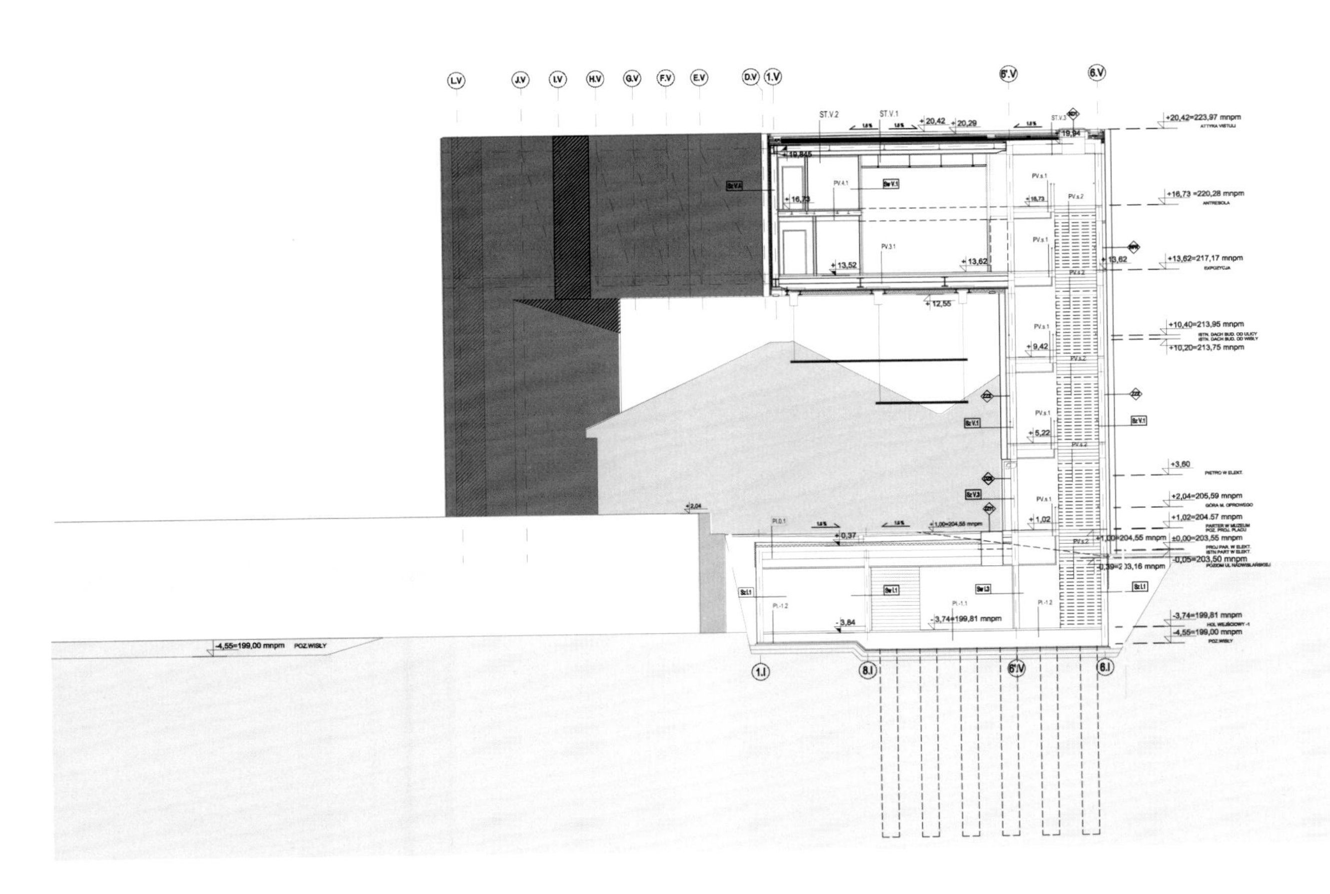

Section 剖面

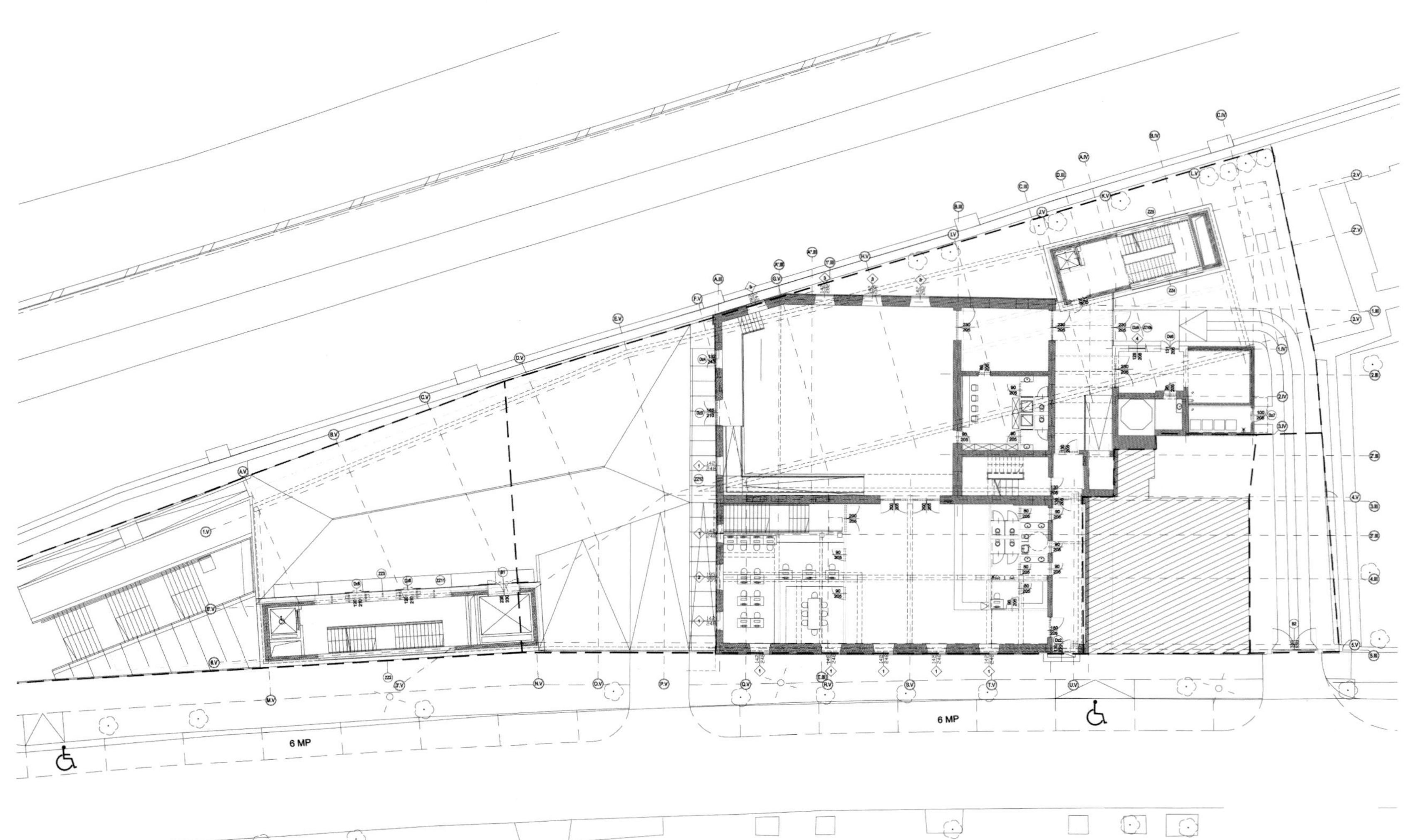

Plan 平面

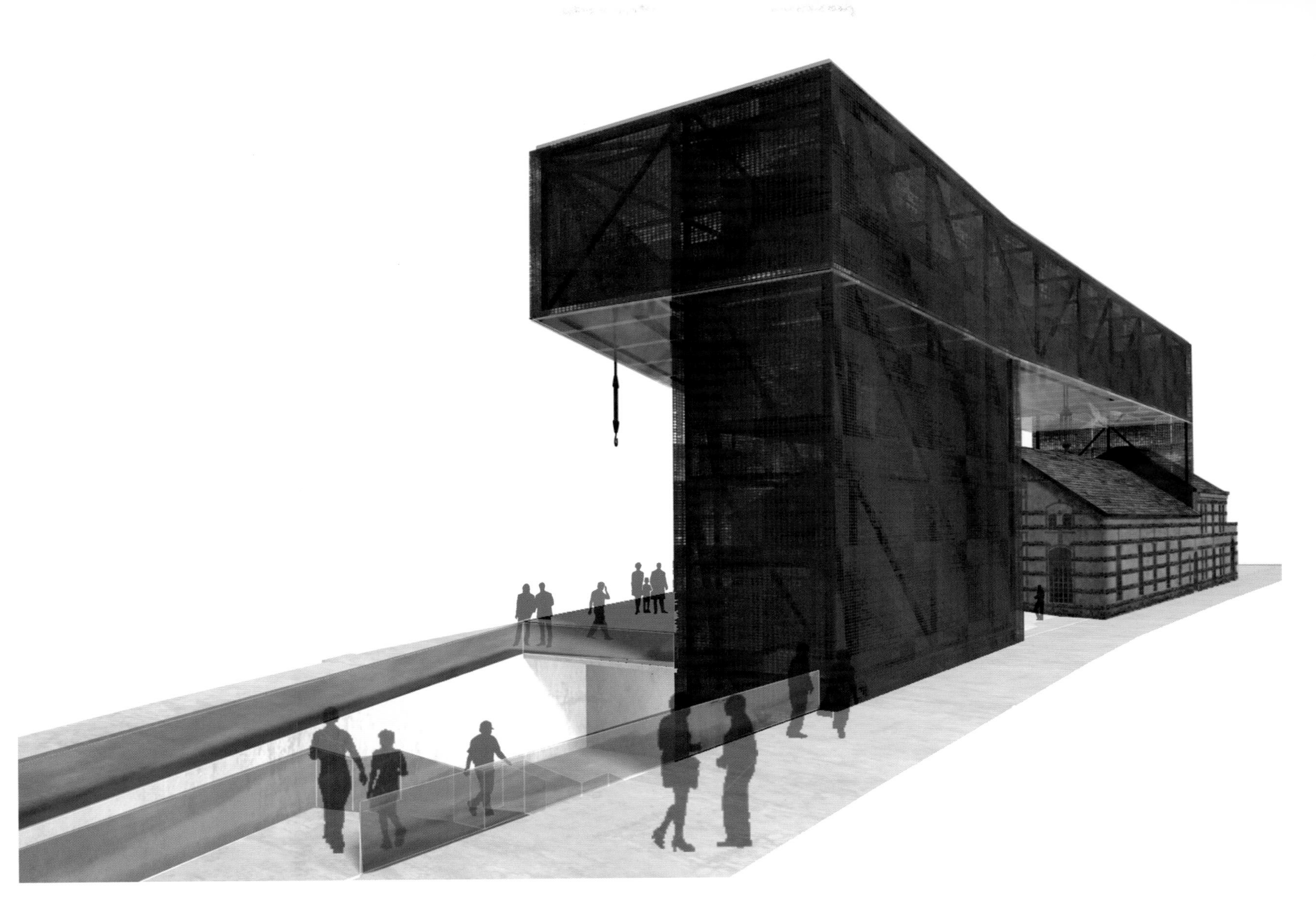

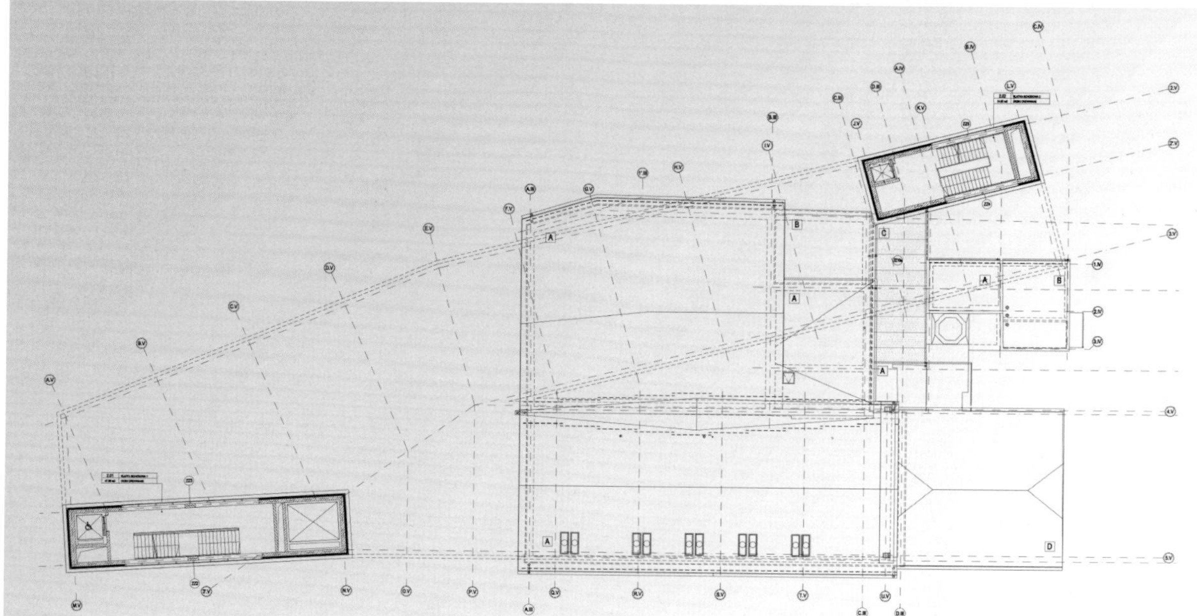

Plan 平面

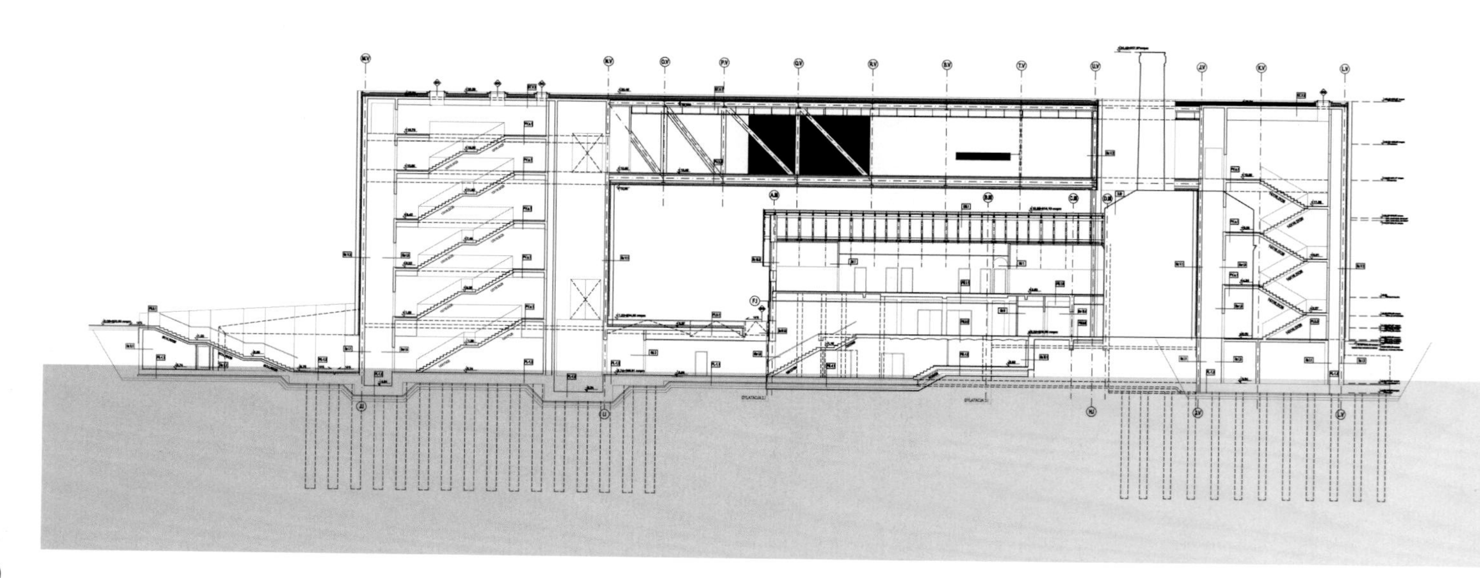

Section 剖面

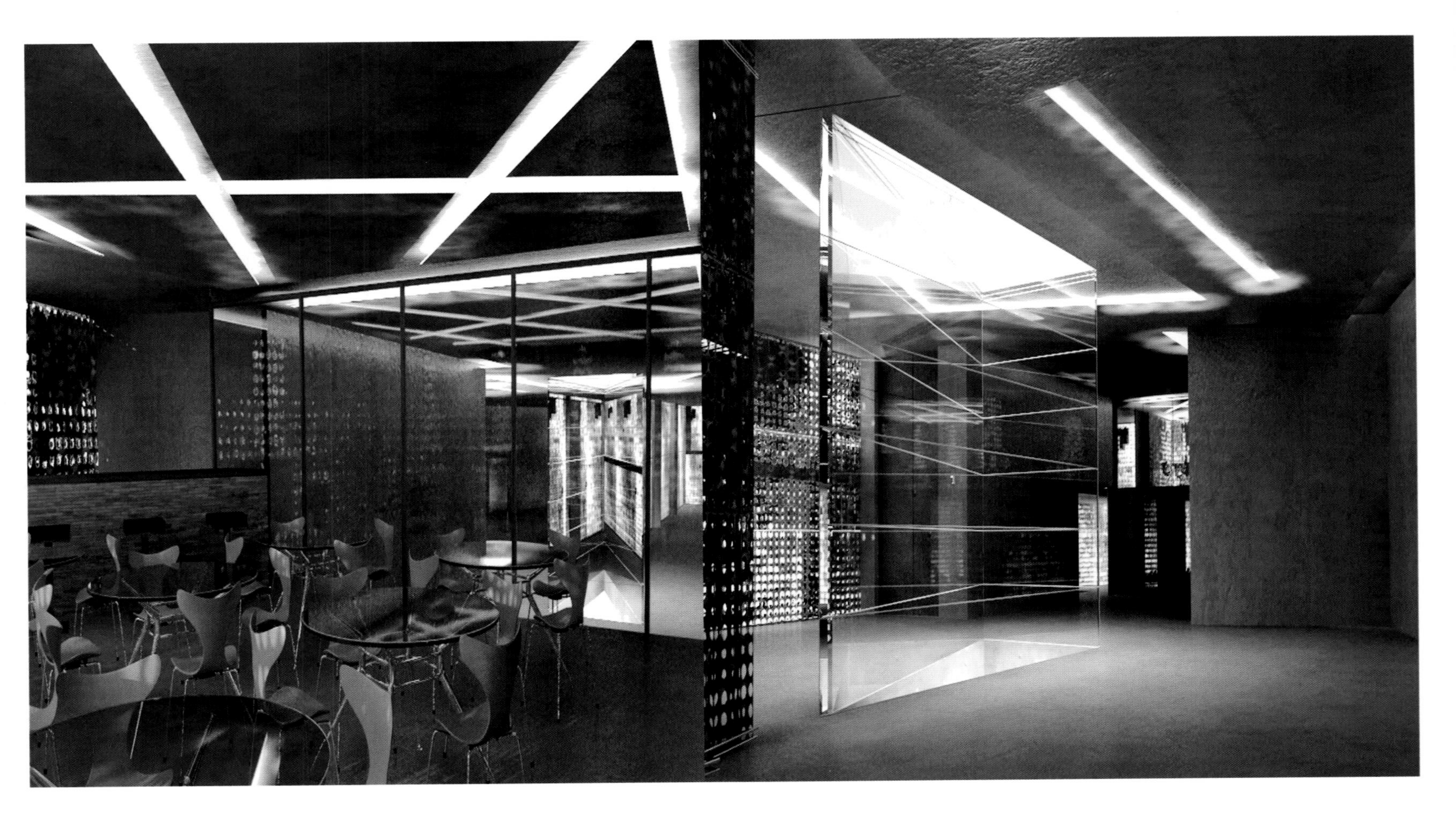

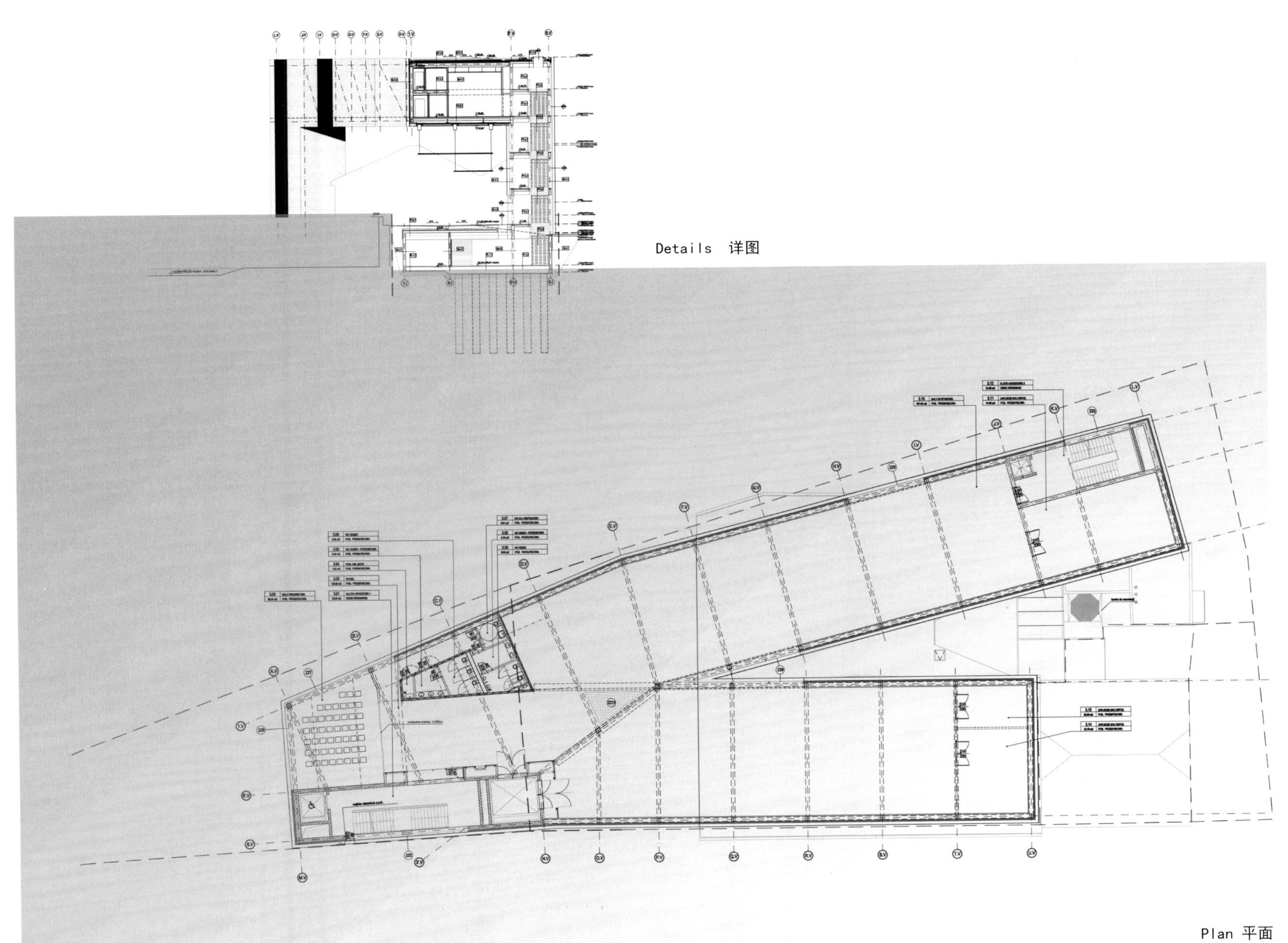

Details 详图

Plan 平面

Tel Aviv Museum of Art

特拉维夫市艺术博物馆

Preston Scott Cohen, Inc. 普雷斯顿·斯科特·科恩公司

Location: Tel Aviv, Israel
Typology: Museum
Image Courtesy: Amit Geron courtesy Tel Aviv Museum of Art.

区位：以色列，特拉维夫市
类型：博物馆
图片提供：Amit Geron，特拉维夫市艺术博物馆

The design for the Amir Building arises directly from the challenge of providing several floors of large, neutral, rectangular galleries within a tight, idiosyncratic, triangular site. The solution is to "square the triangle" by constructing the levels on different axes, which deviate significantly from floor to floor. In essence, the building's levels–three above grade and two below–are structurally independent plans stacked one on top of the other.

These levels are unified by the "Lightfall": an 87-foot-high, spiraling, top-lit atrium, whose form is defined by subtly twisting surfaces that curve and veer up and down through the building. The complex geometry of the Lightfall's surfaces (hyperbolic parabolas) connects the disparate angles of the galleries; the stairs and ramped promenades along them serve as the surprising, continually unfolding vertical circulation system; while the natural light from above is refracted into the deepest recesses of the half-buried building. Cantilevers accommodate the discrepancies between plans and provide overhangs at the perimeter.

In this way, the Amir Bulding combines two seemingly irreconcilable paradigms of the contemporary art museum: the museum of neutral white boxes, which provides optimal, flexible space for the exhibition of art, and the museum of spectacle, which moves visitors and offers a remarkable social experience. The Amir Building's synthesis of radical and conventional geometries produces a new type of museum experience, one that is as rooted in the Baroque as it is in the Modern.

Conceptually, the Amir Building is related to the Museum's Brutalist main building (completed 1971; Dan Eytan, architect). At the same time, it also relates to the larger tradition of Modern architecture in Tel Aviv, as seen in the multiple vocabularies of Mendelsohn, the Bauhaus and the White City. The gleaming white parabolas of the façade are composed of 465 differently shaped flat panels made of pre-cast reinforced concrete. Achieving a combination of form and material that is unprecedented in the city, the façade translates Tel Aviv's existing Modernism into a contemporary and progressive architectural language.

是否能够在一个紧凑、异质的三角形基地上建造出大型、结构中立、矩形的多层博物馆？这是一项挑战。而埃米尔建筑设计方案是为了迎合此项挑战而产生的。解决的方案是通过构造不同层次的轴心“使三角成为矩形”，因此每一层之间都有明显的偏离。在本质上，该建筑物的层高——地上三层地下两层——一层和紧挨着它的另一层之间在结构上都是相互独立的。

这些层次之间是由“垂光”联结起来的：一个 87英尺高、螺旋形、顶部有照明的中庭，整个建筑表面通体有精细的旋转花纹。“垂光”表面的复杂几何形状（双曲抛物面）把艺术馆从各个不同的角度联结起来；当从顶端照射进来的自然光线折射进位于这幢半埋入式建筑的最深处的休息室时，沿着“垂光”安装的楼梯以及倾斜的散步长廊构成了一个令人惊叹的持续展开的垂直循环系统；悬臂调节了规划间的差异，同时在外围提供了悬臂结构。

用这种方法，埃米尔建筑融合了两种当代艺术博物馆看似无法调和的范例：中立的白色的博物馆盒子便于提供最佳的、灵活的空间来履行艺术展览和公开展示的功能，而博物馆壮观的场面容易触动参观者并为其提供一次非凡的社交经历。埃米尔建筑综合了激进的和传统的几何学，创造出一类既扎根于巴洛克派风格也基于现代派风格的新型的博物馆经验。

从概念上来说，埃米尔建筑是与野兽派博物馆的主体建筑(完工于1971年；由建筑师丹·叶檀建造)存在联系。同时，它也与位于特拉维夫市的现代建筑的惯例息息相关，这一点可以在门德尔松多义词汇，包豪斯建筑学派以及白城见到端倪。闪闪发亮的白色抛物线外观是由465 个钢筋混凝土质地的不同形状的扁平面板构成。无论是在形状上还是在材料上，都是这座城市中前所未有的。它的外观将特拉维夫市现存的现代主义转变为与时俱进的建筑语言。

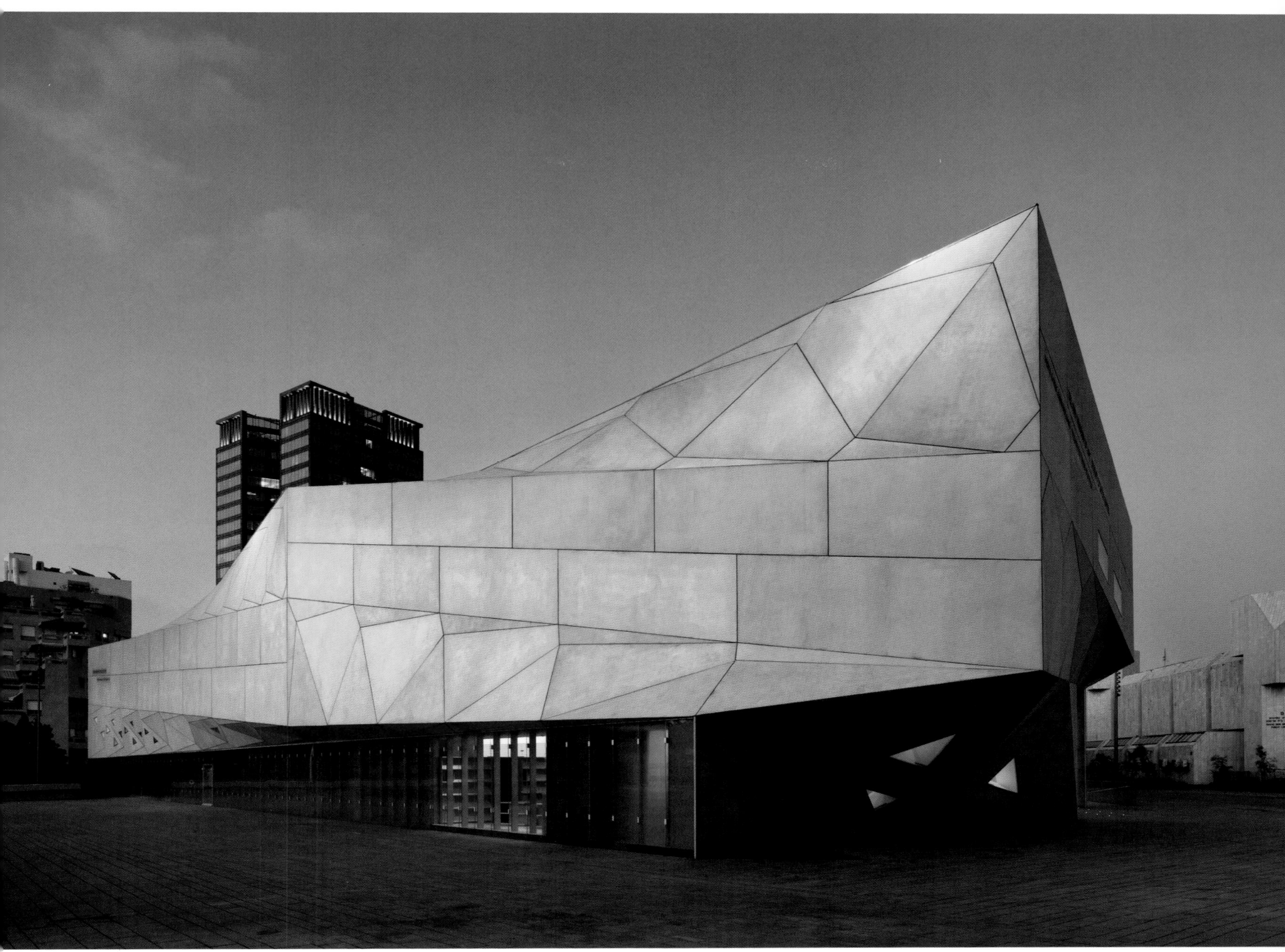

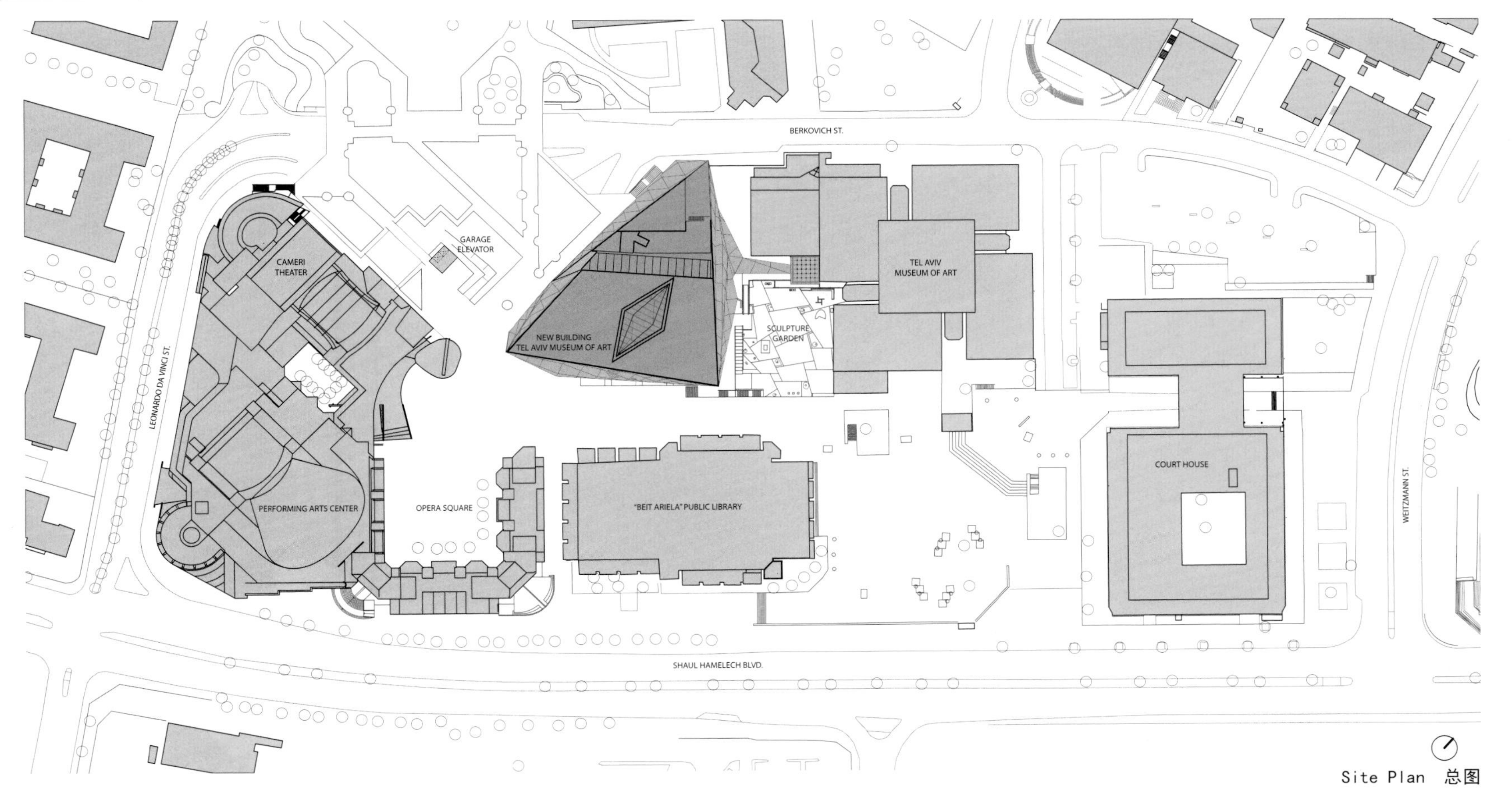
BERKOVICH ST.
GARAGE
ELEVATOR
CAMERI
THEATER
NEW BUILDING
TEL AVIV MUSEUM OF ART
TEL AVIV
MUSEUM OF ART
SCULPTURE
GARDEN
LEONARDO DA VINCI ST.
PERFORMING ARTS CENTER
OPERA SQUARE
"BEIT ARIELA" PUBLIC LIBRARY
COURT HOUSE
WEITZMANN ST.
SHAUL HAMELECH BLVD.

Site Plan 总图

TEL AVIV MUSEUM OF ART
מוזיאון תל אביב לאמנות
HERTA & PAUL AMIR BUILDING
הבניין ע"ש שמואל והרטה עמיר

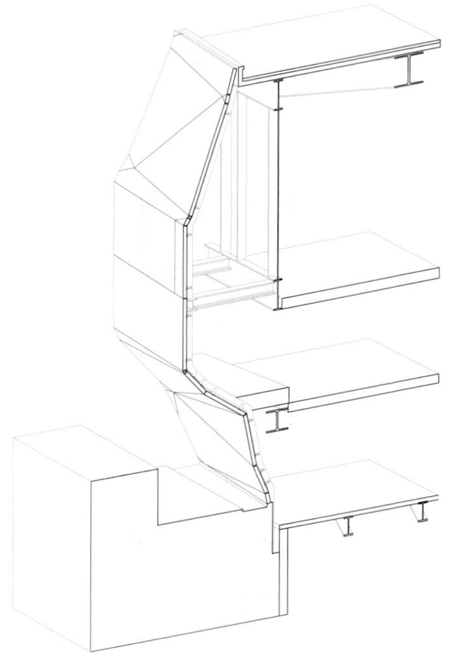

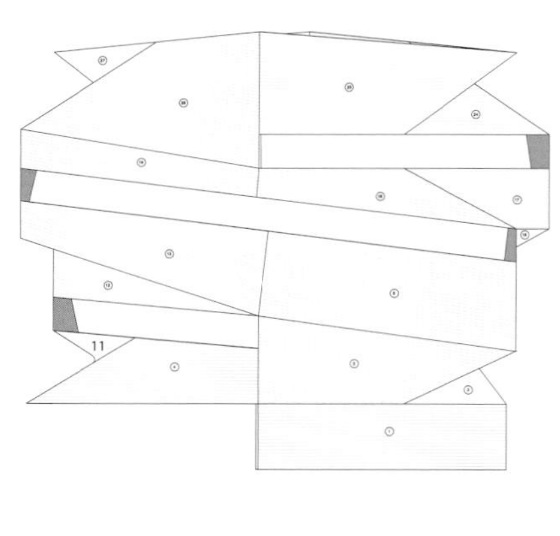
11

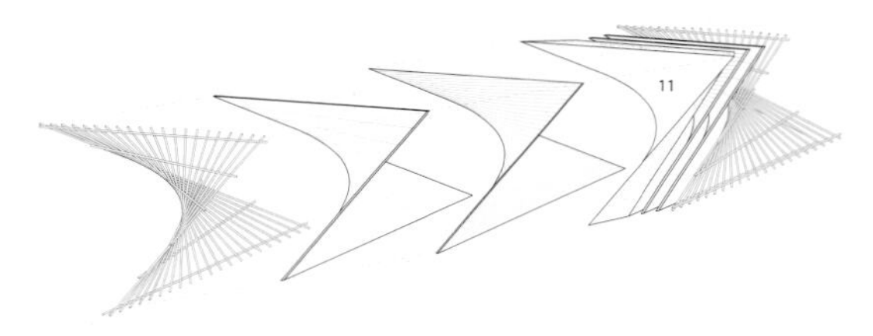
11

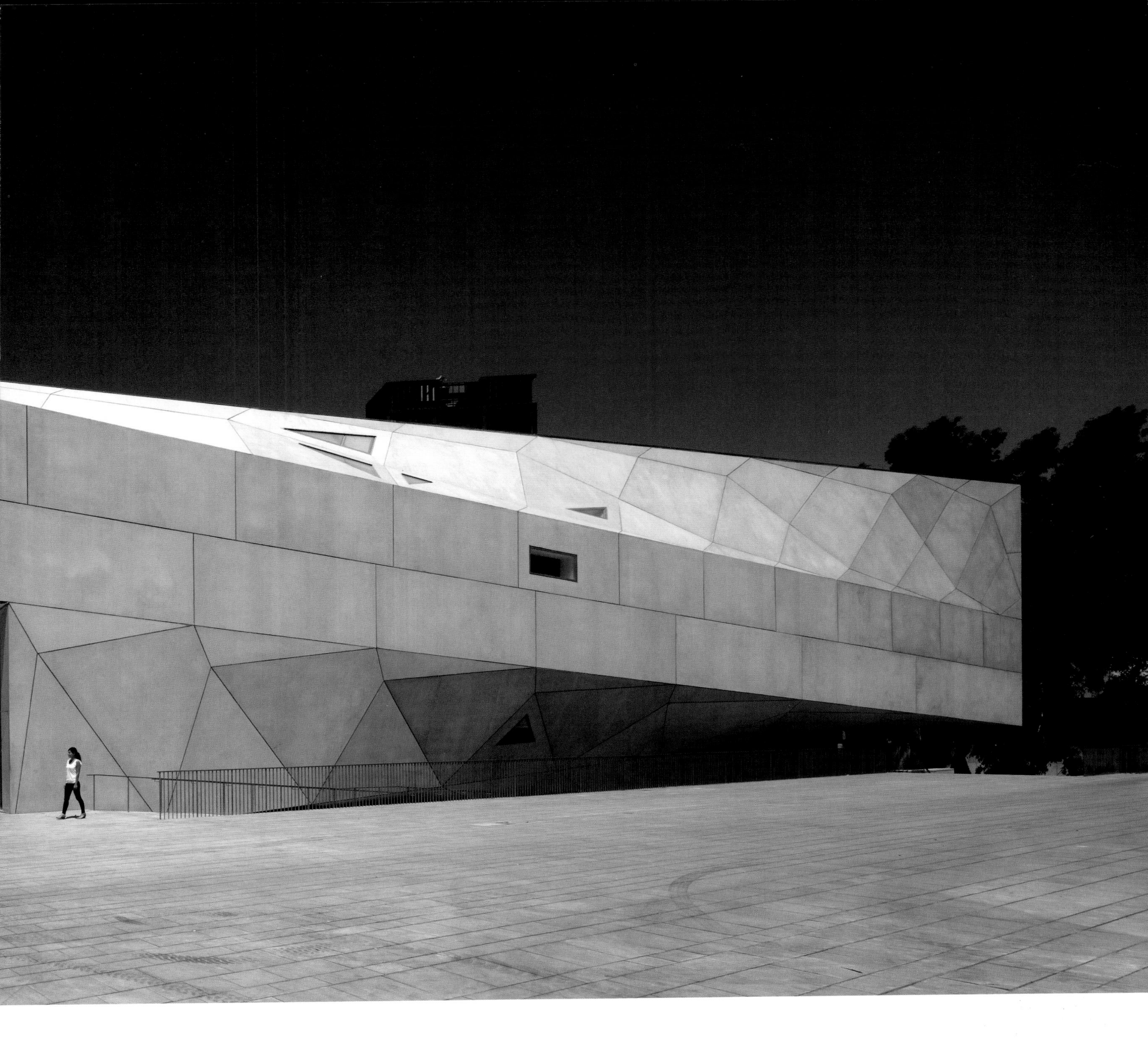

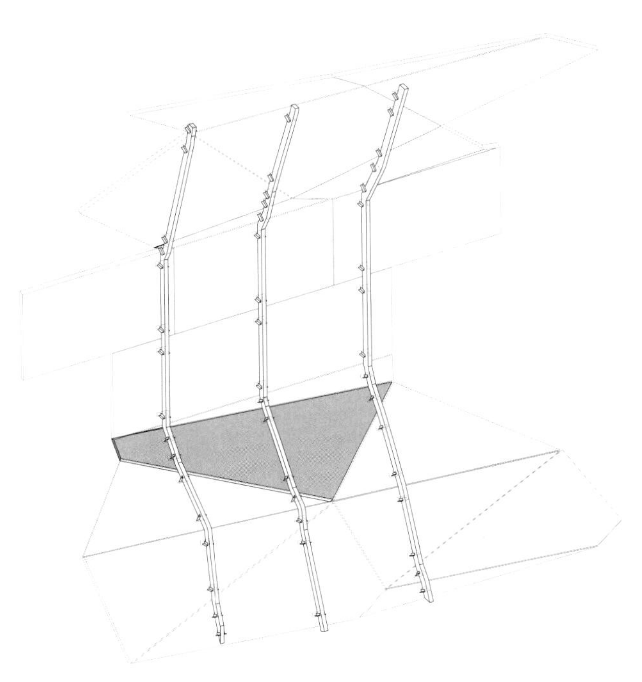

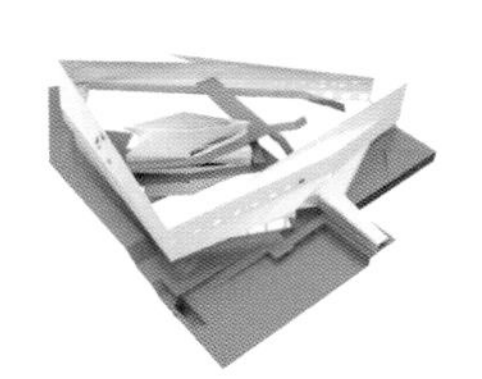

המשכן לאומנויות הבמה
חניון גולדה

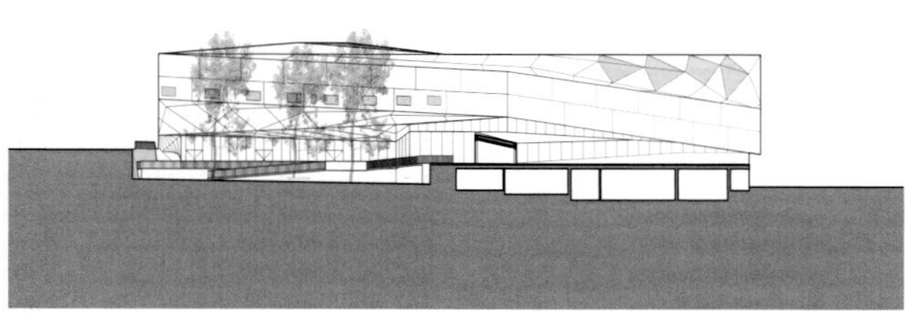

East Elevation 东立面

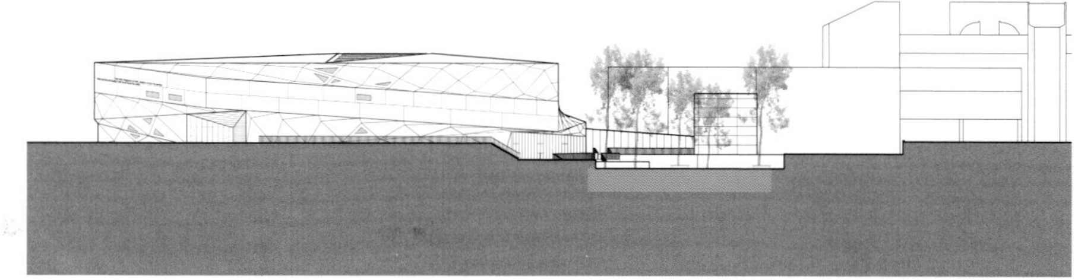

South Elevation 南立面

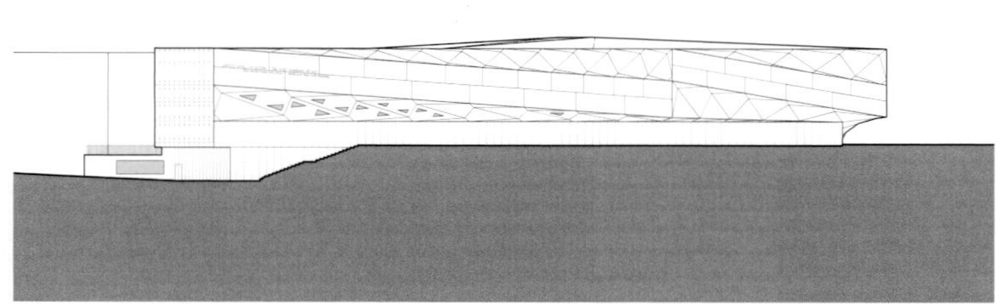

West Elevation 西立面

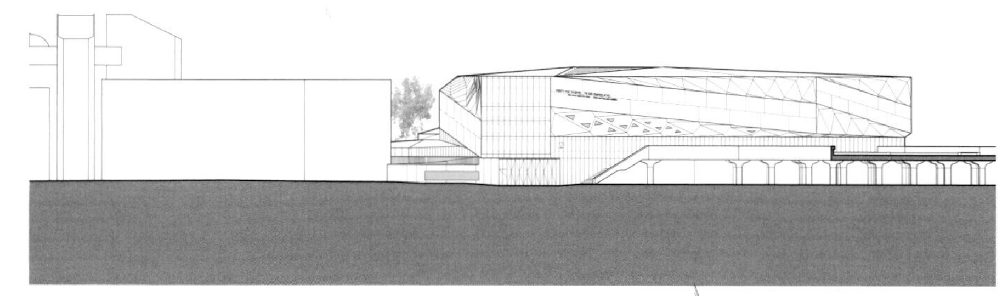

North Elevation 北立面

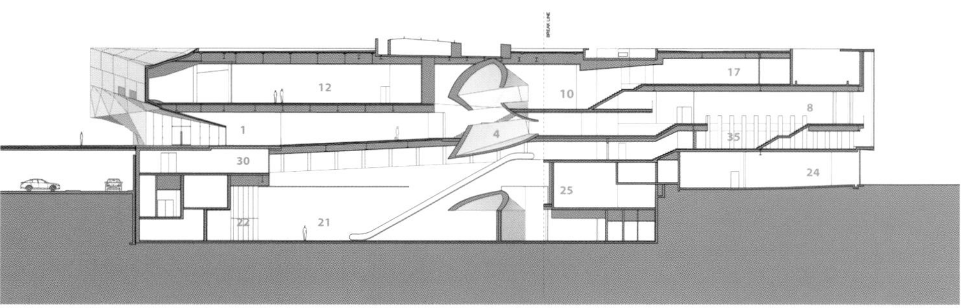

Section A 剖面 A

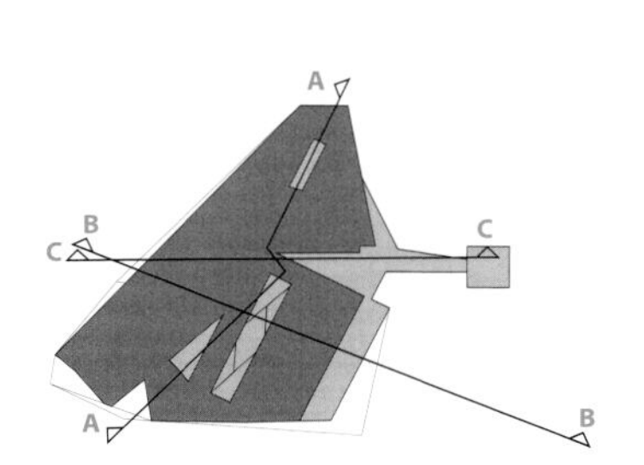

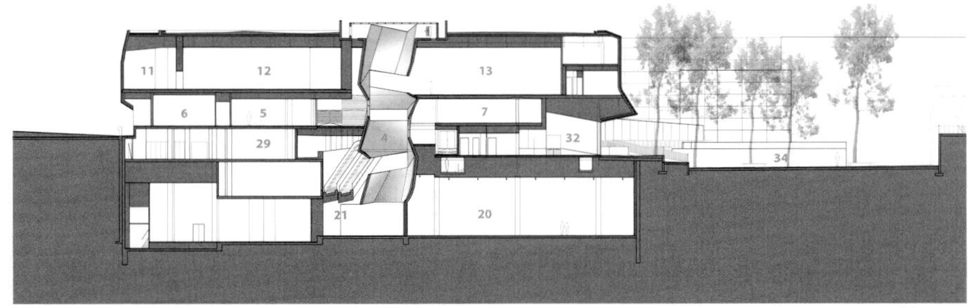

Section B 剖面 B

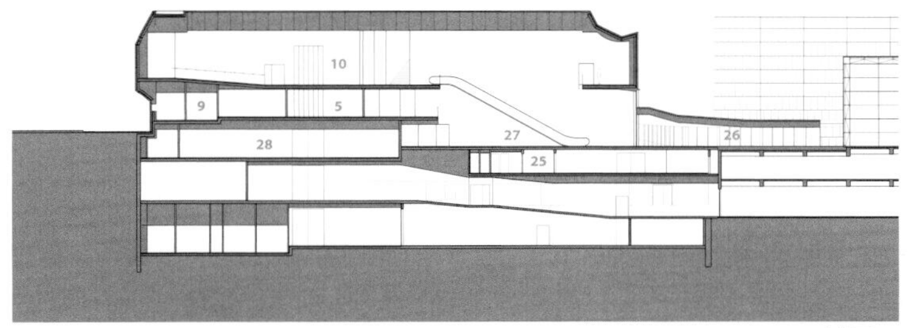

Section C 剖面 C

TEL AVIV MUSEUM OF ART
הבניין ע"ש שמואל והרטה עמיר
HERTA & PAUL AMIR BUILDING

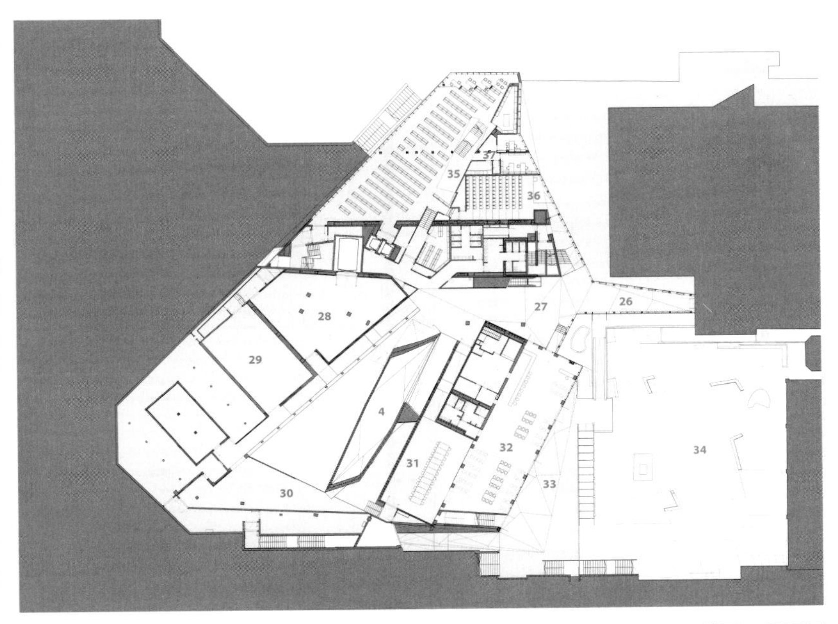

Plan Level -1 地下一层平面

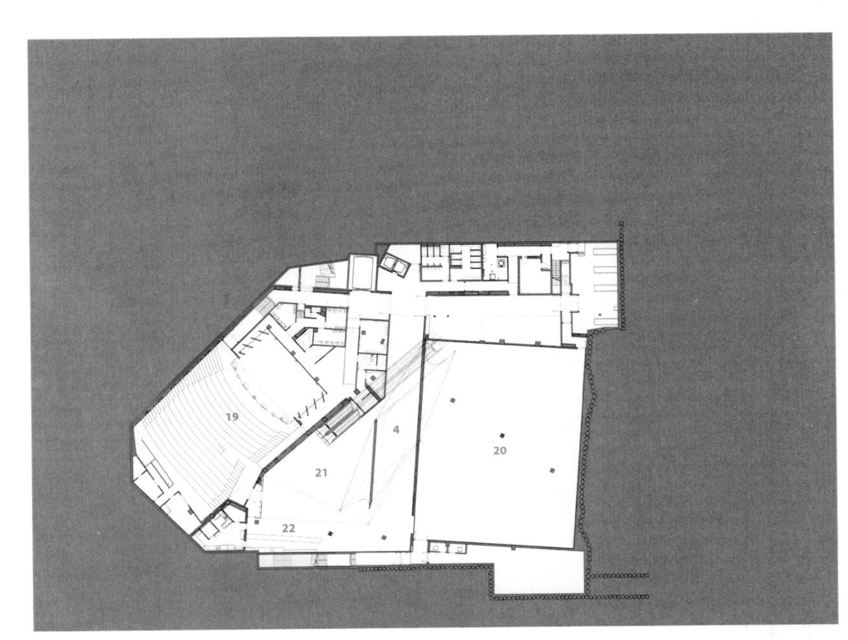

Plan Level0.00 地面层平面

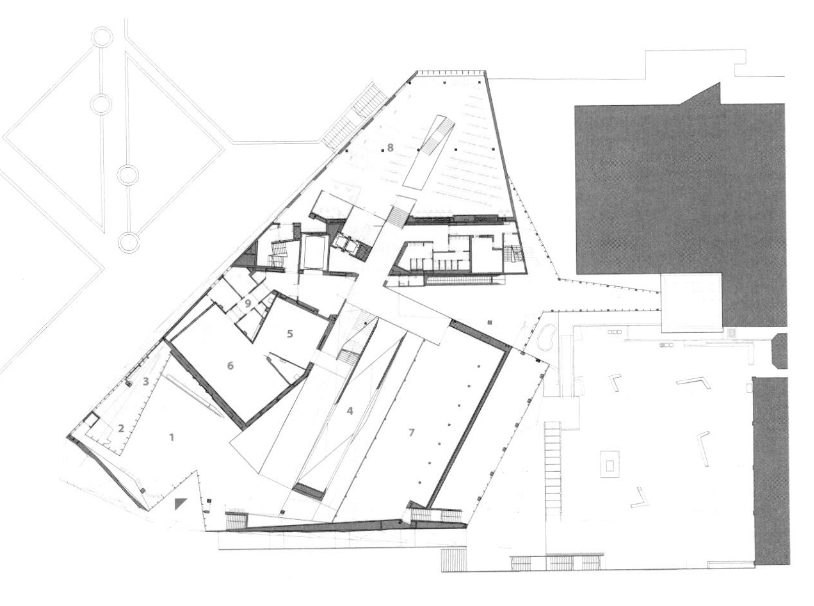

Plan Level 1 一层平面

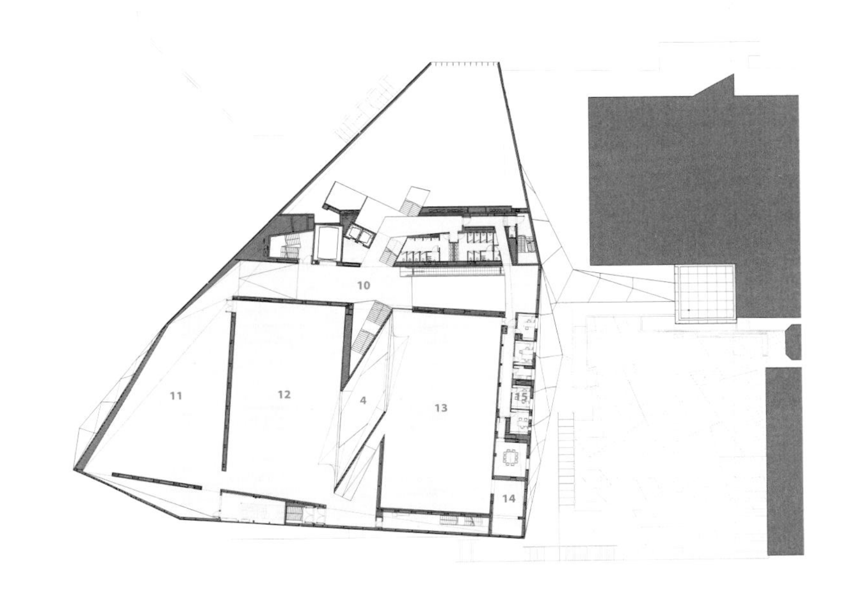

Plan Level 2 二层平面

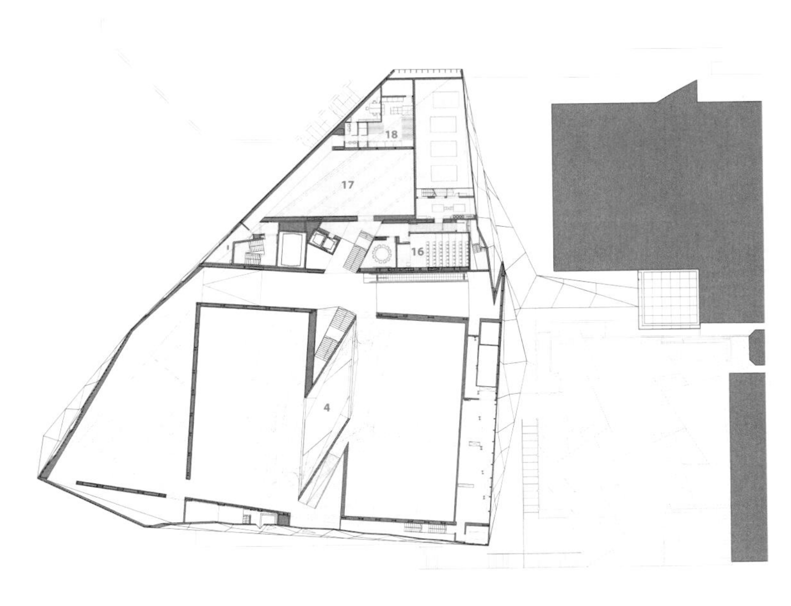

Plan Level 3 三层平面

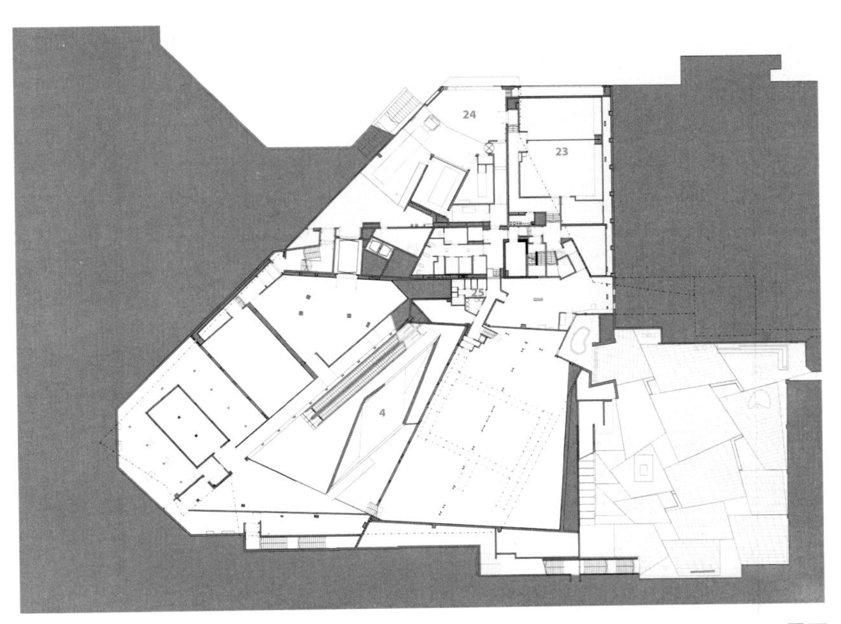

Plan Level -2 地下二层平面

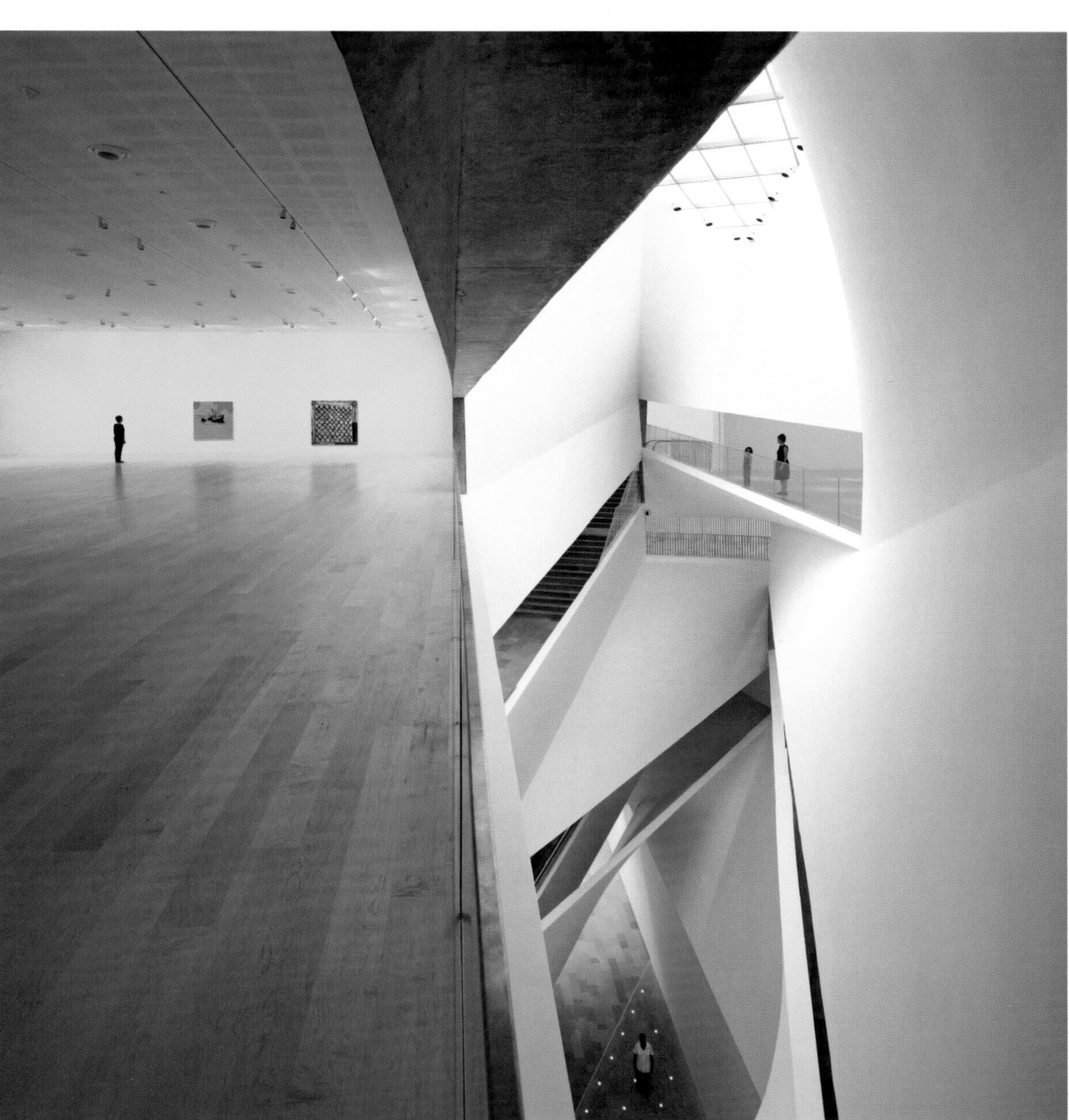

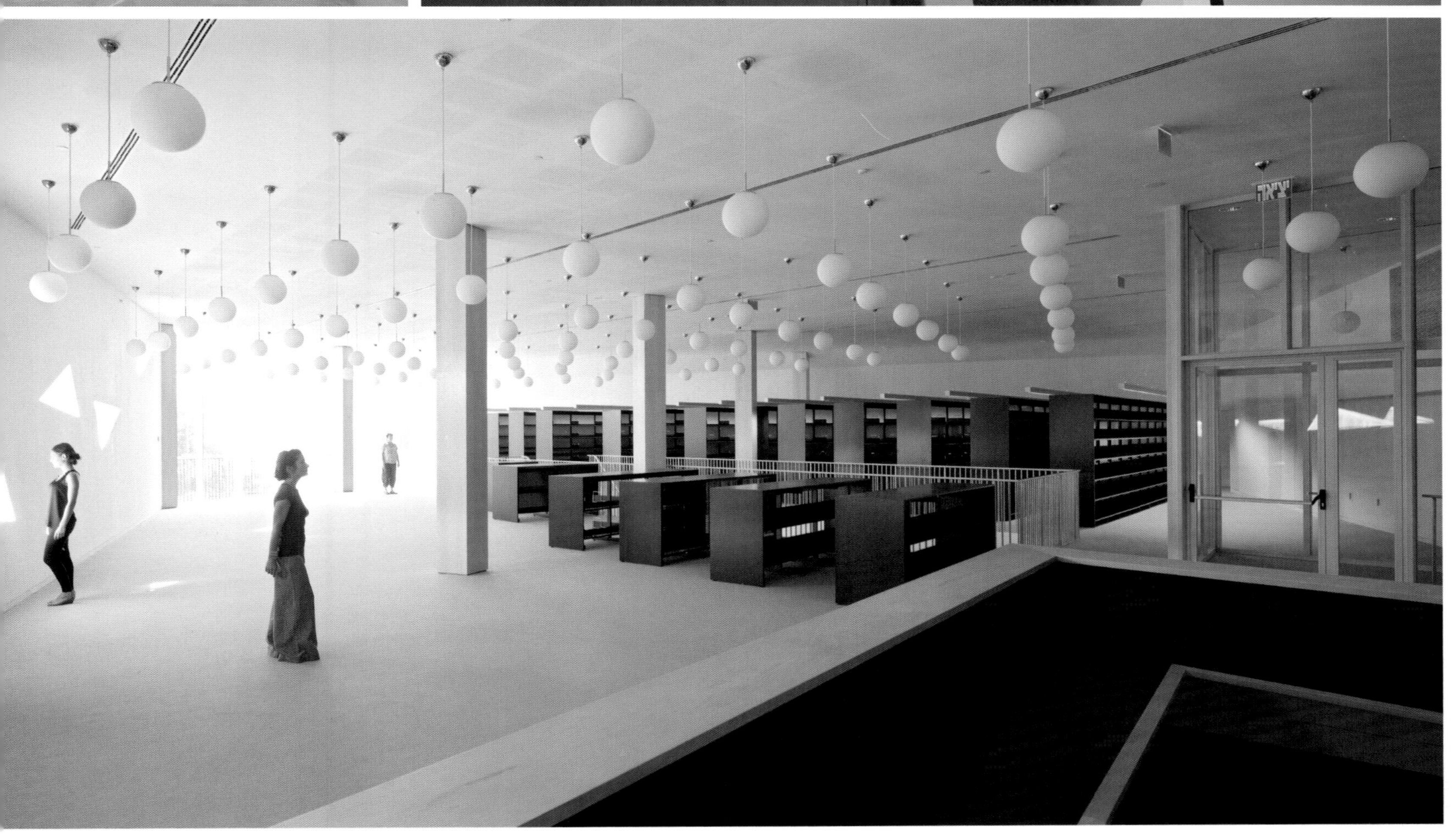

Crystal Nave

水晶殿

Carrilho da Graça Arquitectos　　Carrilho da Graça建筑事务所

Location: Matosinhos. Portugal
Typology: Public
Project Date: 2008
Image Courtesy: Carrilho da Graça Arquitecto

区位：葡萄牙，马托西纽什
类型：公共
项目日期：2008年
图片提供：Carrilho da Graça建筑事务所

The museum is the place where architecture and art commonly intersect.
The project develops itself compactly, in the form of a partially buried longitudinal volume. Its parallelepiped form stretches across the terrain's larger side. It is almost completely covered by a glass pleating, a "skin" that contains a system to filter temperature, light and views. Its zigzagging surface radically changes the use of its materials. More than just a surface, glass is also a kind of building material.
More than transparency, reflections are wanted. As they accumulate, between the facade's folds, the building turns into a visual field. The "iconicity" becomes here pure retinal drive.

博物馆是建筑与艺术普遍交融的地方。

本项目布局紧凑，形成一个局部埋入式的纵向体量。 建筑体的平行形式在基地的较大一侧得到延展。建筑几乎由一层玻璃褶皱覆盖，这个“表皮”包含了用于过滤气温、光线、视野的系统。 曲折的表皮使得材料的使用实现了巨大改变。玻璃不仅仅是表皮，同时也是一种建筑材料。

除了通透性，反射效果也是我们需要的。 当光线在立面折叠片之间聚集时，建筑成为了一个视觉场所。与此同时实现了建筑的“标志性”。

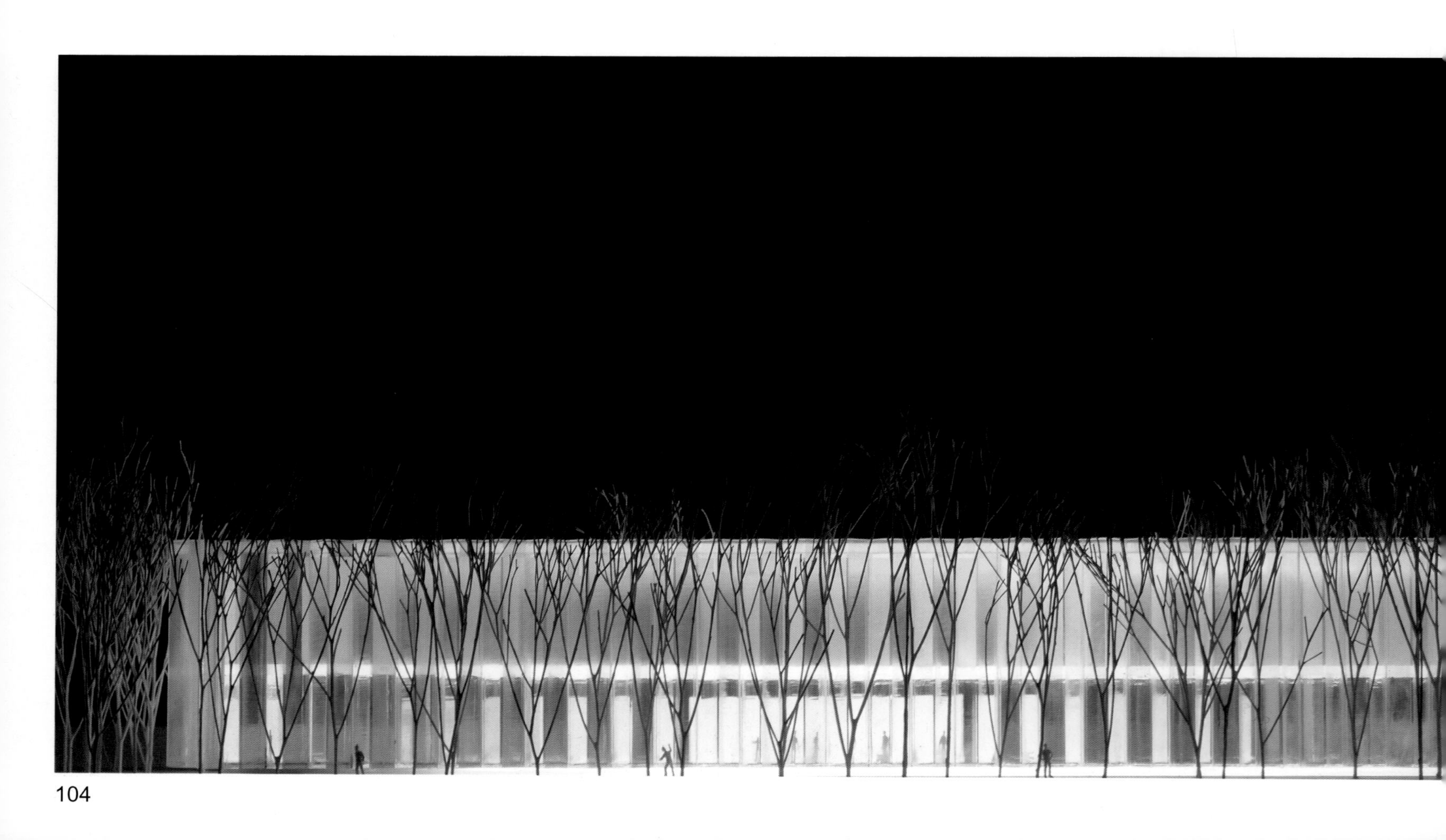

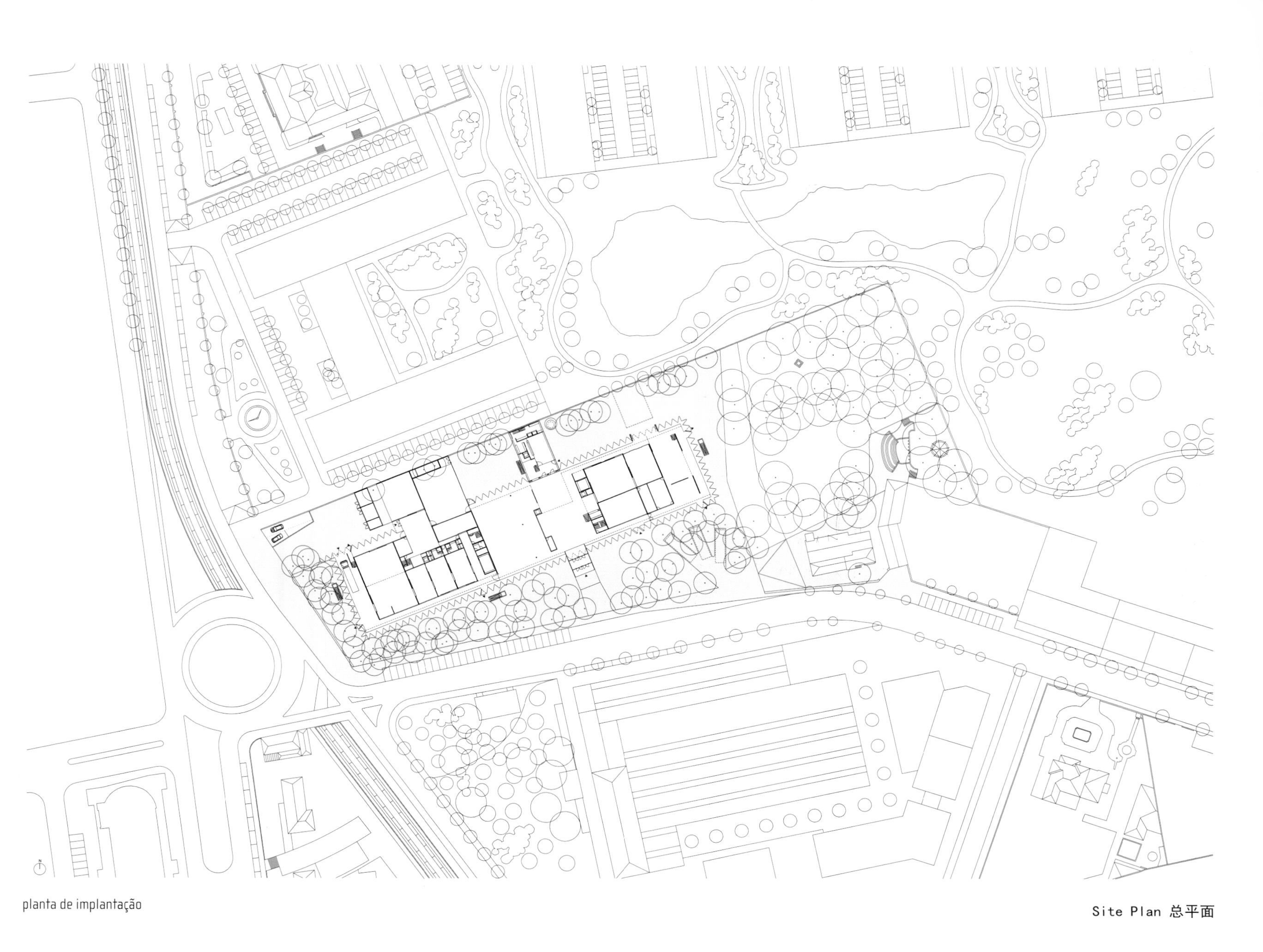

planta de implantação

Site Plan 总平面

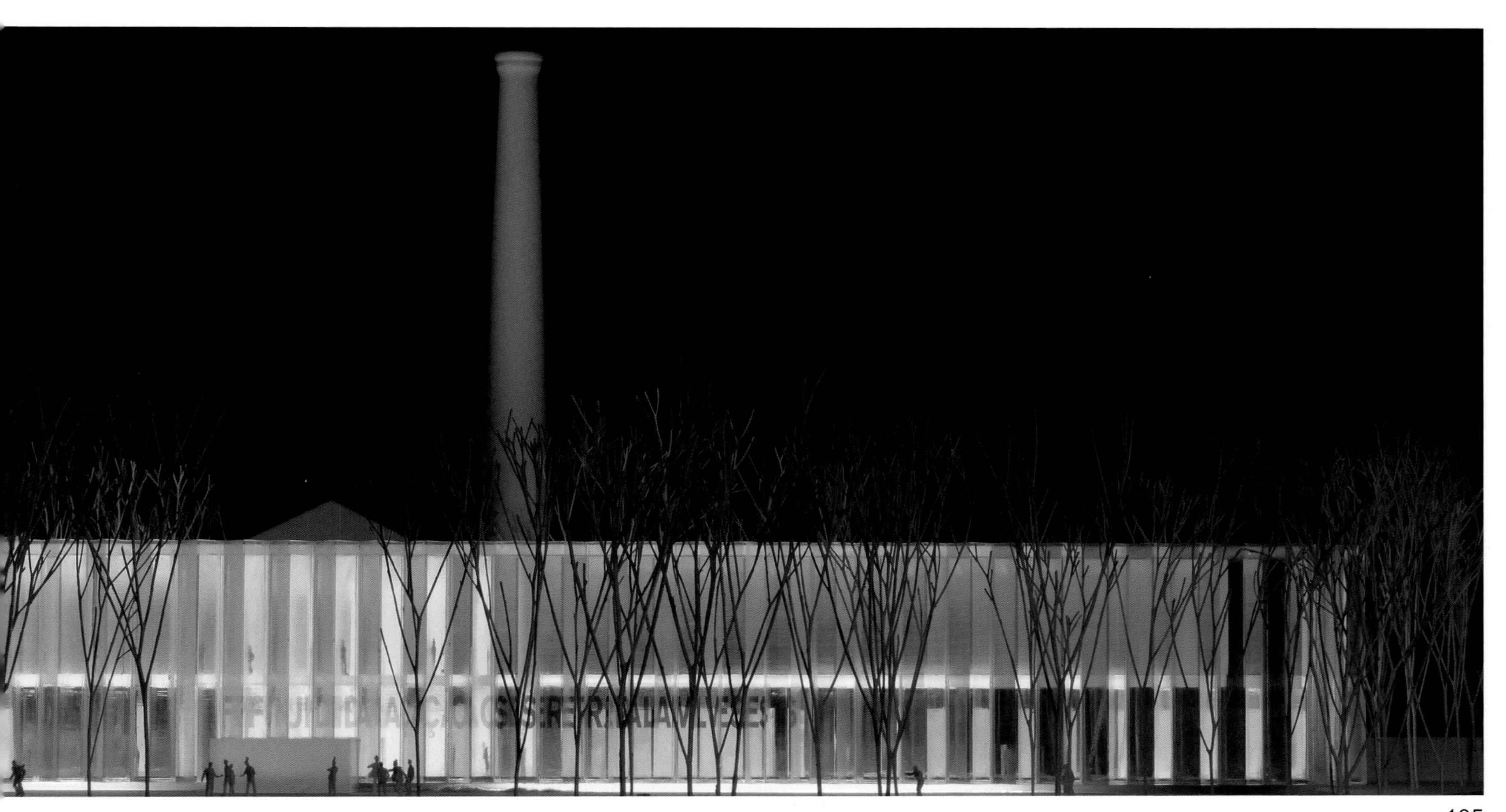

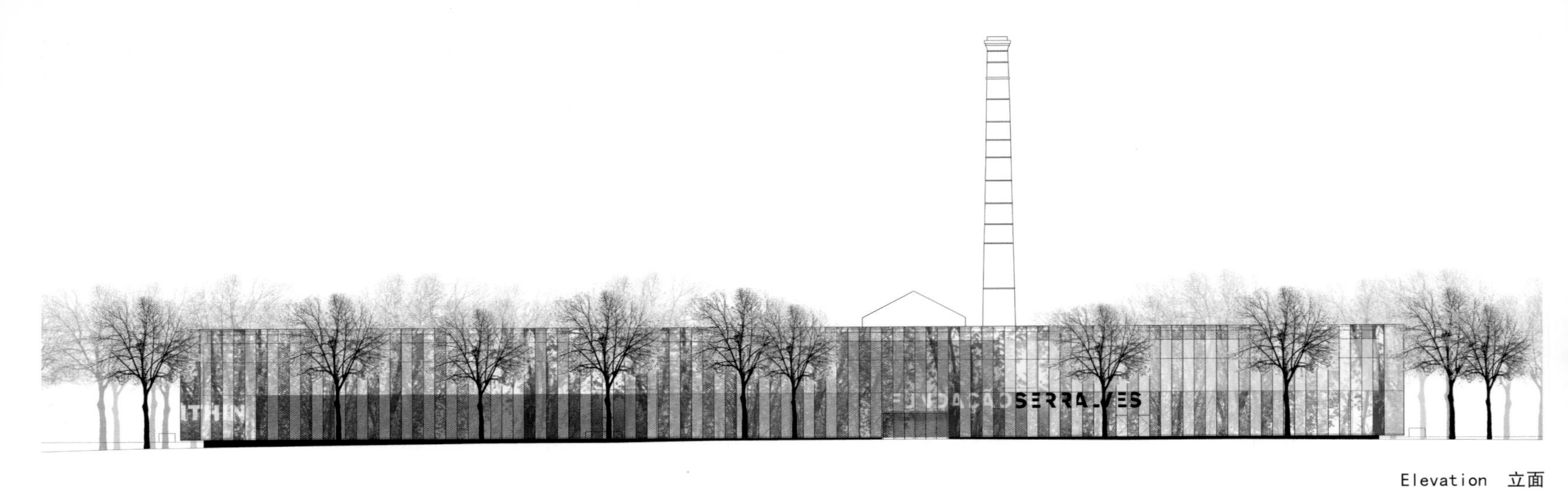

Elevation 立面

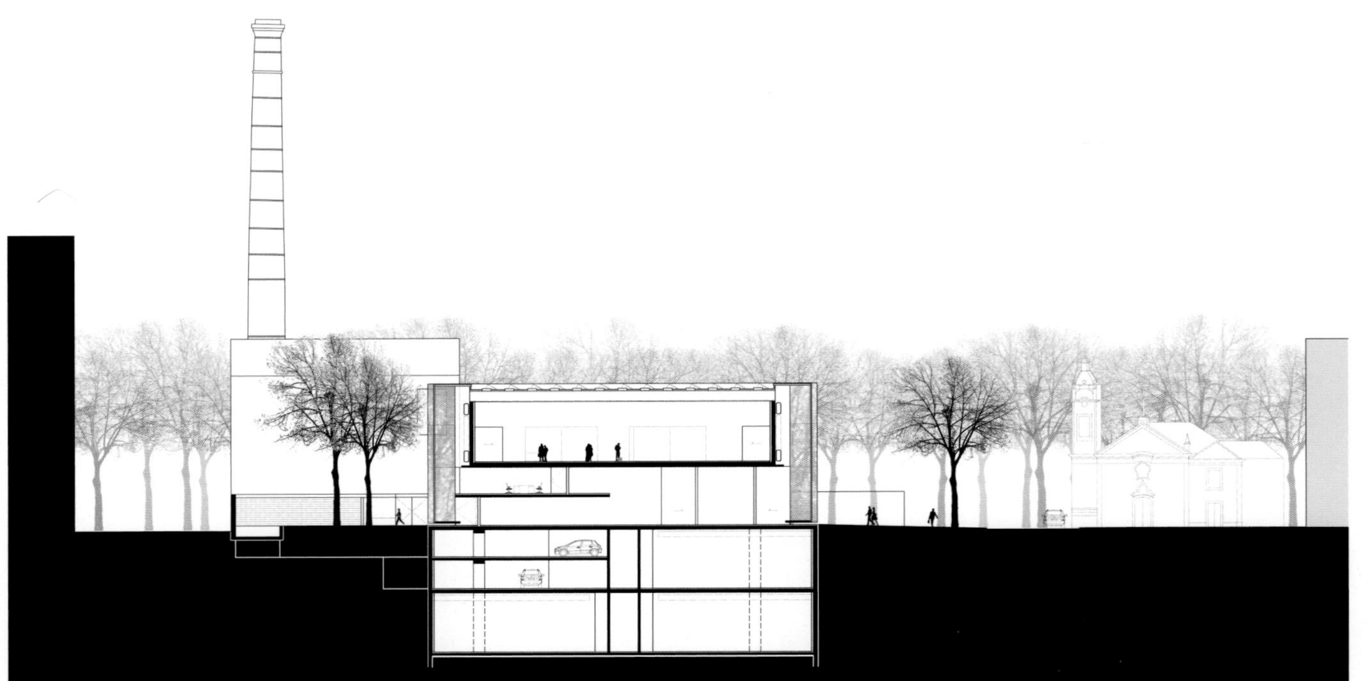

Section 剖面

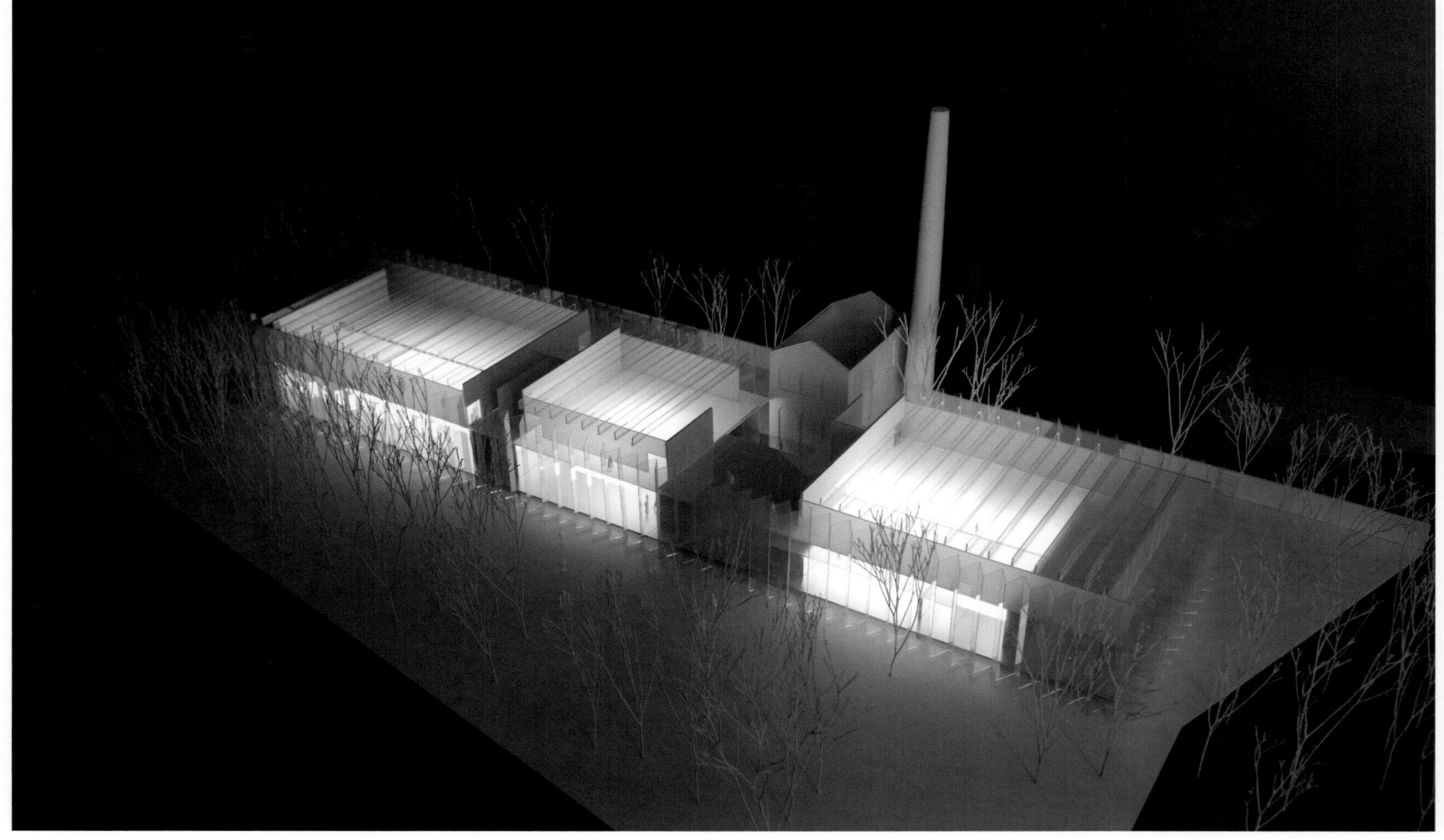

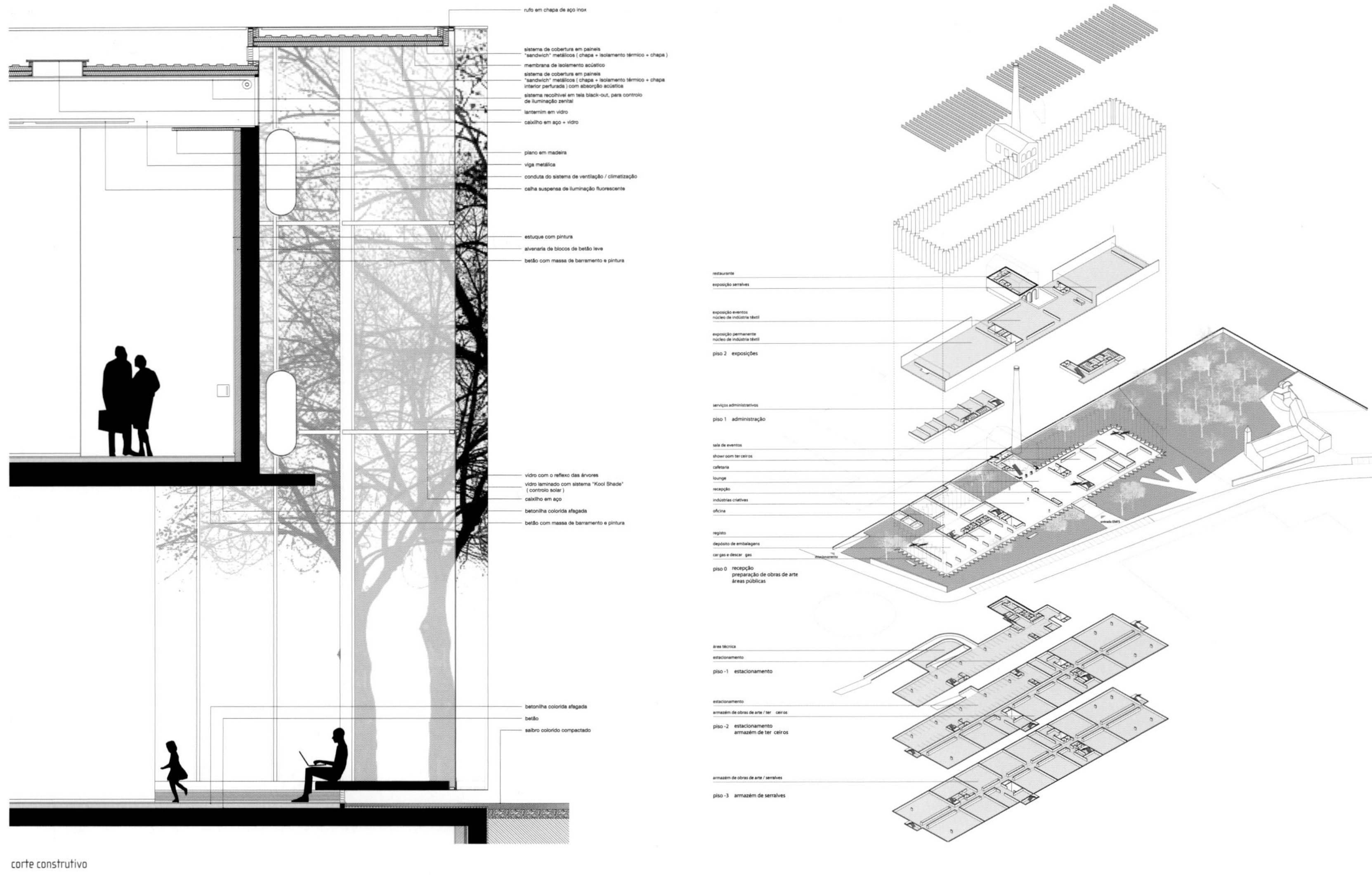

Section Details 剖面详图

Plan 平面

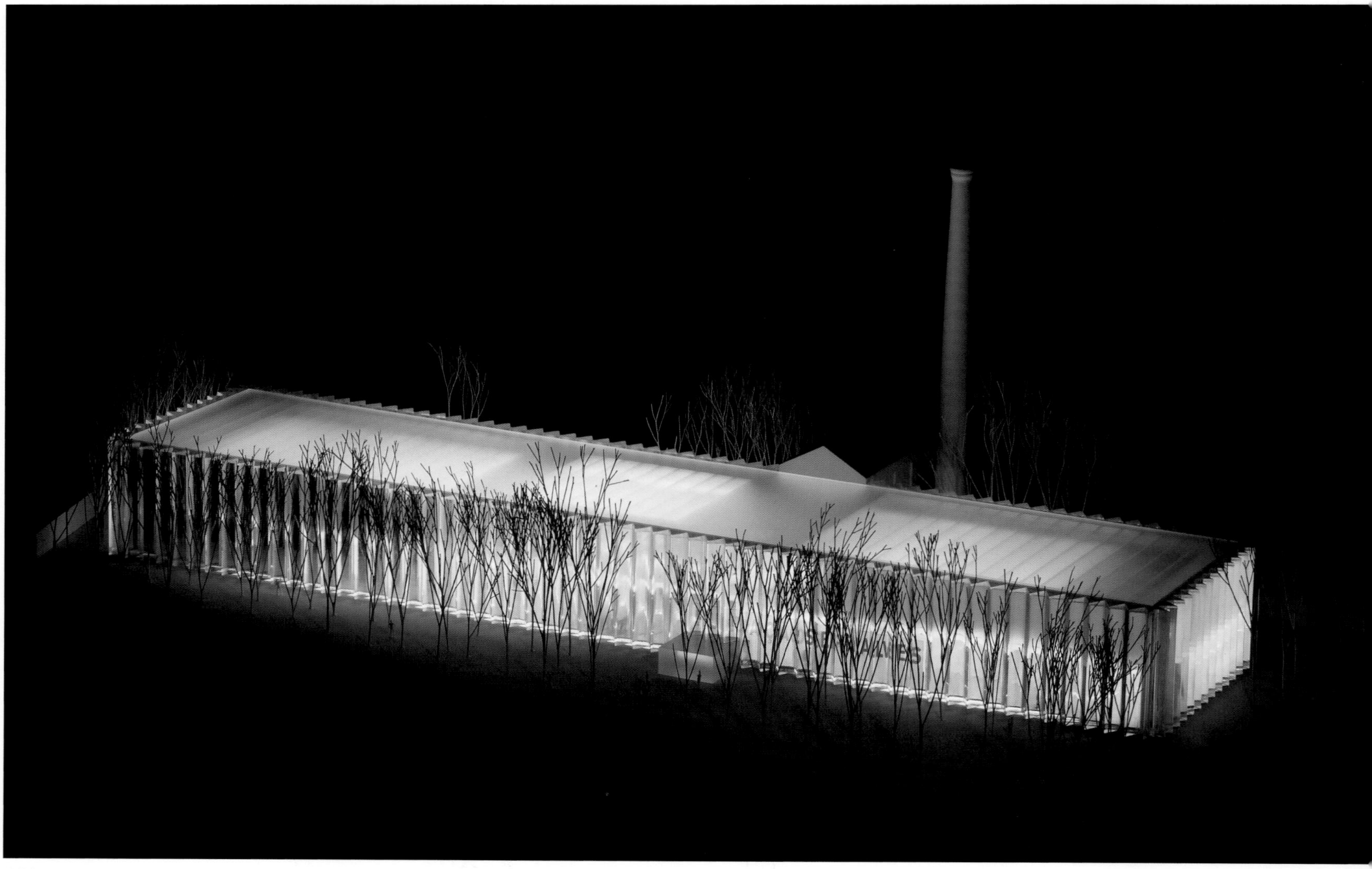

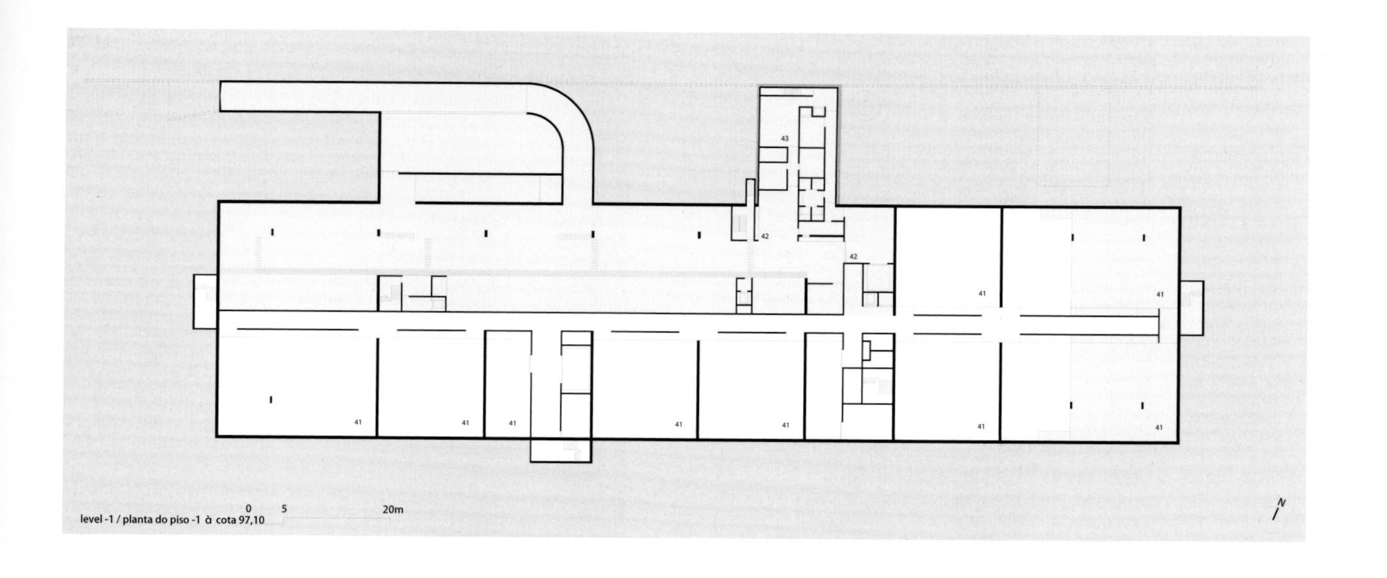

Plan Level -1 地下一层平面

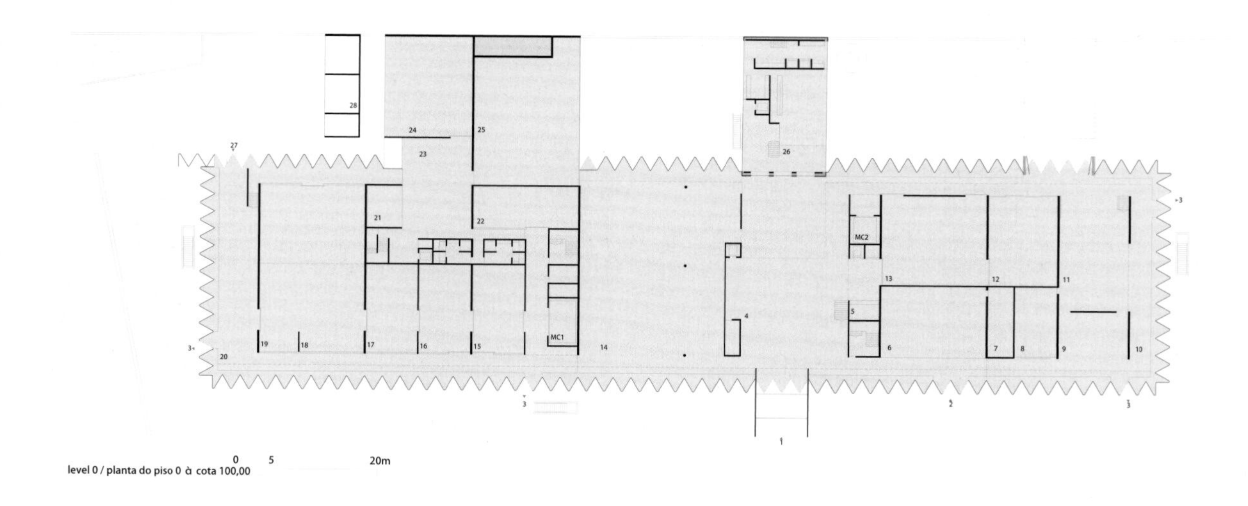

Ground Floor Plan 地面层平面

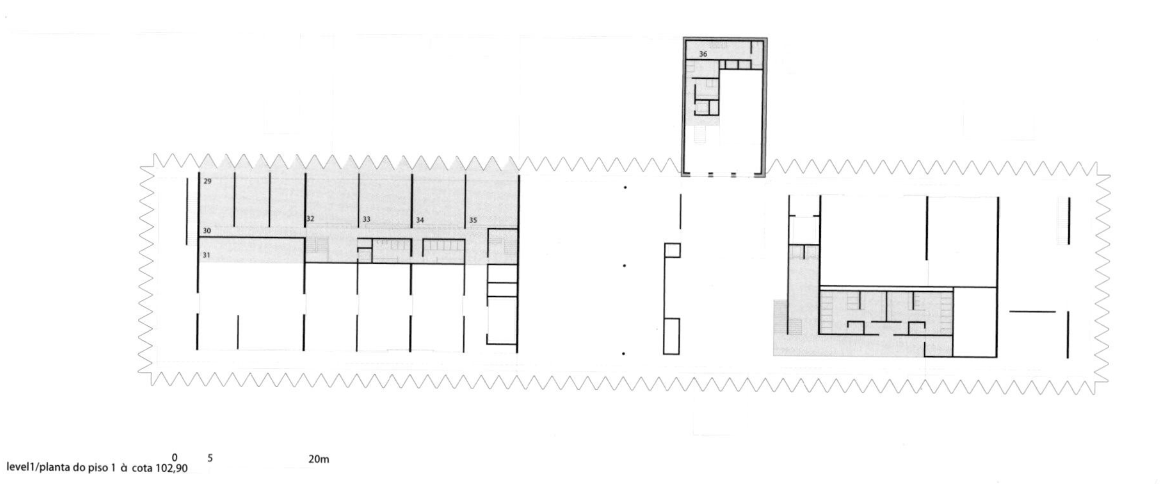

Plan Level 1 一层平面

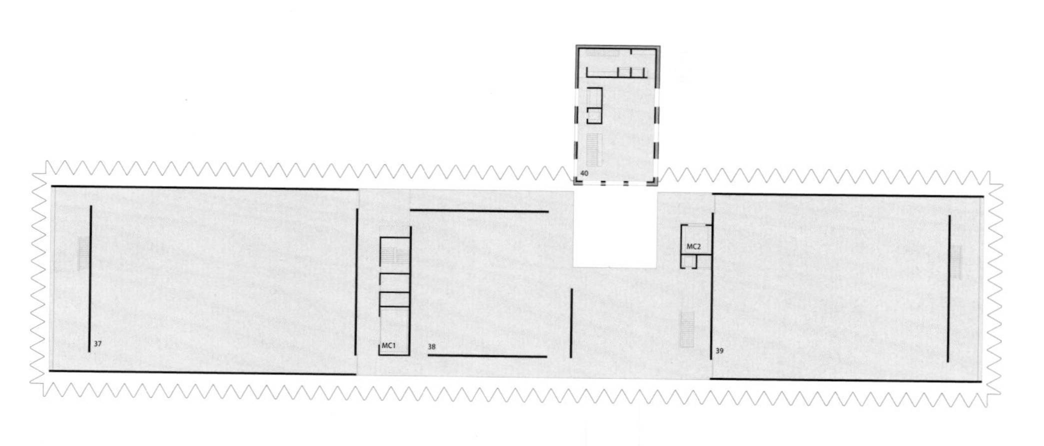

Plan Level 2
二层平面

HIMMEL UND DIE ERSTE
VERGANGEN/
MEHR · ICH SAH DIE
NEUE JERUSALEM/
HIMMEL
WIE EINE BRAUT/
MANN GESCHMUECKT
EINE LAUTE STIMME
SEHT/ DIE
MENSCHEN/
WOHNEN/ UND SIE

espaços
cafetaria
restaurante
elevadores
sala de
wc

Art Museum

美术馆

National Art Museum of China

中国国家美术馆

OMA　　大都会建筑事务所

Location: Beijing, China
Program: 128,000 m^2 for permanent collection and temporary exhibition spaces
Site: 30,000 m^2
Status: Competition first phase, 2010/11
Image Courtesy: OMA

区位：中国，北京
功能：128 000 m^2用于永久性收藏及临时展厅
基地面积：30 000 m^2
状态：2010年11月竞赛第一名
图片提供：大都会建筑事务所

In the past two decades, our museums have become larger and larger; they have now reached a scale at which they can no longer be understood as (large) buildings, but only as (small) cities. Given the area that it covers, the vast number of artworks it will house, the numbers of visitors it will inevitably attract, the turnover of exhibitions it will have to accommodate, NAMOC can be the first museum in the world based on this new paradigm, the first museum conceived as a small city.

In this way, it can incorporate a significant number of breakthroughs, revolutionizing the way in which the museum works today. Like a city, it could mix sectors, "official" and grassroots, it could have a center and a periphery, a Chinese and an international district, modern and historical areas, commercial and "government" neighbourhoods. Like a city, individual sections need not be permanent; areas can be redefined, renovated, or even replaced, without compromising the whole.

To plan NAMOC as a city does not mean that it cannot offer the intimacy that remains the essence of the museum experience: like any city, its individual parts can be small, humane... But like a city, it will offer a degree of variety that will be unique for a single museum. Part of it will be public, other parts could be commercial...

The architecture of the main plinth offers a range of classical, orthogonal museum spaces, to more contemporary, freer forms. Like any city, circulation can be efficient and direct - for larger groups - or meandering and individual. The story of Chinese art can be told, or discovered. The main circulation of the city is based on a five-pointed star that leads from the multiple entry points on the periphery to the center. Here, the star connects to the "lantern", a multistory stack of platforms, wrapped in a red skin, on which temporary exhibitions and events are arranged with the smooth efficiency of a convention center. Although its internal organization is rational, the elastic skin stretched around the metallic frame makes it look like a mystery...

The Lantern is the three-dimensional emblem of NAMOC; a single tangential axis relates the Lantern to the Bird's Nest. In contrast to the intricacy of the city, the six main floors of the Lantern offer wide open spaces, so that the architecture does not interfere with the organization of the exhibitions or events.

In the thickness of the floors, offices, library, research and other services are accommodated; in the "City"they are concentrated on the five arms of the star.

Conceptually, the two halves of NAMOC - "City" and Lantern - are complementary: like the city today, they offer radically different experiences: the small scale, intricate condition of the traditional urban fabric of China, and the contemporary era of radical modernization...

在过去的二十年里，博物馆的规模日益壮大，已达到不再能以(大的)建筑，而要以(小的)城市视之的程度了。因其占地面积之广大，艺术品储藏量之惊人，不可避免会吸引的游客数量之庞大、及其举办展览必需容纳人流量之巨大，中国美术馆成为世界上第一个新典范博物馆，即首个被构想为小城市的博物馆。

通过这种方式，它包含了大量与现今博物馆运作模式截然不同的突破和革新。就像城市一样，它可以混合不同的部门，官方和民间部门相融合。它可以同时设有一个中心和周边区域，一个既民族化又国际化的区域，现代和历史并存的地区，毗邻商业和政府。就像城市一样,独立的部分不需要就此一成不变；不需要对整体结构做丝毫让步，该区域可以被重新界定、重新修复甚至被替代。

将中国美术馆作为城市来规划并不意味着它不能提供舒适感，舒适这一本质一直保留其中：像任何的城市一样，它独立的部分可以是小而具有人文性的……但是像某个城市一样，它将提供一定程度的多

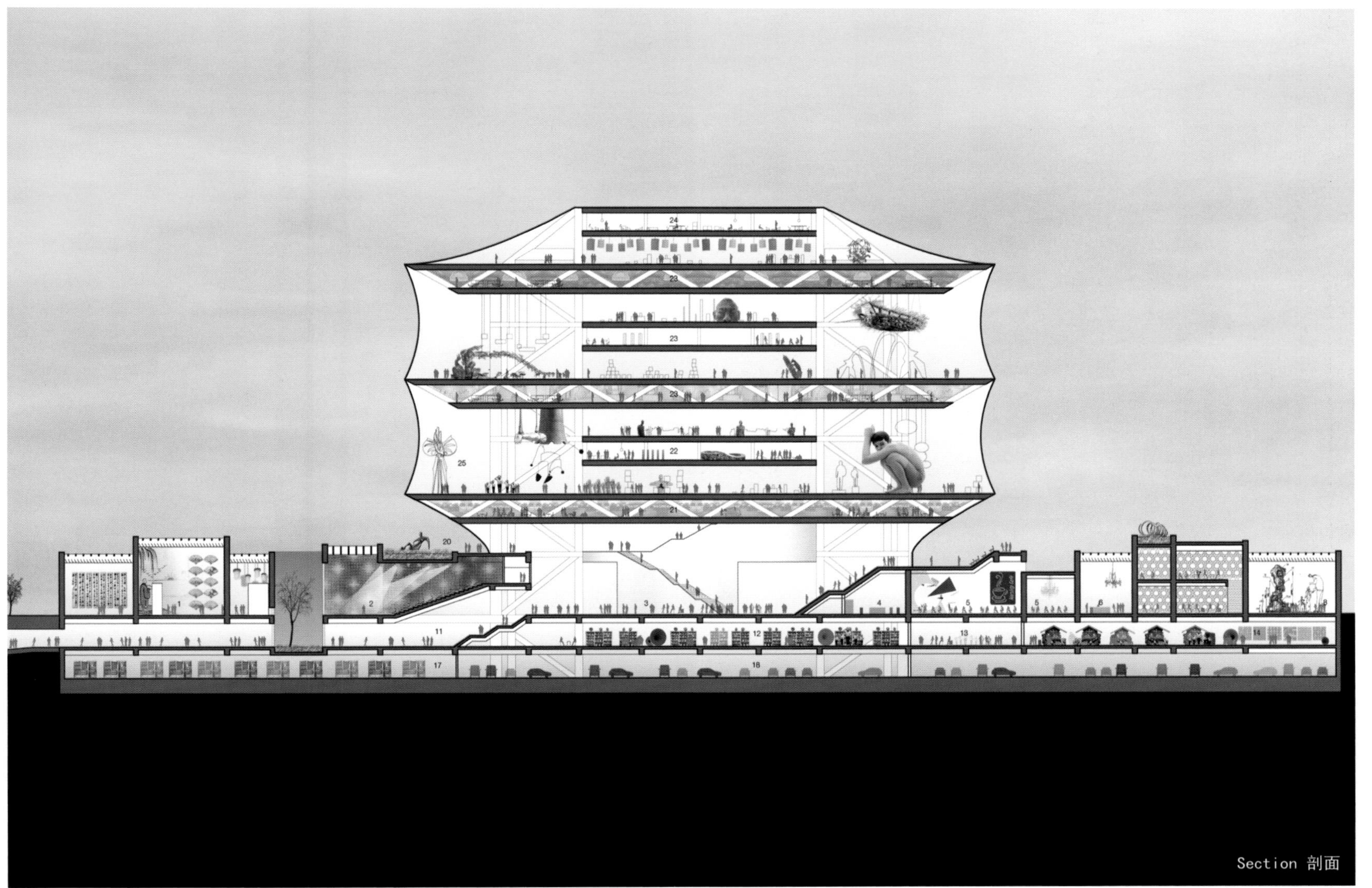

Section 剖面

样化，这对一个独立的博物馆来说是非常独特的。它的一部分是公共性的，其余部分可能商业化……

这座主体底座为方形的建筑提供了各种风格的美术馆空间，从古典的矩形式的设计到更现代、更自由的建筑形式。像任何一个城市一样，美术馆可以实现高效且直接的流通，无论对象是多数群体，还是几个人或者一个人。中国艺术的故事可以被传诵和发现。城市的主要通路是以一个从外围多个入口点向中心汇集的五角星为基础的。在这里，五角星连接着一个“灯笼”——一个裹有红色外壳的多层平台，在高效运作的会议中心的控制下，临时展览和事件就是在这一平台上举办的。虽然它的内部组织结构合理，但围绕金属框架延伸开来的弹性外壳使它看起来像是个谜……

灯笼是中国美术馆的三维象征，一个正切轴把灯笼和鸟巢联系到一起。与错综复杂的城市不同，灯笼的六个主要楼层提供了开阔的空间，这样建筑风格不会影响展览或活动的组织工作。

从地板的厚度，到办公室、图书馆、研究和其他服务都设置完备；在“城市”中，它们集中在五角星的五个角上。

从概念上的来说，作为中国美术馆两大组成部分的“城市”和灯笼是互补的。像如今的“城市”一样，它们能给人们带来截然不同的体验: 传统中国城市建筑规模小而又复杂的情况，及当代中国彻头彻尾的现代化……

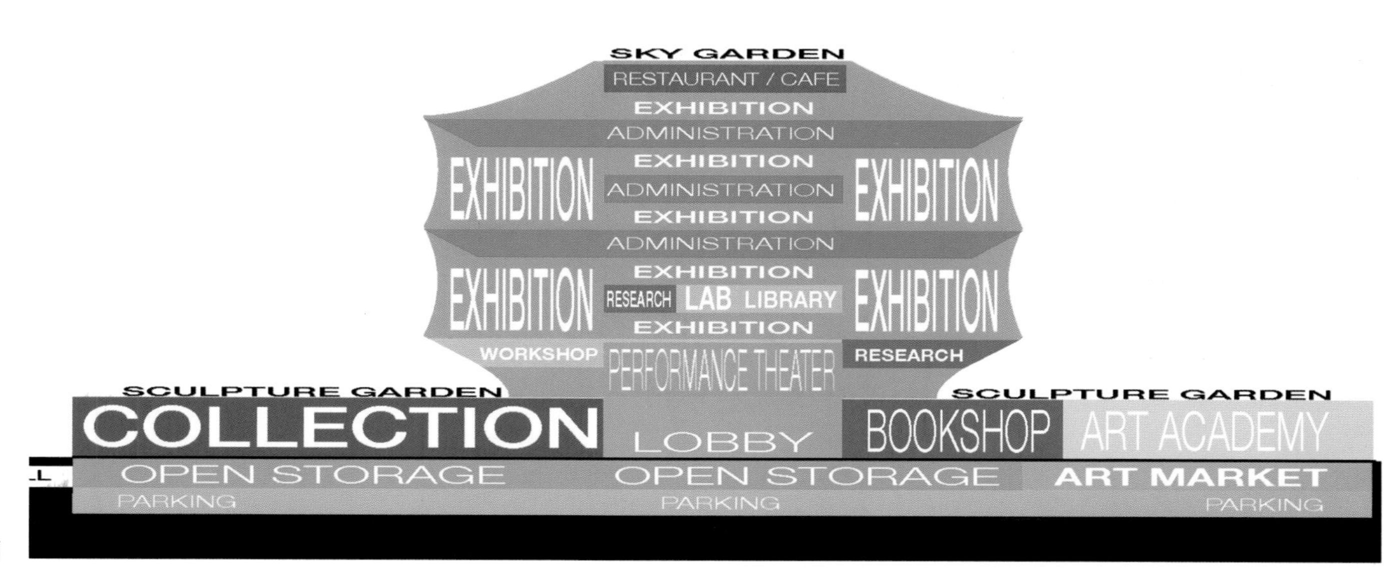
SKY GARDEN
RESTAURANT / CAFE
EXHIBITION
ADMINISTRATION
EXHIBITION
EXHIBITION
ADMINISTRATION
EXHIBITION
EXHIBITION
ADMINISTRATION
EXHIBITION
EXHIBITION
RESEARCH
LAB
LIBRARY
EXHIBITION
EXHIBITION
WORKSHOP
PERFORMANCE THEATER
RESEARCH
SCULPTURE GARDEN
SCULPTURE GARDEN
COLLECTION
LOBBY
BOOKSHOP
ART ACADEMY
OPEN STORAGE
OPEN STORAGE
ART MARKET
PARKING
PARKING
PARKING

Section 剖面

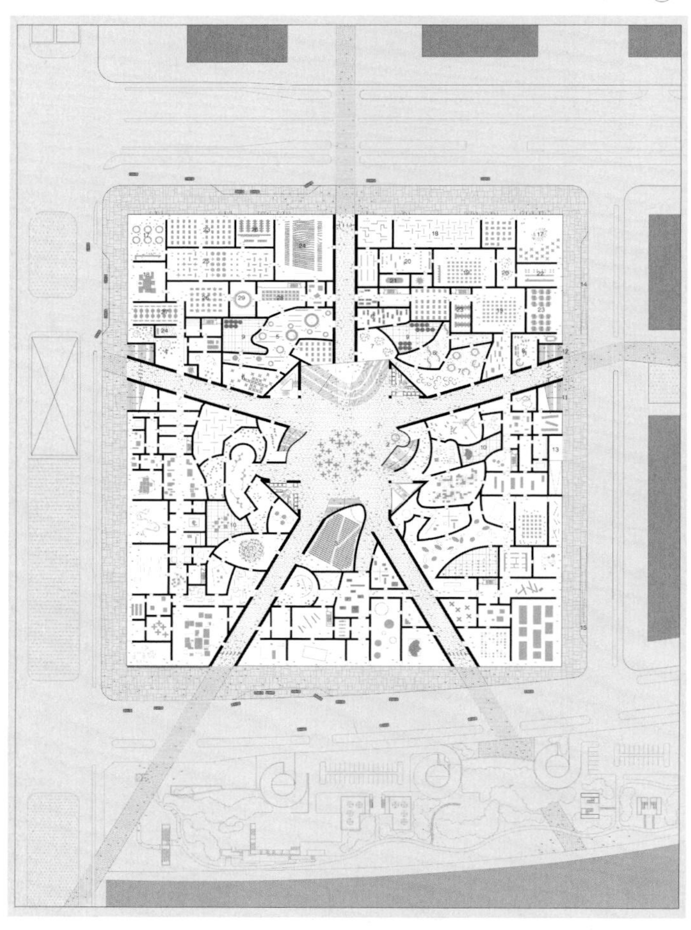

Plan Level 1　一层平面

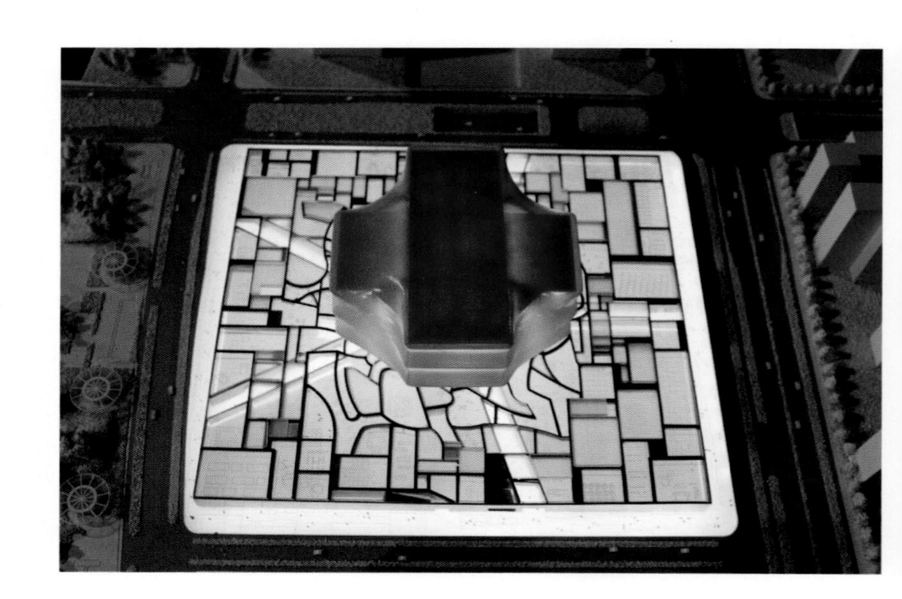

CHINA

Permanent Exhibition:
International Artwork
ASHES OF TIME REDUX
东邪西毒终极版
The Mastermind
M C ESCHER Exhibition
埃舍尔的世界
1/11/2013 - 31/3/2014
中国美术館
售票处
TICKET OFFIC
天上人间
PARADISE
20th Century Chinese Art
Folk Art

The National Gallery of Greenland
格陵兰岛国家美术馆

HEIKKINEN-KOMONEN ARCHITECTS　　HEIKKINEN-KOMONEN建筑事务所

Location: Nuuk, Greenland
Program: Gallery
Status: Competition 2010
Image Courtesy: HEIKKINEN-KOMONEN ARCHITECTS

区位：格陵兰岛，努克
功能：美术馆
状态：2010年竞赛
图片提供：HEIKKINEN–KOMONEN 建筑事务所

The views from the site for the new national gallery of Greenland are breathtaking and powerful, while the surrounding man-made milieu is severe and stark. Located at the edge of the city and in front of the open ocean, the building becomes a mediator between the two contrasting realms.

The proposed building is a freestanding, independent landmark in a context where both natural and artificial environments are dominant in scale and presence. Free from apparent formal references, its introverted and stoic character conceals a monumental interior. Here, art and culture of Greenland can be viewed against the evocative backdrop of the unique local landscape. Inhabitants of the building are invited to enjoy the surrounding seascape and coast, which is brought in through carefully framed views.

The building's program is arranged into eight levels. A tall sky-lit canyon allows for the influence of seasonal changes and daylight conditions on the interior atmosphere. Exhibition spaces, public areas, curation, conservation and storage of the works of art co-exist in and around the vertical central space.

A variety of different exhibition environments is offered, from a sky-lit white box to the dark, dramatic interior of the central void. All levels can be included in the exhibition circulation, or closed for maintenance.

The building is entered on the fourth level through an outdoor exhibition area. Another public level for events, performances and exhibition openings is located at the bottom of the building's central space where a large panoramic opening reveals the ocean horizon in an extraordinary new perspective.

In preparation for the museum's future growth, an extension is proposed beneath the entrance courtyard. It is to be integrated into the existing circulation and services.

从格陵兰岛国家美术馆极目远眺，气势磅礴的视野让人叹为观止，而周边的人造景观则略显肃穆，独自黯淡。坐落于城市边缘，面朝大海的格陵兰岛国家美术馆成为了两大对比鲜明领域间的调和者。

于自然及人工环境在规模及效果方面都占据主导地位的景致之中拟建建筑物，以独立式的地标性建筑为其定位。该建筑设计摆脱了外观稍显拘谨的参考方案，其内敛的斯多葛式特色在其经典不朽的内部设计中展现得淋漓尽致。在这里，在呼之即出且独具特色的当地景观背景的映衬下，格陵兰岛的艺术与文化一览无余。建筑内的居民应邀观赏周边海洋及海岸的海景，在细致构架的视野范围内，人们真是大饱眼福了。

建筑项目依照八种层次进行规划安排。高大的天窗采光峡谷设计使四季更替所带来的影响与白昼的光照条件共同作用于内部氛围。艺术品的展览空间、公共区域、屏模、维护及存储在垂直的中央空间内部及周围共存。

从天窗采光的白盒到中央区域黑暗却激动人心的内部设计，各式各样的展览环境层出不穷。所有的设计层次均囊括在展览循环之内，或是关闭以便维护。

格陵兰岛国家美术馆的入口设于第四层次，需要穿过户外展览区域。另一个用于举办活动、表演及公共展览的场馆设于美术馆基层的中央区域，宽广海域的壮美景象以非凡全新的视角全部呈现在人们面前。

在博物馆未来发展的筹备期间，提出于入口庭院处进行地下延伸，而这也将成为现有流线及服务的整合部分。

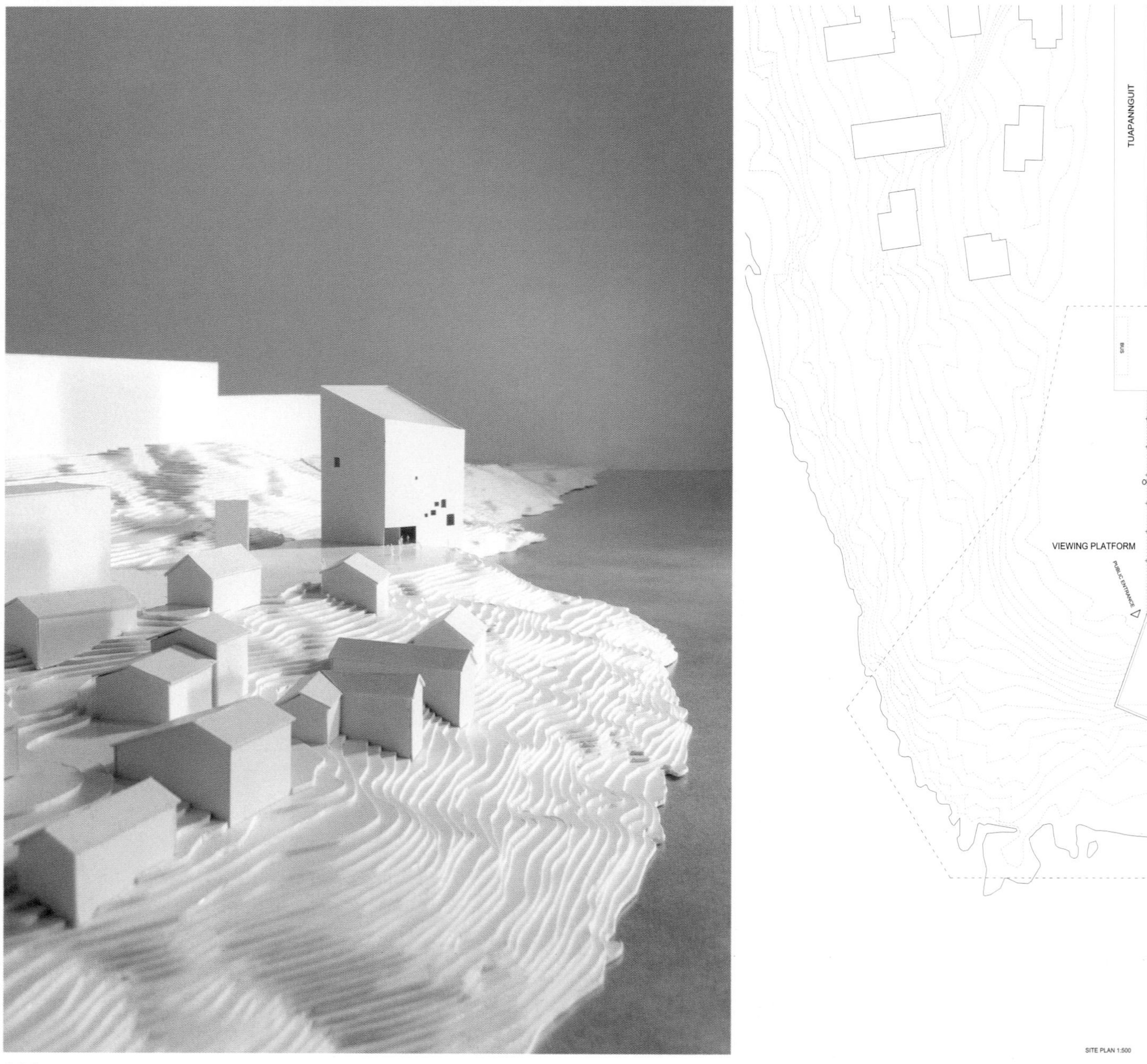

SITE PLAN 1:500　　VIEW FROM THE BAY

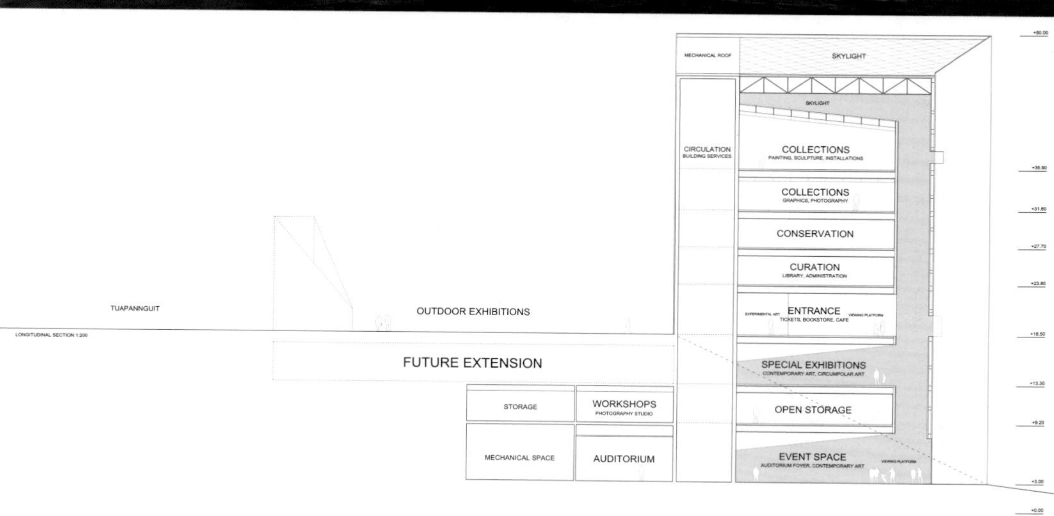
TUAPANNGUIT
OUTDOOR EXHIBITIONS
FUTURE EXTENSION
MECHANICAL ROOF
SKYLIGHT
SKYLIGHT
CIRCULATION
BUILDING SERVICES
COLLECTIONS
PAINTING, SCULPTURE, INSTALLATIONS
COLLECTIONS
GRAPHICS, PHOTOGRAPHY
CONSERVATION
CURATION
LIBRARY, ADMINISTRATION
EXPERIMENTAL ART
ENTRANCE
TICKETS, BOOKSTORE, CAFE
VIEWING PLATFORM
SPECIAL EXHIBITIONS
CONTEMPORARY ART, CIRCUMPOLAR ART
OPEN STORAGE
EVENT SPACE
AUDITORIUM FOYER, CONTEMPORARY ART
VIEWING PLATFORM
STORAGE
WORKSHOPS
PHOTOGRAPHY STUDIO
MECHANICAL SPACE
AUDITORIUM
+50.00
+35.90
+31.60
+27.70
+23.80
+18.50
+13.30
+9.20
+3.00
+0.00

Section 剖面

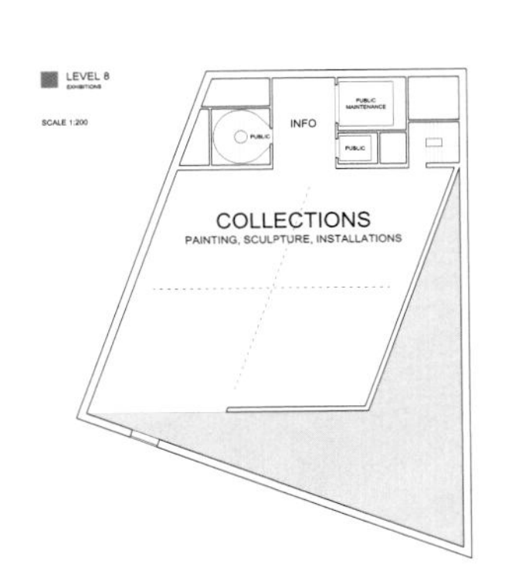

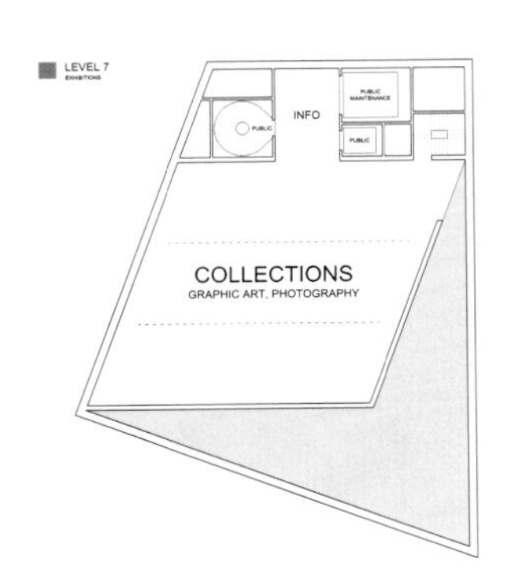

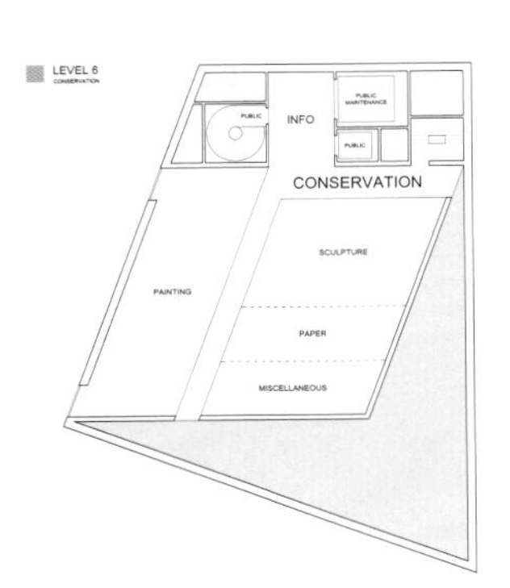

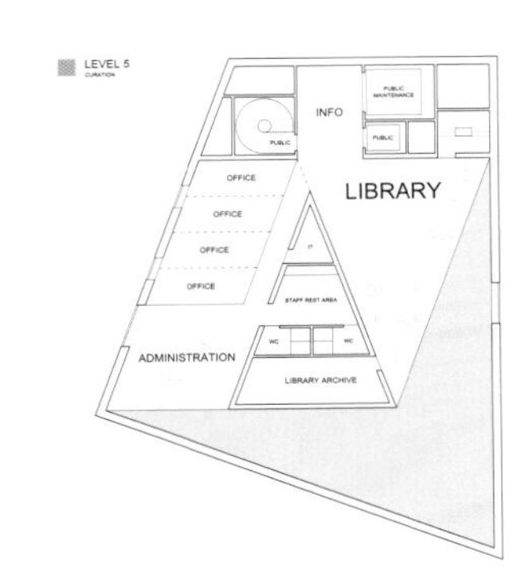

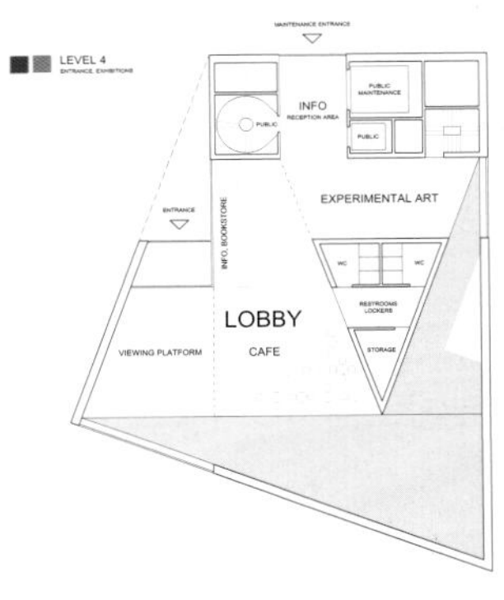

Plans 平面

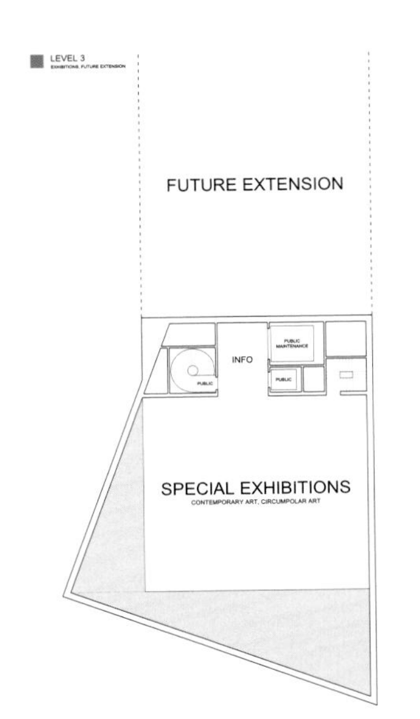

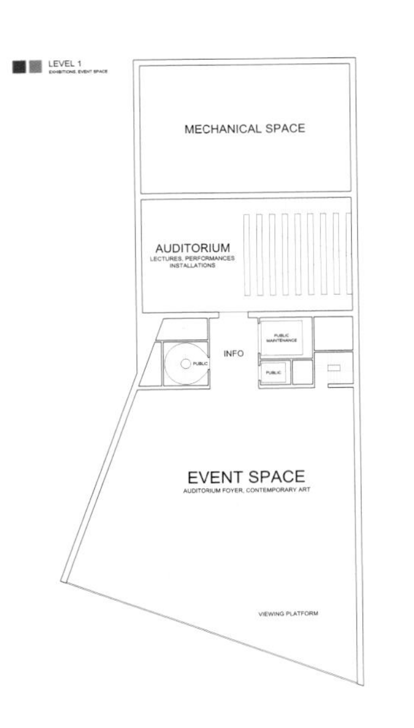

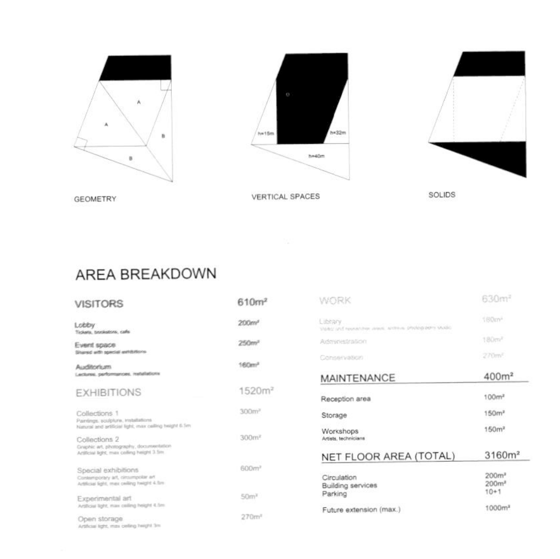

AREA BREAKDOWN

VISITORS	610m²
Lobby Tickets, bookstore, cafe	200m²
Event space Shared with special exhibitions	250m²
Auditorium Lectures, performances, installations	160m²
EXHIBITIONS	1520m²
Collections 1 Paintings, sculpture, installations Natural and artificial light, max ceiling height 6.5m	300m²
Collections 2 Graphic art, photography, documentation Artificial light, max ceiling height 3.5m	300m²
Special exhibitions Contemporary art, circumpolar art Artificial light, max ceiling height 4.5m	600m²
Experimental art Artificial light, max ceiling height 4.5m	50m²
Open storage Artificial light, max ceiling height 3m	270m²
WORK	630m²
Library	180m²
Administration	180m²
Conservation	270m²
MAINTENANCE	400m²
Reception area	100m²
Storage	150m²
Workshops Artists, technicians	150m²
NET FLOOR AREA (TOTAL)	3160m²
Circulation	200m²
Building services	200m²
Parking	10+1
Future extension (max.)	1000m²

Details 详图

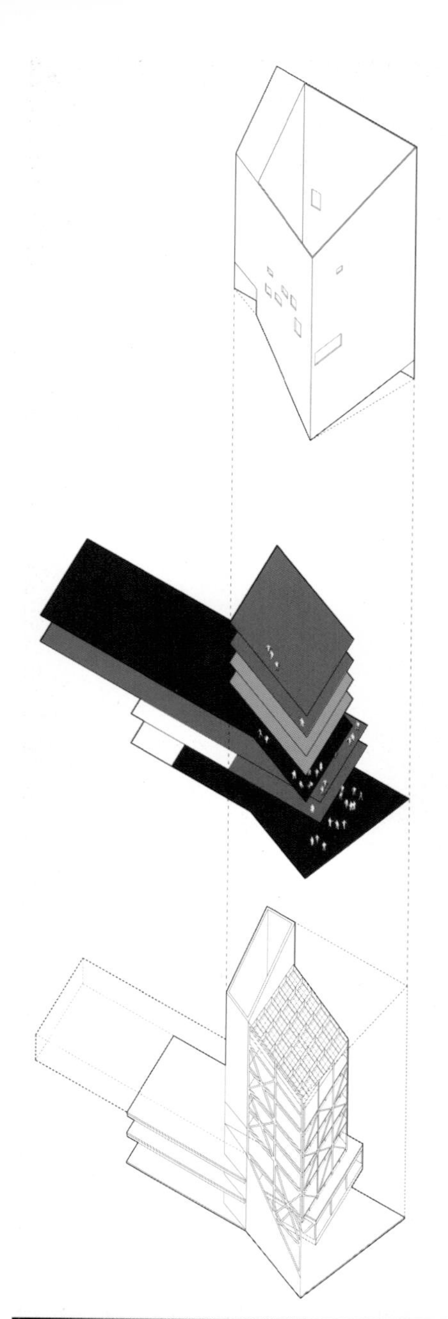

ENVELOPE

FUNCTIONAL CONCEPT

- LEVEL 8
  EXHIBITIONS
- LEVEL 7
  EXHIBITIONS
- LEVEL 6
  CONSERVATION
- LEVEL 5
  CURATION
- LEVEL 4
  ENTRANCE, BOOKSTORE, CAFE, EXHIBITIONS
- LEVEL 3
  EXHIBITIONS, FUTURE EXTENSION
- LEVEL 2
  EXHIBITIONS, MAINTENANCE
- LEVEL 1
  EXHIBITIONS, EVENT SPACE, MECHANICAL SPACES

STRUCTURAL CONCEPT

The building is made of cast-in-situ reinforced concrete and prefabricated steel parts that allow for straightforward construction and minimized material consumption. The assembly of the new national gallery refers to the culture of shipbuilding.

The building is stabilized by its cast-in-situ reinforced concrete core. The load-bearing envelope is composed of steel trusses and large rectangular steel tubes for lateral load-resistance. The facade is made of hot-rolled steel sheets welded in-situ, lending the monolithic appearance of the building's facade. The uniform skin is attached to the structural frame by springs that facilitate thermal movements.

The cantilevered interior spaces are supported by steel trusses and anchored to the cast-in-situ reinforced concrete foundation. The slabs are made of cast-in-situ reinforced concrete with corrugated steel sheets as formwork.

The large planar truss of the roof allows for a non-structural ceiling membrane, and thus the use of natural light in the entire building.

Conceptual Diagram 概念分析图

Onassis Cultural Center

奥纳西斯文学美术馆

AS.Architecture Studio　　法国AS建筑工作室

Location : Athens, Greece
Floor Area: 20,000 m^2
Status :Completed in 2010
Image Courtesy : AS.Architecture Studio

区位：希腊，雅典
建筑面积：20000 m^2
状态：2010年建成
图片提供：法国AS建筑工作室

Foundation includes an opera house - theater of 900 seats, an auditorium conference hall-cinema of 200 seats, an open air amphitheatre of 200 seats, a library, a restaurant and an exhibition hall.

The building is designed as a simple, diaphanous volume built of Thassos marble, elevated over a glass base. The facades are simultaneously opaque and transparent depending on whether one perceives them from near or afar. The project is revealed in the movement, the approach towards the building. The opacity of the stone is balanced by transparency, rhythm and matter.

This urban scenography is based on the treatment of the building's facades as a living, responsive membrane, reflecting the activity of the Foundation itself and its openness onto the city and the world. The simplicity of the volume and the abstraction of architectural expression create the monumental aspect that would have seemed almost impossible because of the small size of the plot.

Beyond the facades, a precious and surprising object that brings together the three auditoriums crosses the entire building. This is the heart of the project. Magnified by the void surrounding, its envelope forms the ideal stage for the Cultural Center's events, whose presence is marked at the scale of the building and that of the city.

基地包括一座可容纳900人的歌剧院、一座可容纳200人的音乐厅、一间可容纳200人的露天阶梯教室、一个图书馆、一个餐厅、一个展览厅。

奥纳西斯文学美术馆由达索大理石建在玻璃底座上，外形简洁大方、呈半透明状。墙体立面的近景与远景分别呈现出不透明与透明的透视效果，随着逐渐的靠近，整个建筑充满动感。不规则的实体石材与规则透明的玻璃体取得了和谐的平衡。

建筑表面的处理透射出文学美术馆本身的活动，形成勃勃生机，使建筑风格体现出城市的特色。由于建筑本身体量的限制，简约的条纹石材表面和抽象的建筑表达造就的美术馆如同不可能完成的任务，让人叹为观止。

在内部，贯穿整个建筑的巨大主体容纳了三个视听空间。这是建筑的核心部分。四周空间衬托出主体的华丽壮观，其外壳包含的先进舞台装置代表了奥纳西斯文学美术馆以及整个城市的品味。

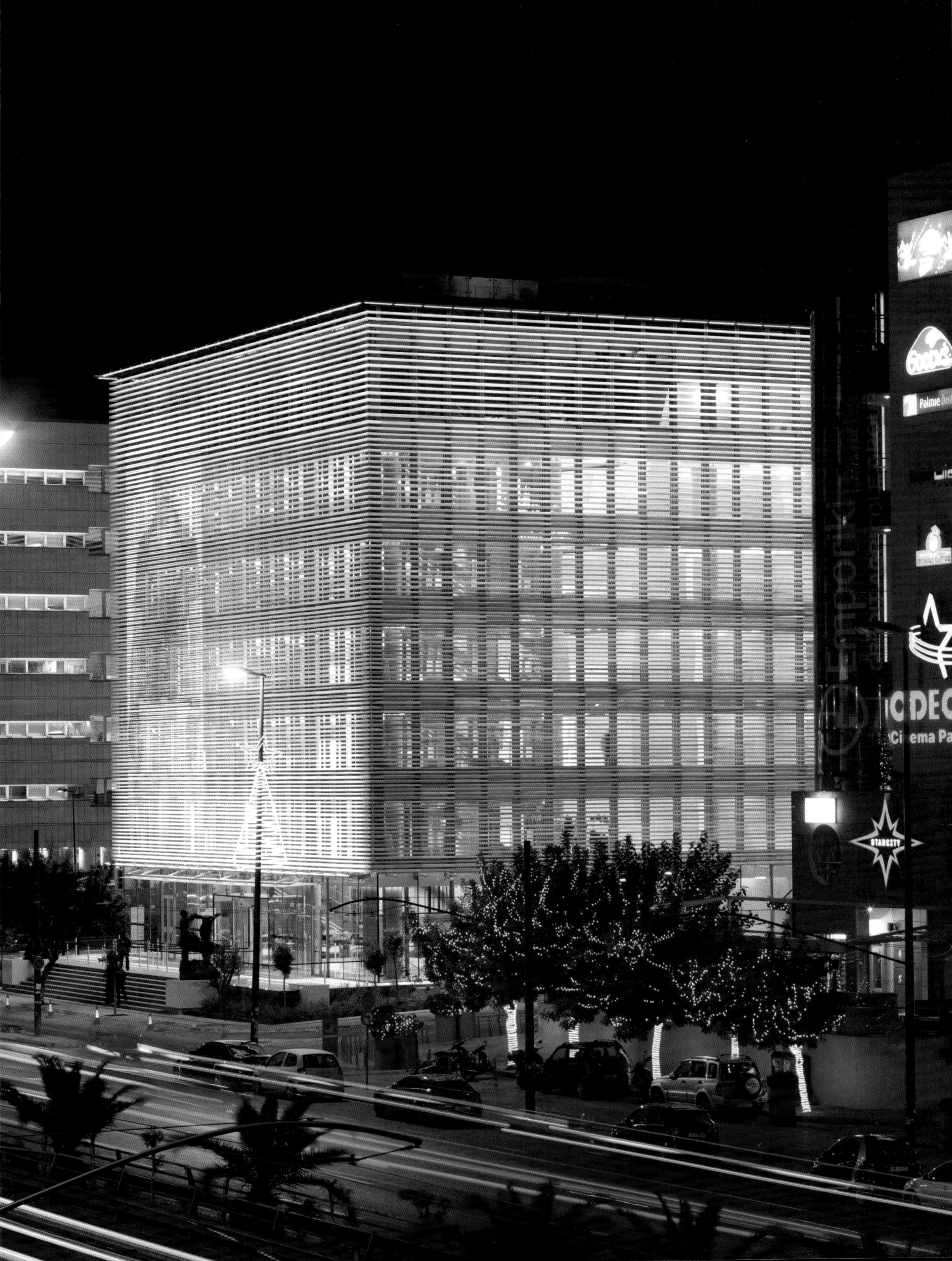
STARCITY

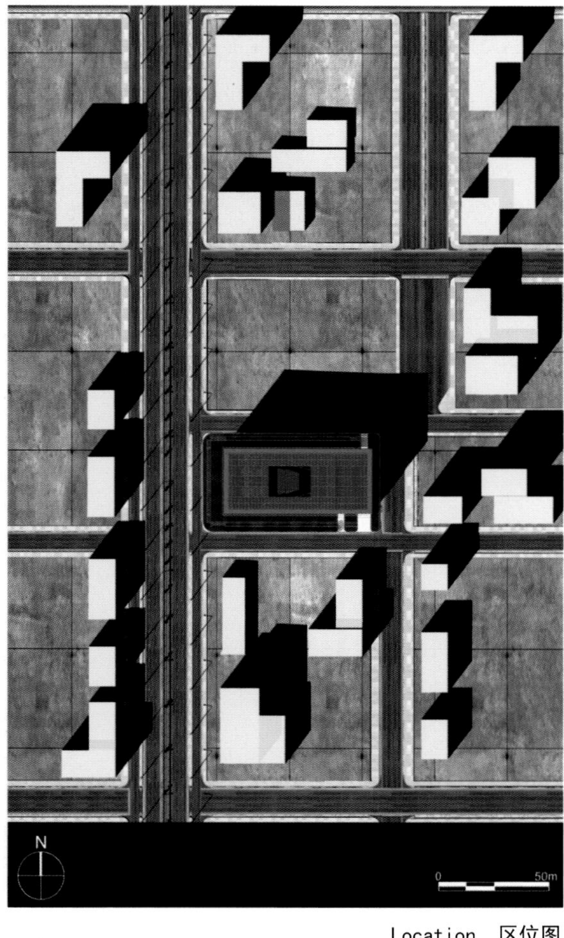

Location 区位图

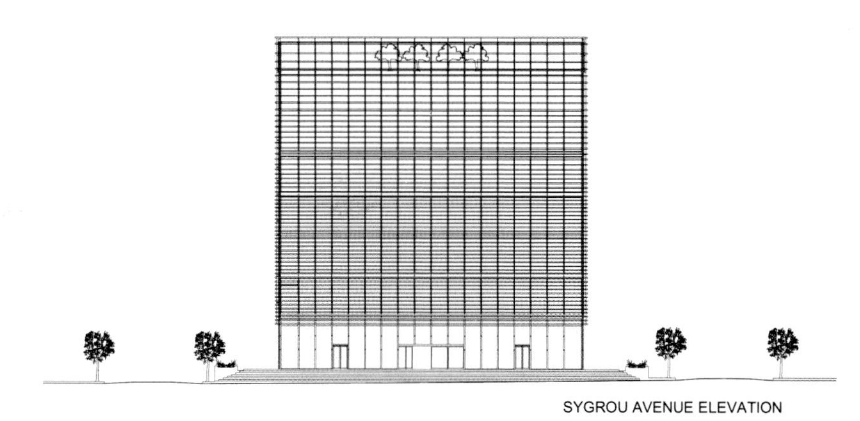

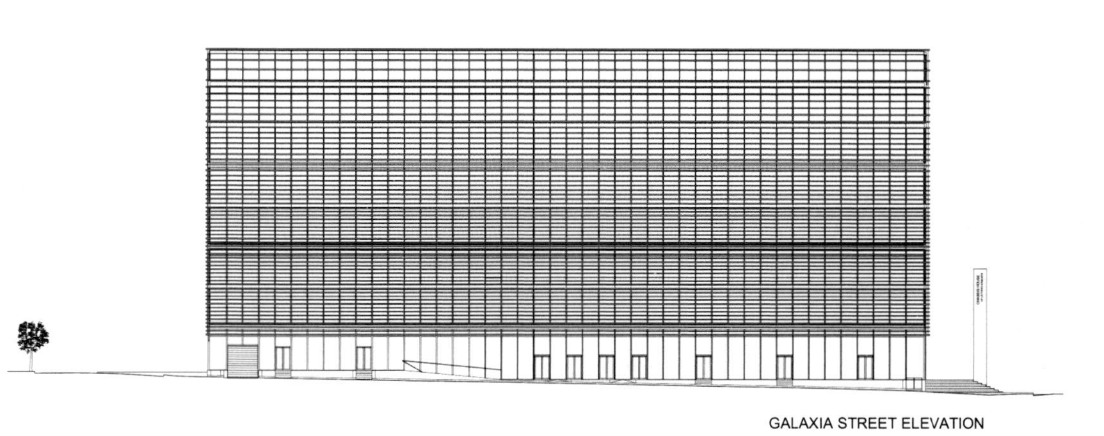

Elevations 立面

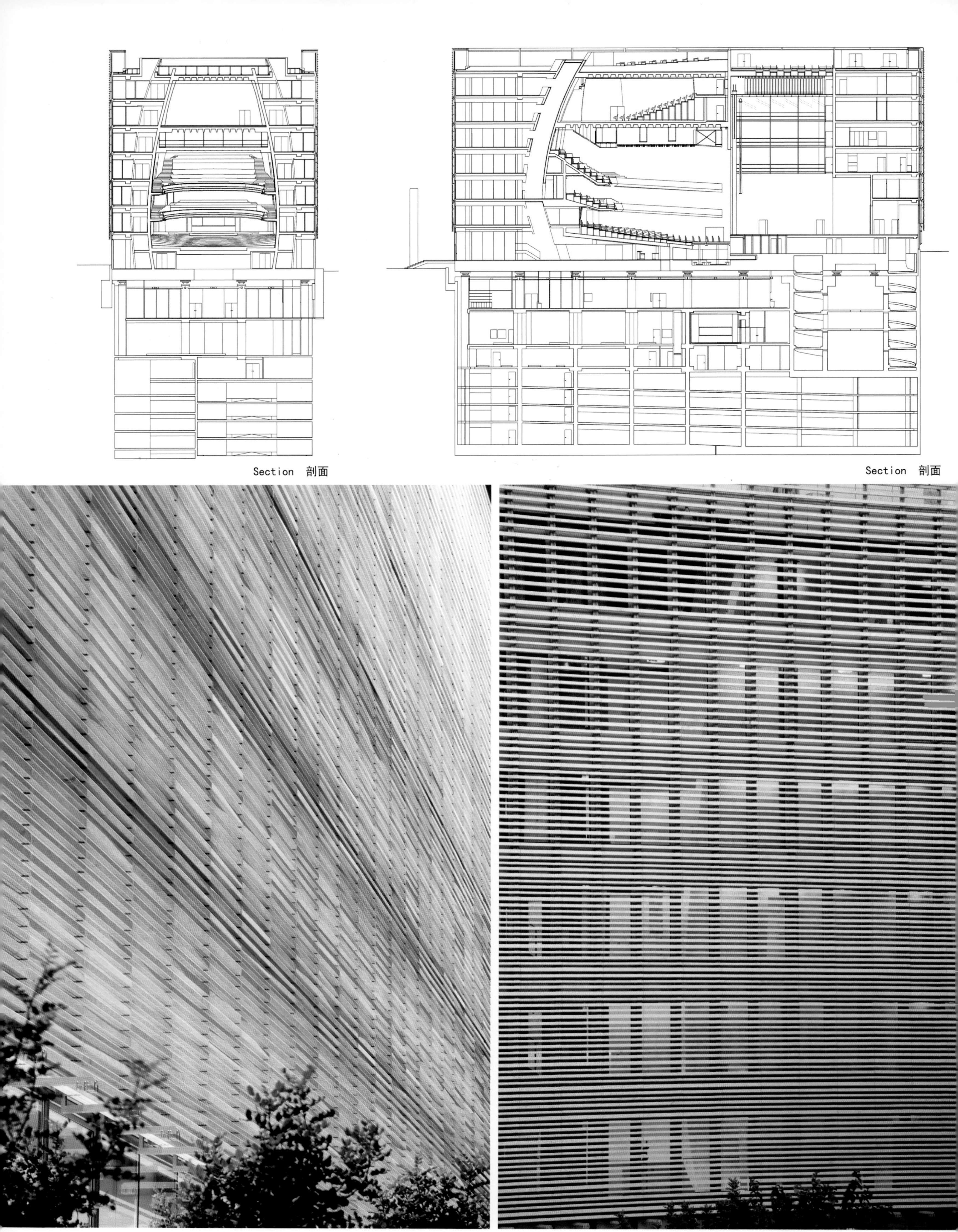

Section 剖面

Section 剖面

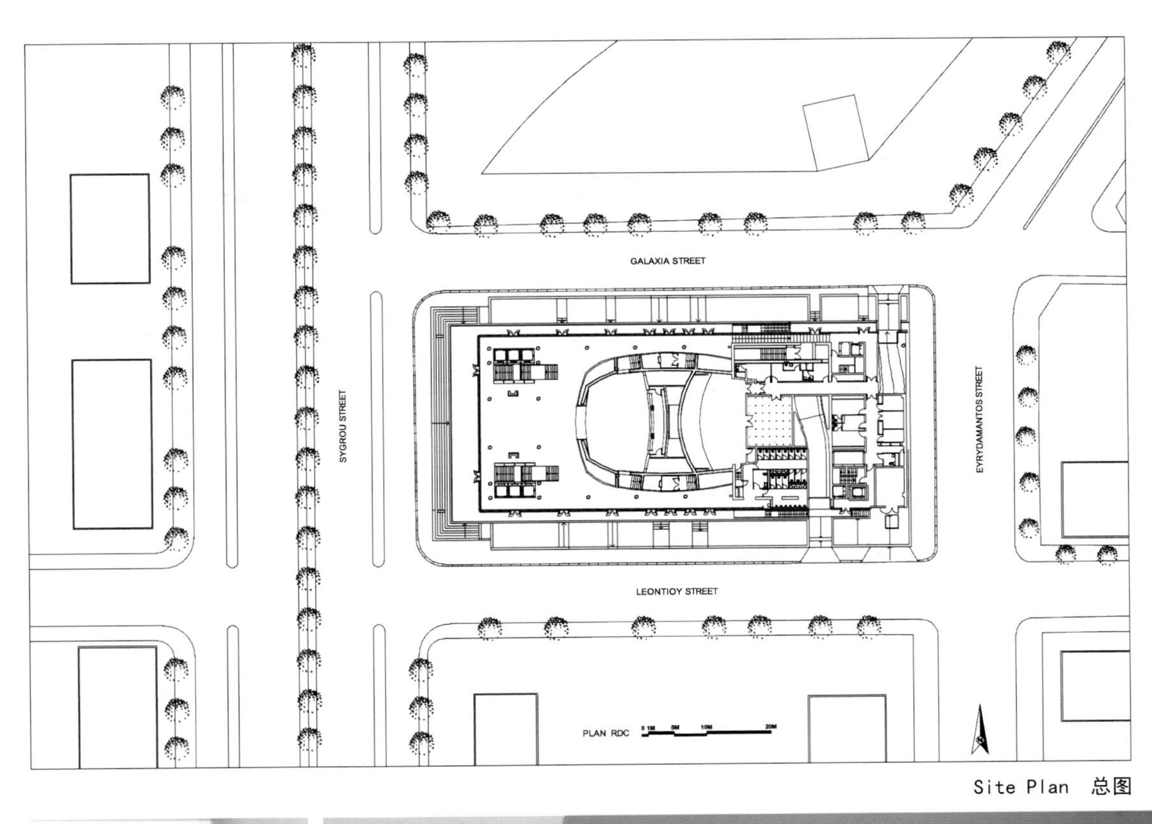
GALAXIA STREET
SYGROU STREET
EYRYDAMANTOS STREET
LEONTIOY STREET
PLAN RDC

Site Plan 总图

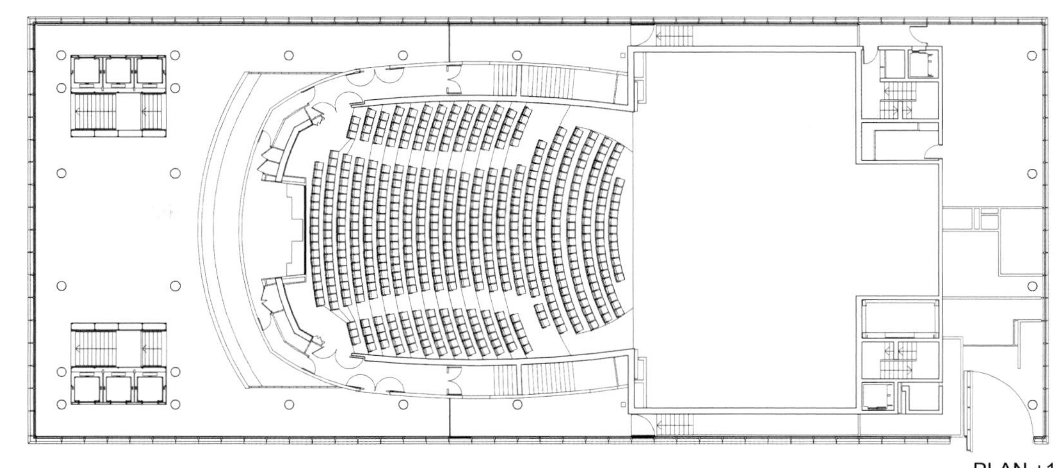

PLAN +1
THEATRE 1
一层平面　剧场

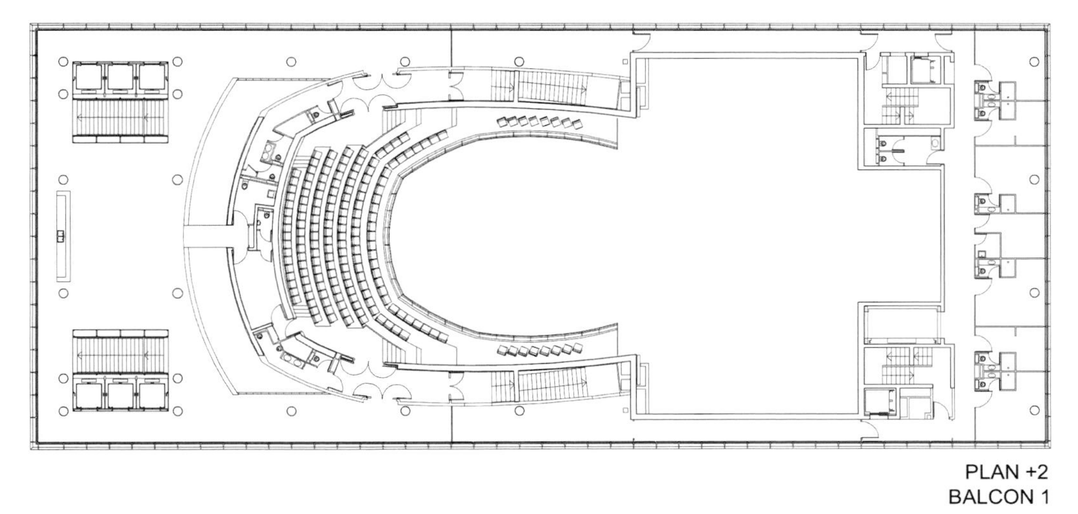

PLAN +2
BALCON 1
二层平面　楼座

Jamia Art Gallery of JMI

JMI Jamia画廊

Romikhosla Design Studio   Romikhosla设计工作室

Location: JMI University, New Delhi
Program: Art Gallery
Area: 810 m^2
Status: Completed in 2008
Image Courtesy: Saurabh Pandey

区位：新德里，JMI大学
功能：画廊
面积：810 m^2
状态：2008年建成
图片提供：Saurabh Pandey

The university of Jamia was established in the 1930's. As the university evolved, it introduced a wide range of contemporary academic disciplines such as media studies and central Asian studies. Jamia University is popularly regarded as a progressive avant garde campus. In 2008, the vice chancellor proposed a new cultural hub for the university that would have as its core a contemporary students' canteen, a unique art gallery and landscaped lawns.

The architects chose white marble in the canteen and white metal louvers in the art gallery to express this contemporary identity. The art gallery has become a community space for gathering alternative expressions of culture and identity. This role signaled the canteen and the art gallery as iconic models of architectural expression in contemporary Indian academic institutions.

The art gallery has three main parts to it. The front gallery that is naturally lit is primarily designed for the display of popular art and student exhibitions. The second space is the main internal gallery which is lit by controlled light and can be divided into two smaller galleries with the help of the central pivoting wall. This gallery is designed for the great university art collection, as well as for external artists who want to exhibit their work here. The third exhibition space is the open air sculpture court at the rear of the building.

Other than this, the art gallery also has two artist studios adjacent to the sculpture court which are designed for short term stay of visiting artists.

JMI大学建于20世纪30年代。作为一所逐渐演变而来的大学，JMI大学引进了诸如媒体研究及中东研究等众多领域的现代学科。大学被广泛视作一个进步的前锋大学。2008年，副校长提议为该大学建造一个新的文化中心，该中心以当代的学生食堂、独特的画廊及景观草坪为核心。

建筑师选择使用白色大理石装修食堂，用白合金百叶窗装饰画廊，用以展示出该建筑的当代身份。画廊已成为集合文化及个性交替展现的社区空间了。这一角色赋予餐厅及美术画廊以当代印度学院机构式的标志性建筑表现力。

画廊由三大部分组成。自然采光照明的前部画廊是专为流行艺术品和学生作品展览而设计的。第二大空间构成则是主内部画廊，可控灯光负责室内照面，且可在中央旋转墙的帮助下分成两个较小的画廊。该画廊是为展示名牌大学艺术收藏品而设计的，也可用于外界艺术家举办其作品展。第三大展览空间是位于建筑物后方的露天雕塑场。

除此之外，该画廊也设有两大艺术家工作室，它们与雕塑庭院相连，主要提供来访艺术家短期停留时使用。

M.F.Husain Art Gallery

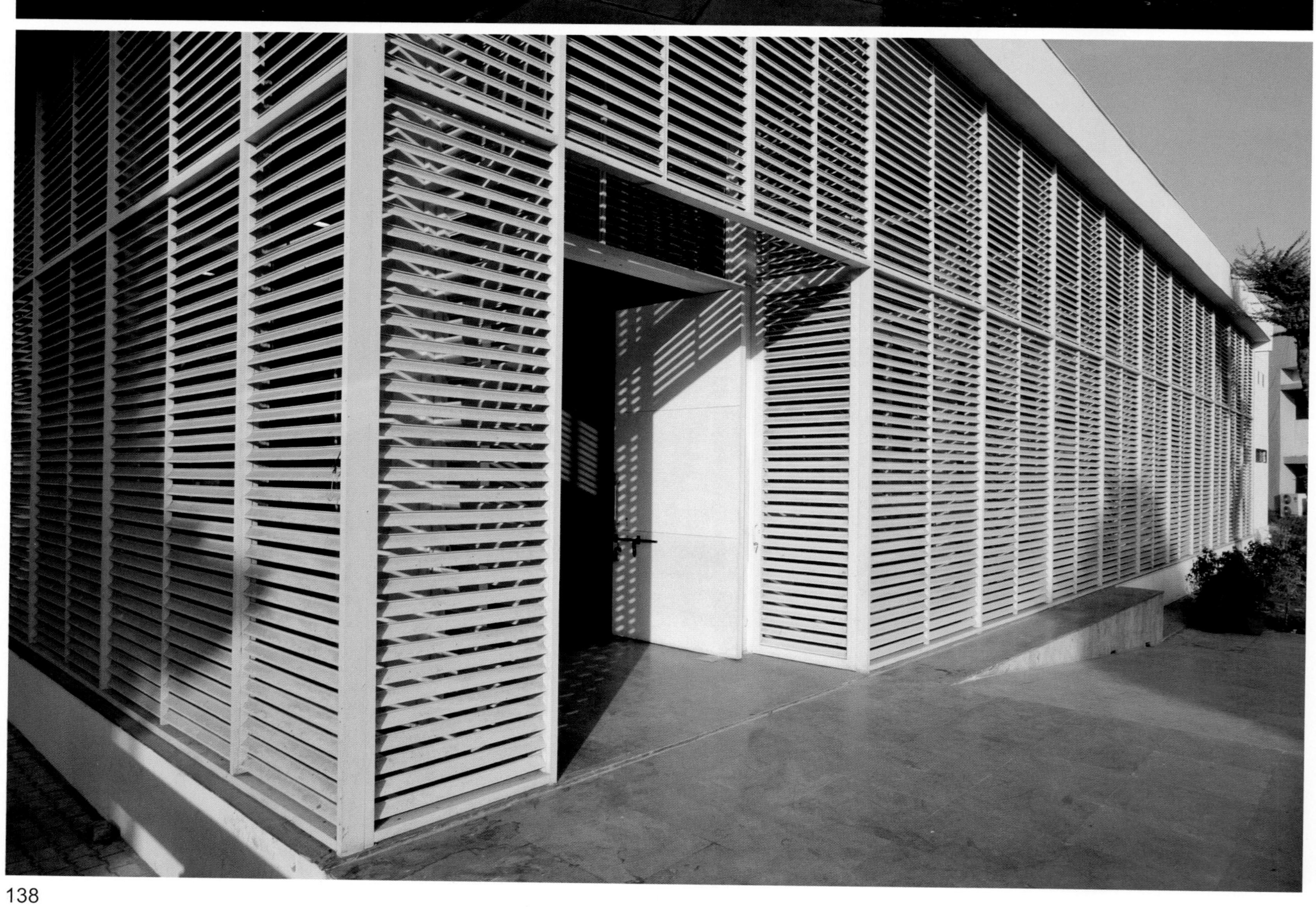

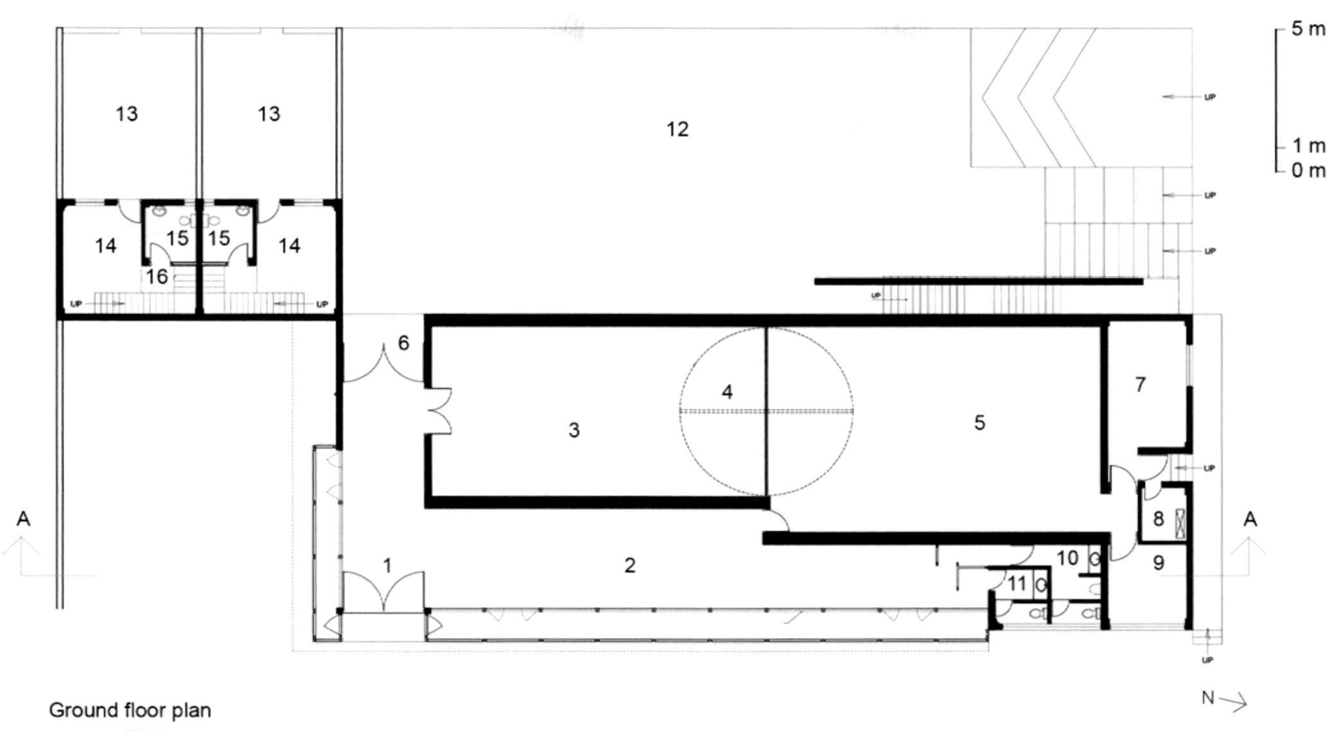

Ground floor plan
Plans 平面

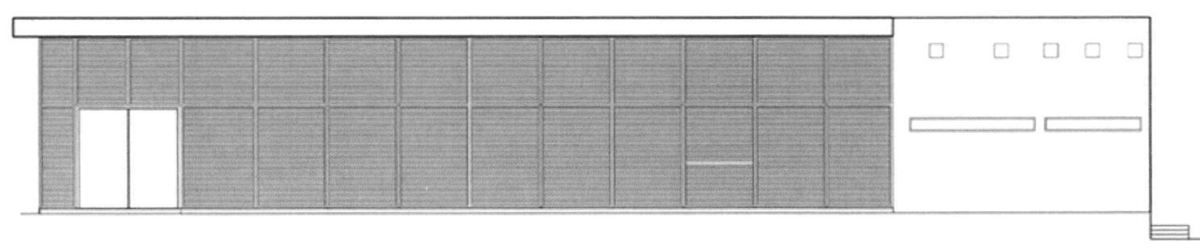

East elevation- With screen
带有百叶的立面

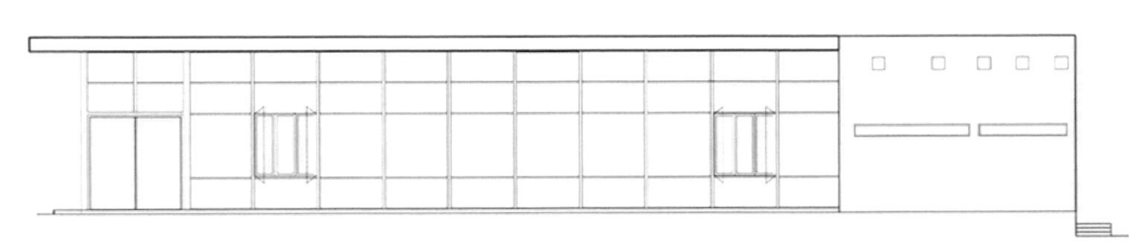

East elevation- Without screen
无百叶的立面

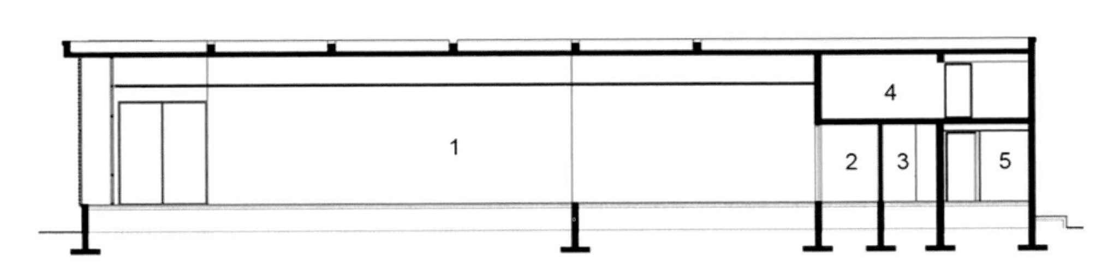

Section AA
剖面 AA

LEGEND

Ground Floor Plan:

1 - Main entrance
2 - Gallery lobby
3 - Gallery I
4 - Centrally pivoted wood partition door
5 - Gallery II
6 - Back entrance
7 - Manager's office
8 - Elec. room
9 - Store
10 - Gents toilet
11 - Ladies toilet
12 - Sculpture Court
13 - Open Courtyard
14 - Studio
15 - Toilet
16 - Pantry

Section:

1 - Gallery Lobby
2 - Ladies Toilet
3 - Gents Toilet
4 - AC Plant
5 - Store

Exhibition

展览馆

PJCC – The POD Exhibition Hall

PJCC – POD展厅

Studio Nicoletti Associati Italy and Hijjas Kasturi Associates Sdn Malaysia
意大利Nicoletti联合工作室，马来西亚
Hijjas Kasturi联合工作室

Location: Petaling Jaya, Kuala Lumpur, Malaysia
Program: Exhibition Hall
Dimension: 800 m^2
Status: 2010 (built)
Image Courtesy : © H. Lim Ho

区位：马来西亚，吉隆坡，八打灵再也
功能：展厅
规模：800 m^2
状态：2010竣工
图片提供：H. Lim Ho

In the area of Petaling Jaya, west of Kuala Lumpur, a great urban development is under way for the establishment of a new urban center. As a landmark for this area, the developer wanted to host his on-site offices and sale's showroom in an iconic pavilion that would reflect the spirit and the architectural style of the whole development.

Water droplets in nature was the inspiration for 'The Pod' pavilion structure creating a dynamic spherical form resulting in a primitive building archetype with a modern twist. The round and soft shape of The Pod is formed as a series of elliptical sections of variable widths and heights. Slithers of windows bring natural daylight into the spaces below

Internally the pavilion is divided into two parts: one zone is dedicated to the office area and the other contains the main showroom. The building appears to be sliced diagonally into a series of ribbons which wrap up and over the building creating a dynamic space within, forming a layered protective shell.

The structure is fabricated from tubular steel members with the exterior skin made of spectrally reflective aluminium panels. The exterior skin colour shades change's depending on the reflection of the sun, dynamically.

The Pod is surrounded by greenery with a reflective water pool wrapping around the edges that will contribute to the bioclimatic behaviour and wellness of its inhabitants.

为了在吉隆坡西部的八打灵再也打造一个新的城市中心，城市建设正在筹备之中。作为这一全新中心的地标建筑，开发商希望能够将他们的现场办公室和销售展示大厅放在一个标志性的建筑里。这个建筑能够反映出整个地区开发的精神和建筑风格。

这座名为“茧”的展示馆，其建筑灵感来源于自然界的水滴。这个动感十足的球形建筑是在原始建筑体的基础上赋予了现代感的扭动。看上去浑圆而柔软的茧是由一系列椭圆状结构以不同的宽度和高度组合而成，而自然之光则通过能够滑动的窗户进入到馆内。

馆内分为两个部分：一个区域用来办公，另一个则包含了主要的销售展示大厅。这个建筑看上去像是一条条斜放着的丝带组合在一起，在其中创造了一方动态的空间而其外则形成坚固的保护外壳。

整个建筑是由管状钢结构组件构成，外面覆盖一层反射灵敏度高的铝板材。这个外层的“皮肤”能够根据太阳光的反射自动地改变颜色。

“茧”周围环绕着绿地，在整个建筑物的最边缘还围绕着一汪水池，能够对建筑物内居民的生物气候行为和健康作出反射性调整。

hotel
DESARIA
PJCC

PJCC

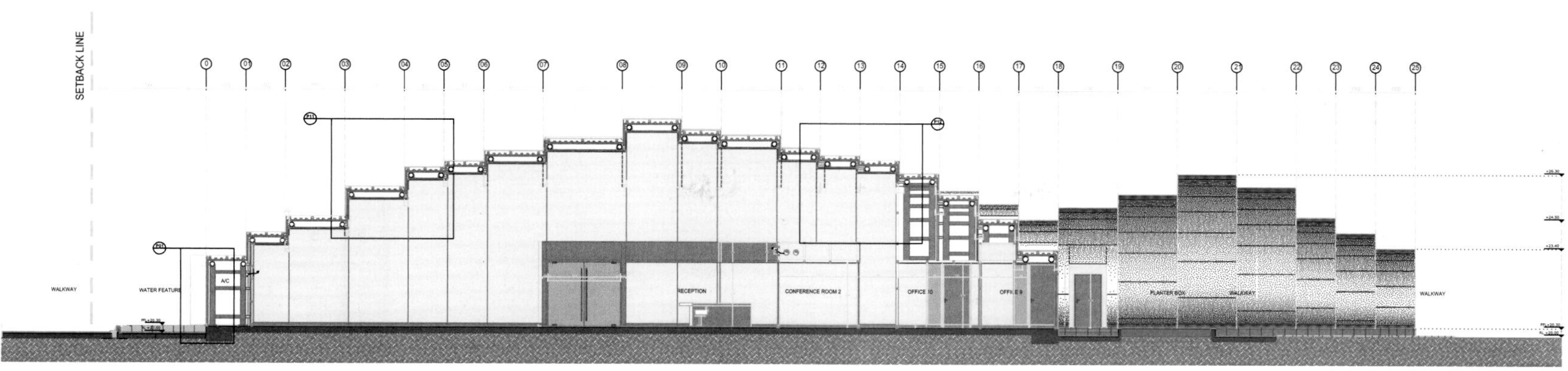

Section A-A 剖面A-A

Plan Level 01 一层平面

PJCC

PJCC

PJCC

Shanghai Expo 2010 Italian Pavilion

2010上海世博会意大利馆

Studio Nicoletti Associati NICOLETTI联合工作室

Location: Shanghai, China
Typology: Culture
Project Area: 6,000 m²
Status: 2008 International competition, honourable mention
Image Courtesy: Studio Nicoletti Associati

区位：中国，上海
类型：文化
面积：6000m²
状态：2008年国际竞赛荣誉奖
图片提供：NICOLETTI联合工作室

Pavilion's dynamic shape represents Italian innovation and fantasy, meanwhile the better city - better life theme is represented through the main features of Italian urban quality. Pavilion's structure, raised up from the ground, is the metaphor of a flying bird. The surface is shaped in accord with a warped geometry, the same of many living creatures as leafs and animals. The pavilion is suspended upon a white platform surrounded by water pools, and represents a specific Italian quality: lightness, intended by Italo Calvino as a agility, elegance, soar.

The structure is in steel to grant the precision of a prefabricated assembling system which will have as result an easy and economic construction. Three columns with stairs and services are going to support a horizontal structure with suspended floors. There will be no other structural parts, in order to obtain an internal space entirely opened and free.

Under the pavilion a space similar to the Italian square is provided, protected by the sun, refreshed by the water, opened to the surroundings. Here, a circular island will host different kind of shows.

Inside the pavilion nature and city are integrated as it uniquely happens in Italian cities. The first floor is covered with a grass surface and the three columns are the metaphor of city buildings.

外形动感的意大利馆阐述着意大利的创新精神与魔幻色彩，同时也将“城市，让生活更美好”的世博主题融入了意大利的都市品质。这个馆体结构拔地而起，象征着一只展翅欲飞的鸟儿。其平面形状为扭曲的几何图形，形似生物界中的树叶和多种动物。整个场馆悬于一块白色的平台之上，平台之下绿水环绕，寓意意大利的民族品质：轻。意大利著名作家伊塔罗・卡尔维诺则将这品质总结为体态轻盈，气质优雅，气度不凡。

整个馆体为钢结构，这样就保证了馆体预制构件组装系统的分毫不差。这样的设计做到了简便而又经济。主体为悬空平台的水平结构由三根支柱支撑。柱体包括台阶与一些辅助功能。除此之外，没有其他结构零件，以实现内在空间完全的开放与自由。

馆体正下方有一个类似意大利广场的空间。这里阳光充足，清水常新，四周毫无阻挡。其中有一处圆形小岛用以举办各类演出。

馆内自然与城市的完美组合唯有在意大利城市中得以窥见。一层由绿草覆盖，而穿插其间的那三根支撑的柱体则象征城市里的高楼大厦。

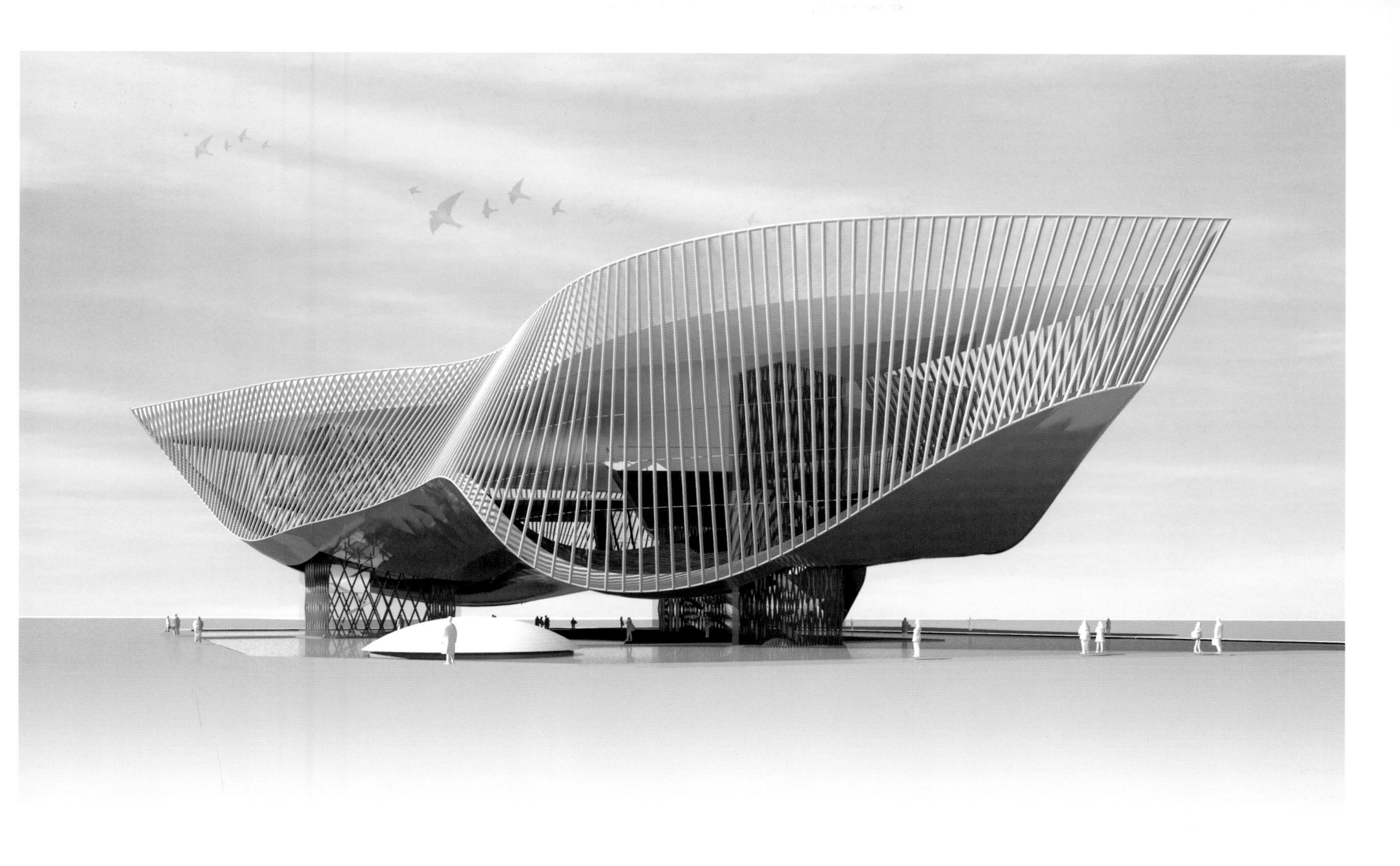

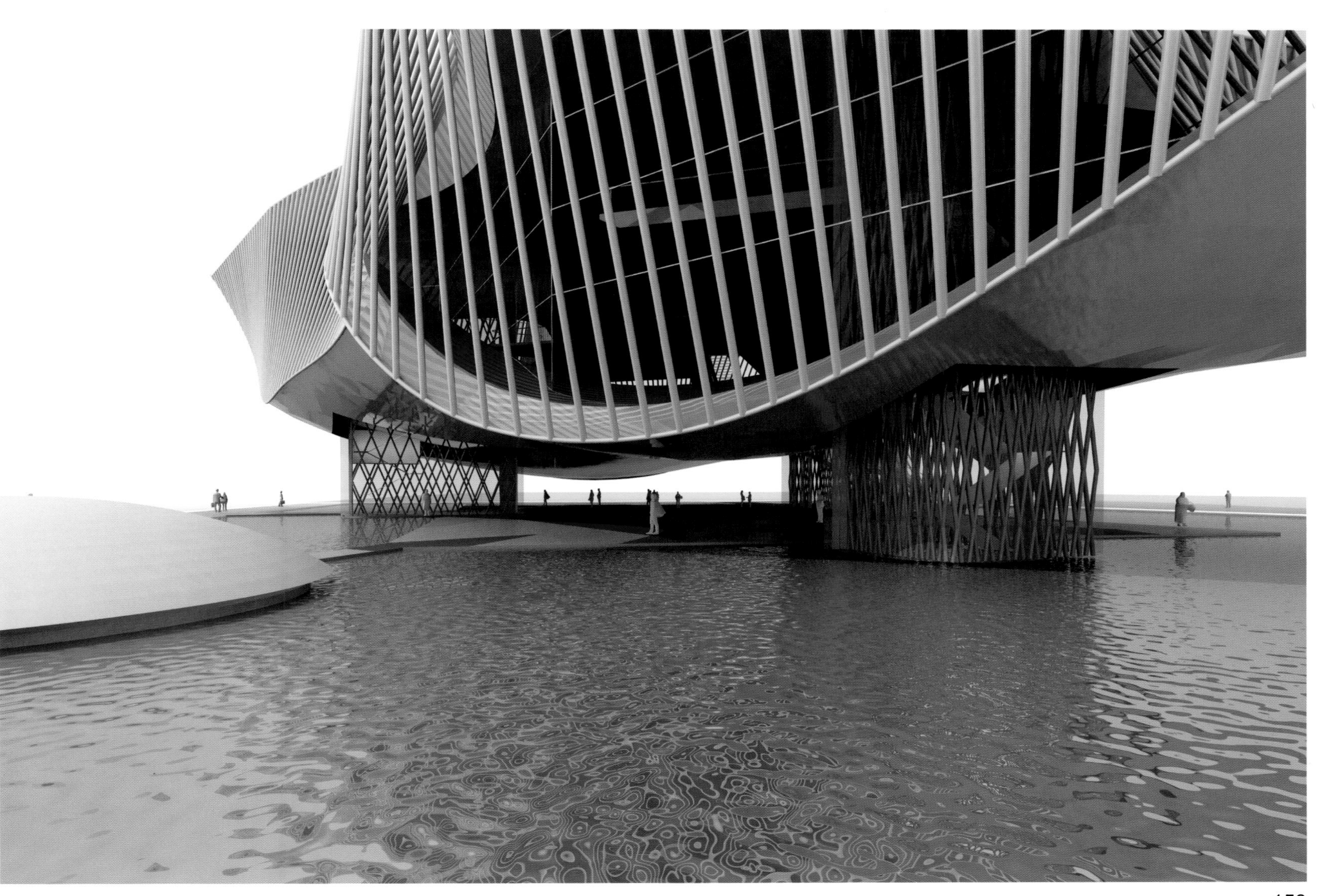

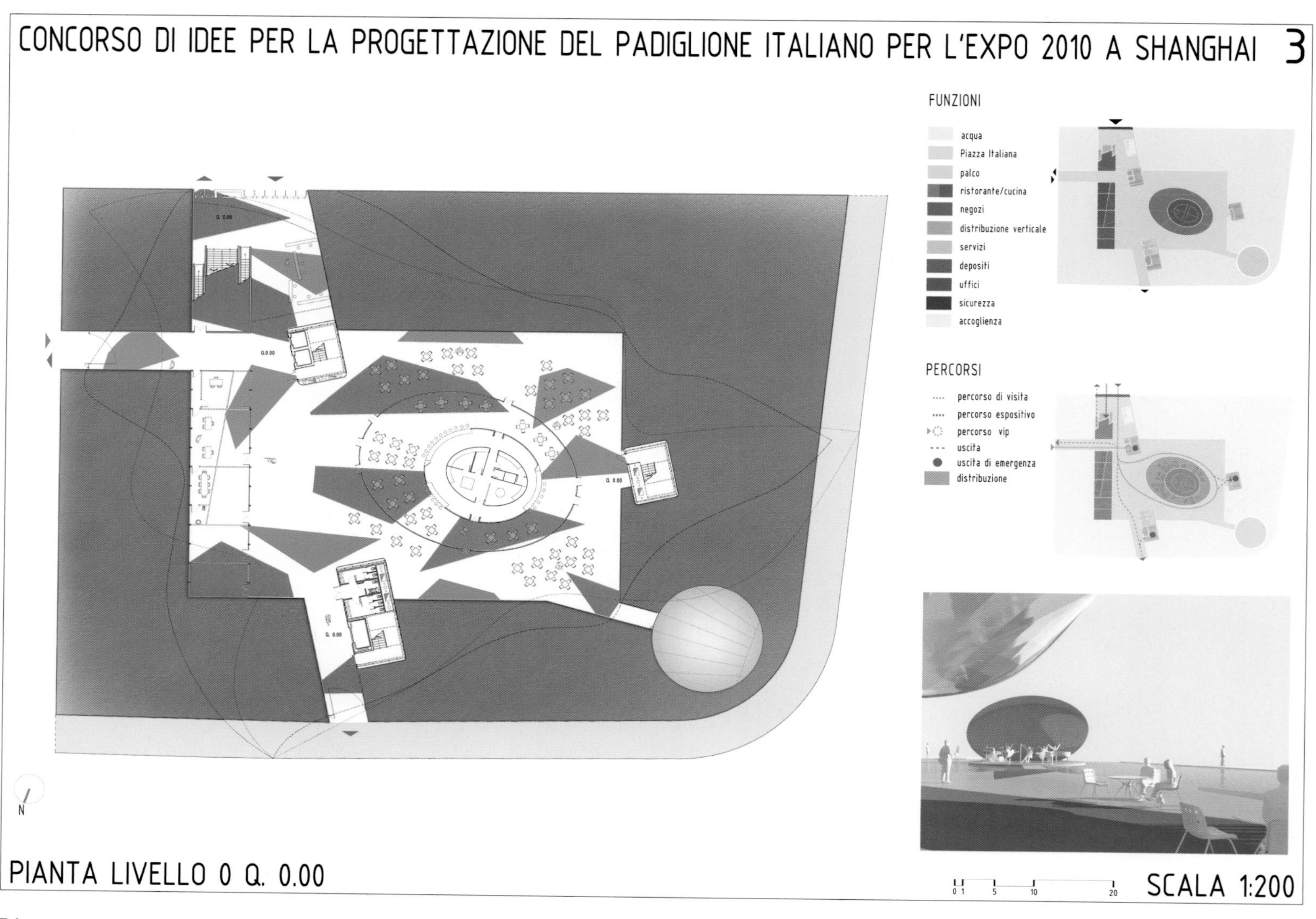

CONCORSO DI IDEE PER LA PROGETTAZIONE DEL PADIGLIONE ITALIANO PER L'EXPO 2010 A SHANGHAI 3
FUNZIONI
acqua
Piazza Italiana
palco
ristorante/cucina
negozi
distribuzione verticale
servizi
depositi
uffici
sicurezza
accoglienza
PERCORSI
percorso di visita
percorso espositivo
percorso vip
uscita
uscita di emergenza
distribuzione
Q. 0.00
N
PIANTA LIVELLO 0 Q. 0.00
0 1 5 10 20
SCALA 1:200

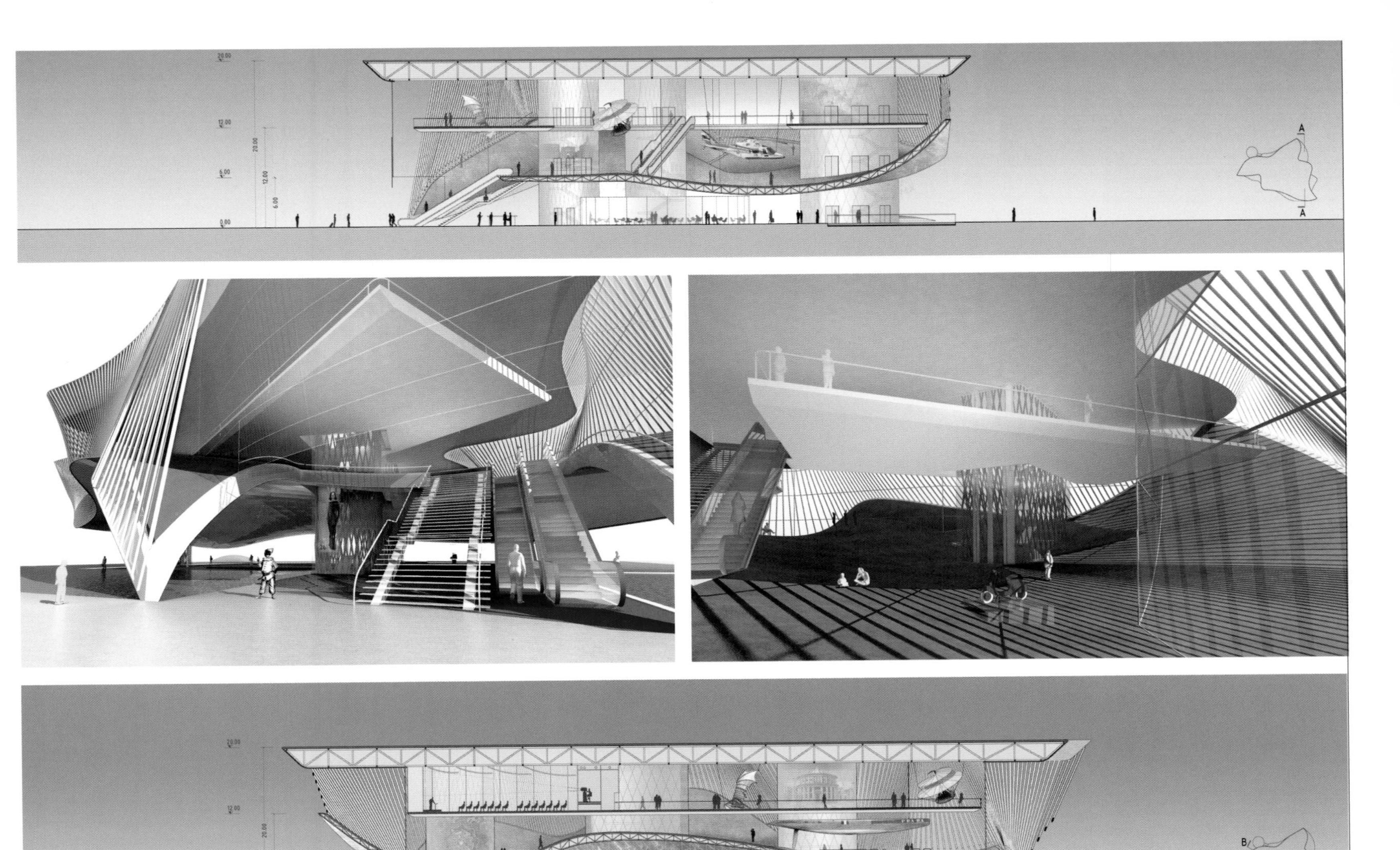

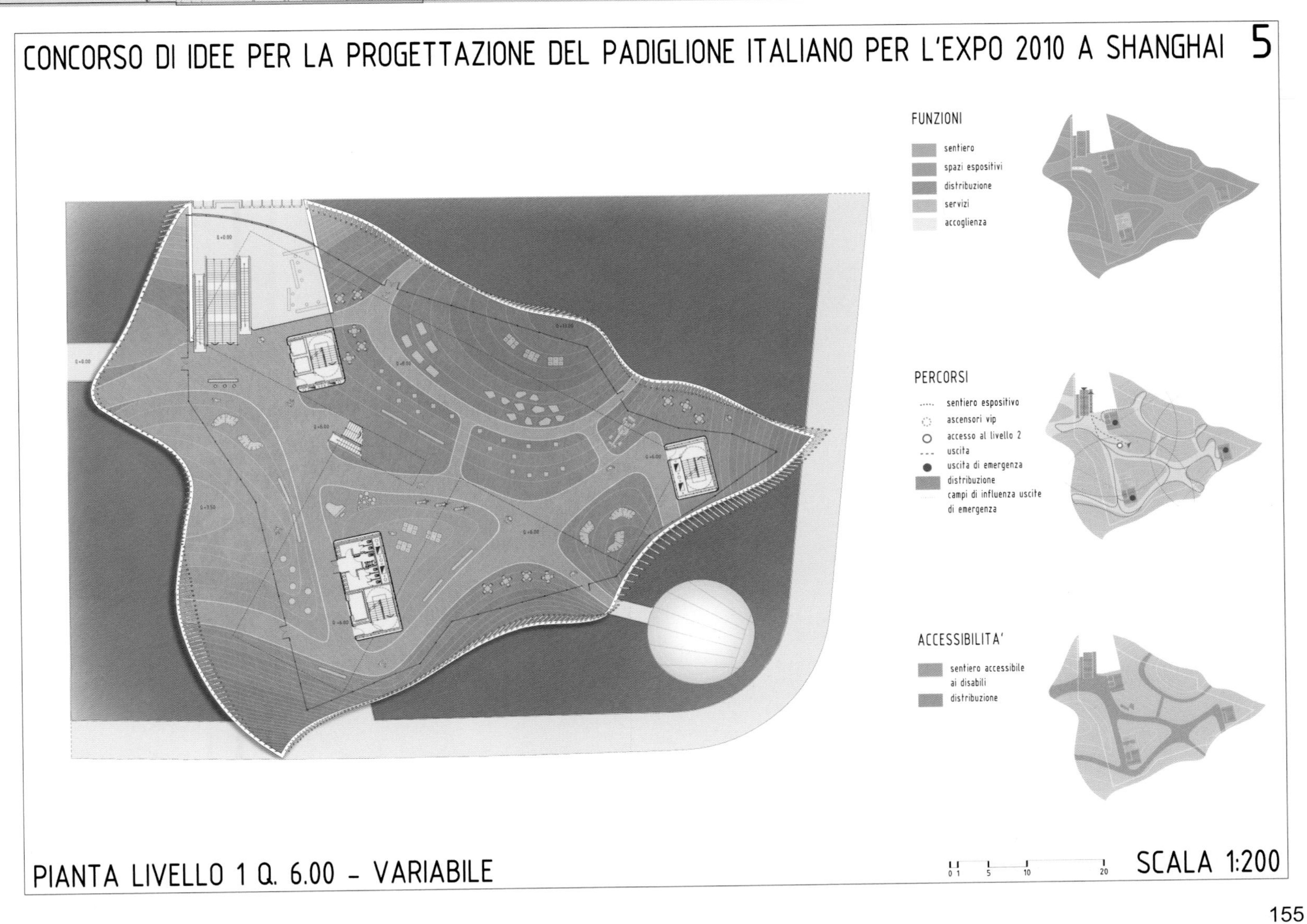

CONCORSO DI IDEE PER LA PROGETTAZIONE DEL PADIGLIONE ITALIANO PER L'EXPO 2010 A SHANGHAI 5
FUNZIONI
sentiero
spazi espositivi
distribuzione
servizi
accoglienza
PERCORSI
sentiero espositivo
ascensori vip
accesso al livello 2
uscita
uscita di emergenza
distribuzione
campi di influenza uscite di emergenza
ACCESSIBILITA'
sentiero accessibile ai disabili
distribuzione
PIANTA LIVELLO 1 Q. 6.00 - VARIABILE
SCALA 1:200

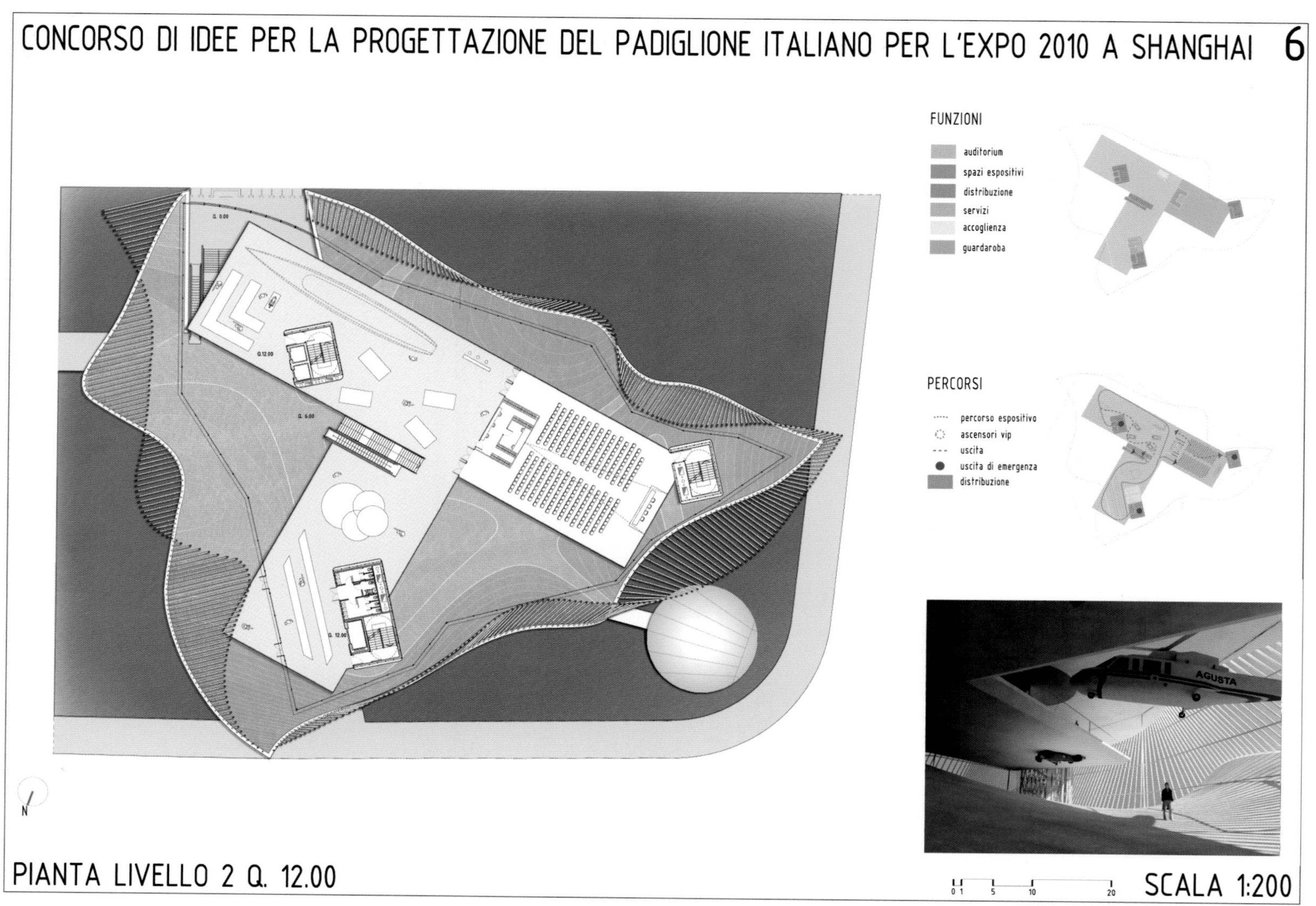

立面 Elevation

ITALIA

prospetto nord

立面 Elevation

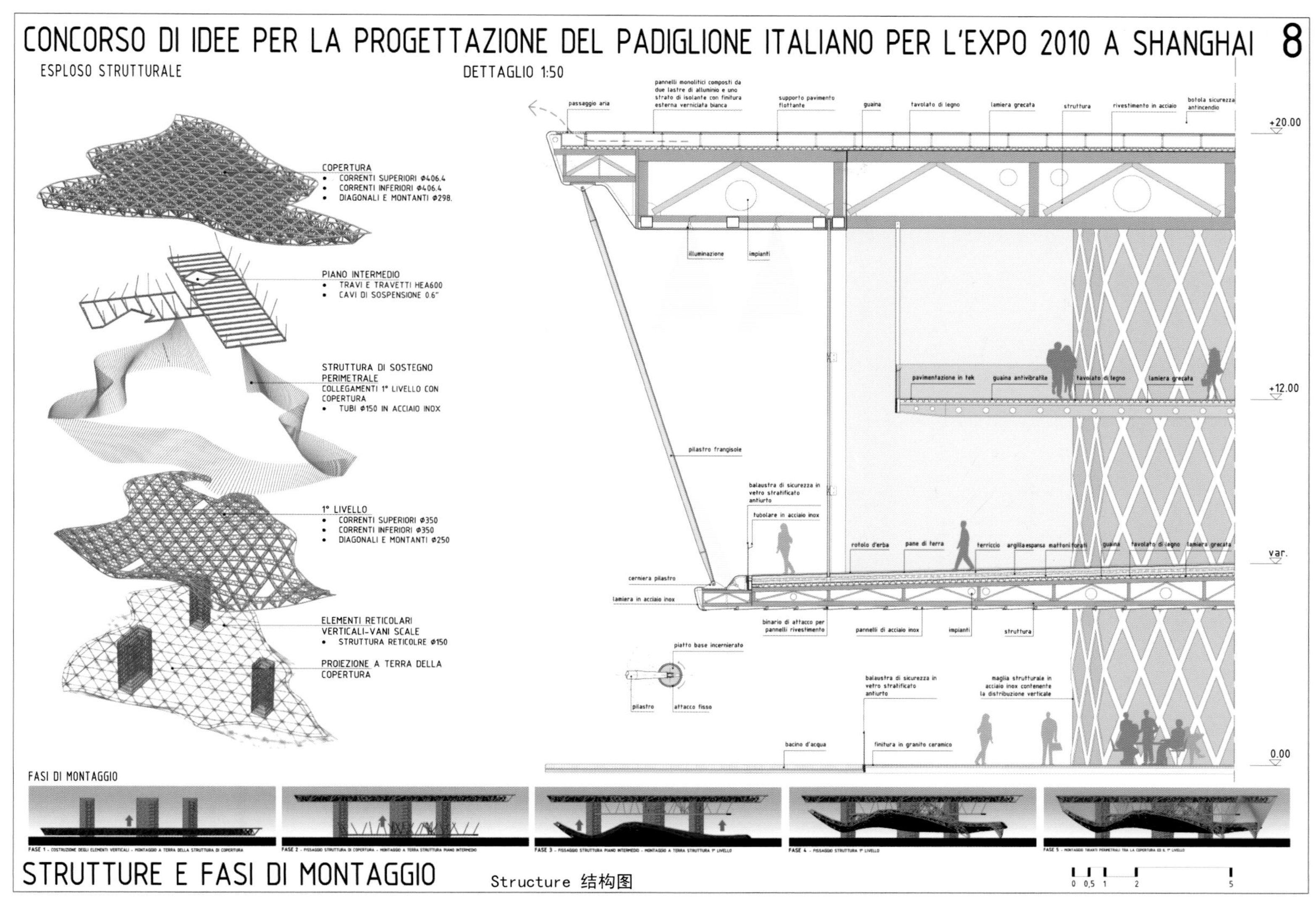

CONCORSO DI IDEE PER LA PROGETTAZIONE DEL PADIGLIONE ITALIANO PER L'EXPO 2010 A SHANGHAI 8
ESPLOSO STRUTTURALE
DETTAGLIO 1:50
COPERTURA
• CORRENTI SUPERIORI ø406.4
• CORRENTI INFERIORI ø406.4
• DIAGONALI E MONTANTI ø298.
PIANO INTERMEDIO
• TRAVI E TRAVETTI HEA600
• CAVI DI SOSPENSIONE 0.6"
STRUTTURA DI SOSTEGNO PERIMETRALE
COLLEGAMENTI 1° LIVELLO CON COPERTURA
• TUBI ø150 IN ACCIAIO INOX
1° LIVELLO
• CORRENTI SUPERIORI ø350
• CORRENTI INFERIORI ø350
• DIAGONALI E MONTANTI ø250
ELEMENTI RETICOLARI VERTICALI-VANI SCALE
• STRUTTURA RETICOLRE ø150
PROIEZIONE A TERRA DELLA COPERTURA
FASI DI MONTAGGIO
+20.00
+12.00
var.
0.00
passaggio aria
supporto pavimento flottante
guaina
favolato di legno
lamiera grecata
struttura
rivestimento in acciaio
botola sicurezza antincendio
illuminazione
impianti
pavimentazione in tek
guaina antivibratile
favolato di legno
lamiera grecata
pilastro frangisole
balaustra di sicurezza in vetro stratificato antiurto
tubolare in acciaio inox
rotolo d'erba
pane di terra
terriccio
argilla espansa
cerniera pilastro
lamiera in acciaio inox
binario di attacco per pannelli rivestimento
pannelli di acciaio inox
impianti
struttura
piatto base incernierato
pilastro
attacco fisso
balaustra di sicurezza in vetro stratificato antiurto
maglia strutturale in acciaio inox contenente la distribuzione verticale
bacino d'acqua
finitura in granito ceramico
STRUTTURE E FASI DI MONTAGGIO
Structure 结构图
0 0,5 1 2 5

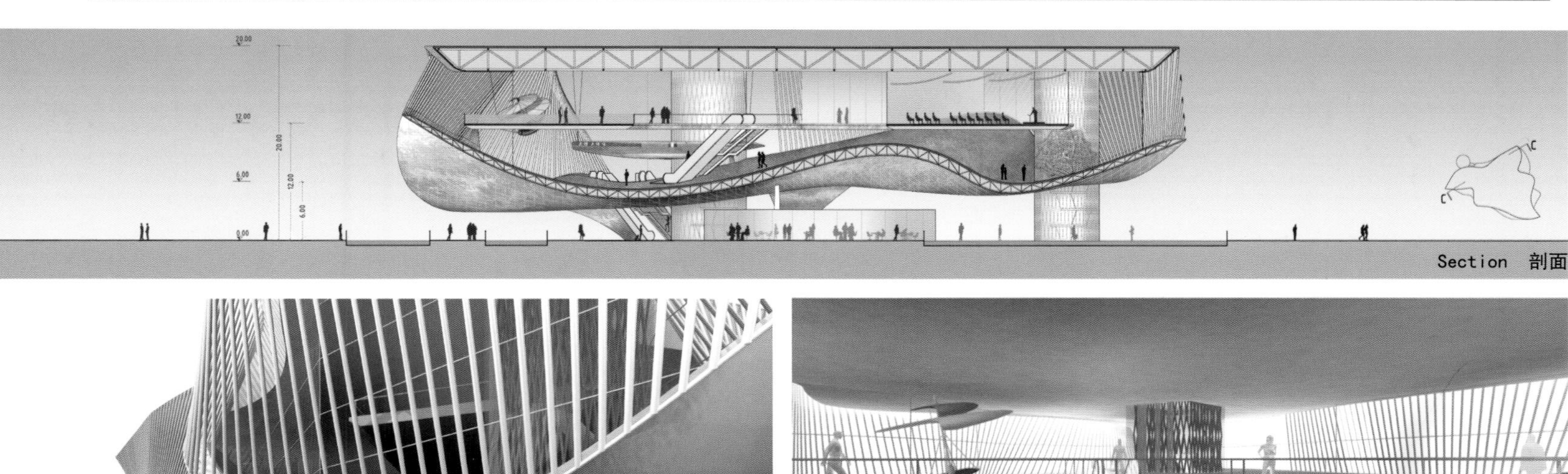

20.00
12.00
6.00
0.00
Section 剖面

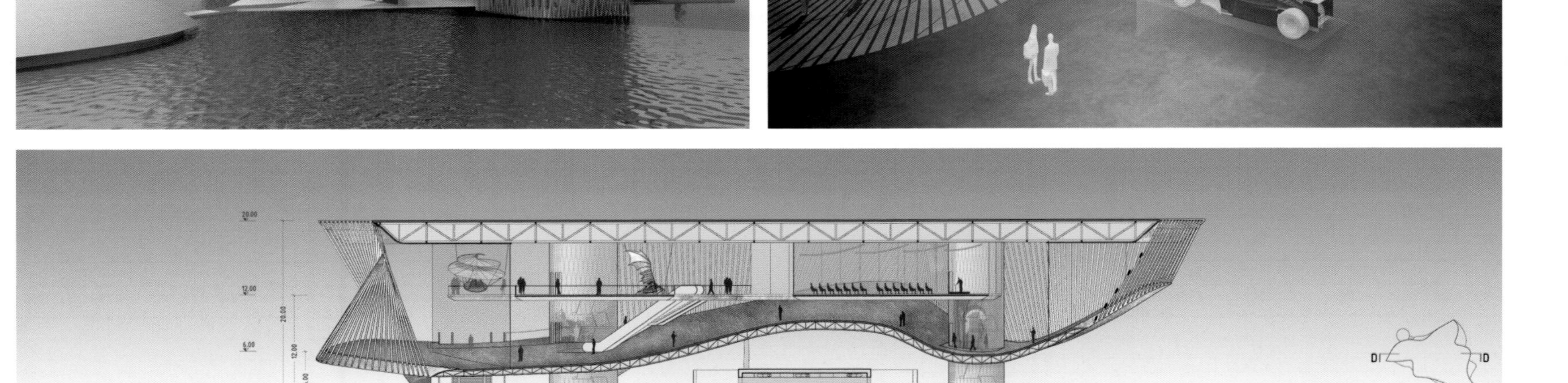

20.00
12.00
6.00
0.00
Section 剖面

Yeosu Expo 2012 Thematic Pavilion - South Korea

2012韩国丽水世博会主题馆

Studio Nicoletti Associati NICOLETTI联合工作室

Location: Yeosu, South Korea
Total Area: 5,000 m^2
Gross Floor Area: 4,000 m^2
Image Courtesy: Studio Nicoletti Associati

区位：韩国，丽水
总面积：5 000 m^2
总建筑面积：4 000 m^2
图片提供：NICOLETTI联合工作室

The Thematic Pavilion emerges from Yeosu's waters as a Great Blue Whale hurling itself out of the port becoming the Messenger and the Symbol of Yeosu Expo 2012 - The Living Ocean and Coast. Its Iconic and Symbolic Design holds the messages that Yeosu Expo 2012 wants to transmit to the world: oceans are the origin of life on earth and we have the responsibility to protect them. The metaphor spread from the Thematic Pavilion wants to draw the attention to how fundamental are to the planet's oceans and coasts resources, and how dangerous it is, for such a fragile ecosystem, an irresponsible behaviour. It is significant that such a message is entrusted by the largest living mammal on earth and which is, because of men, in danger of extinction. In Yeosu, Nature itself will make its voice heard through its most mysterious and majestic representative, for the protection and the survival of the fragile planetary ecosystems. The Great Blue Whale will thus become the symbol not only of Yeosu Expo 2012 but the symbol of the movement for the preservation of Oceans and Coasts and their treasures.

跃然于丽水河之上的主题馆犹如一只巨型蓝鲸跃出港口，成为2012丽水博览会的标志。这届博览会的主题即为海洋生命与海岸。它标志性与象征性的设计表达了本届博览会旨在向世界传递的讯息：海洋是地球生命的起源地，因此保护海洋匹夫有责。由主题馆传达出来的寓意想要让人们重视保护海洋和海岸资源意义多么重大，以及不负责任的行为对于如此脆弱的生态环境是多么的危险。令人引以深思的是这样的信息正由地球上最大的，也正因为人类而濒临灭绝的哺乳动物传达出来。在这里，自然将通过其最神秘、最庞大的代表发出呼吁：让脆弱的生态环境得以保护与生存。巨型蓝鲸将不仅成为2012丽水博览会的标志，更成为保护海洋资源以及海岸资源运动的标志。

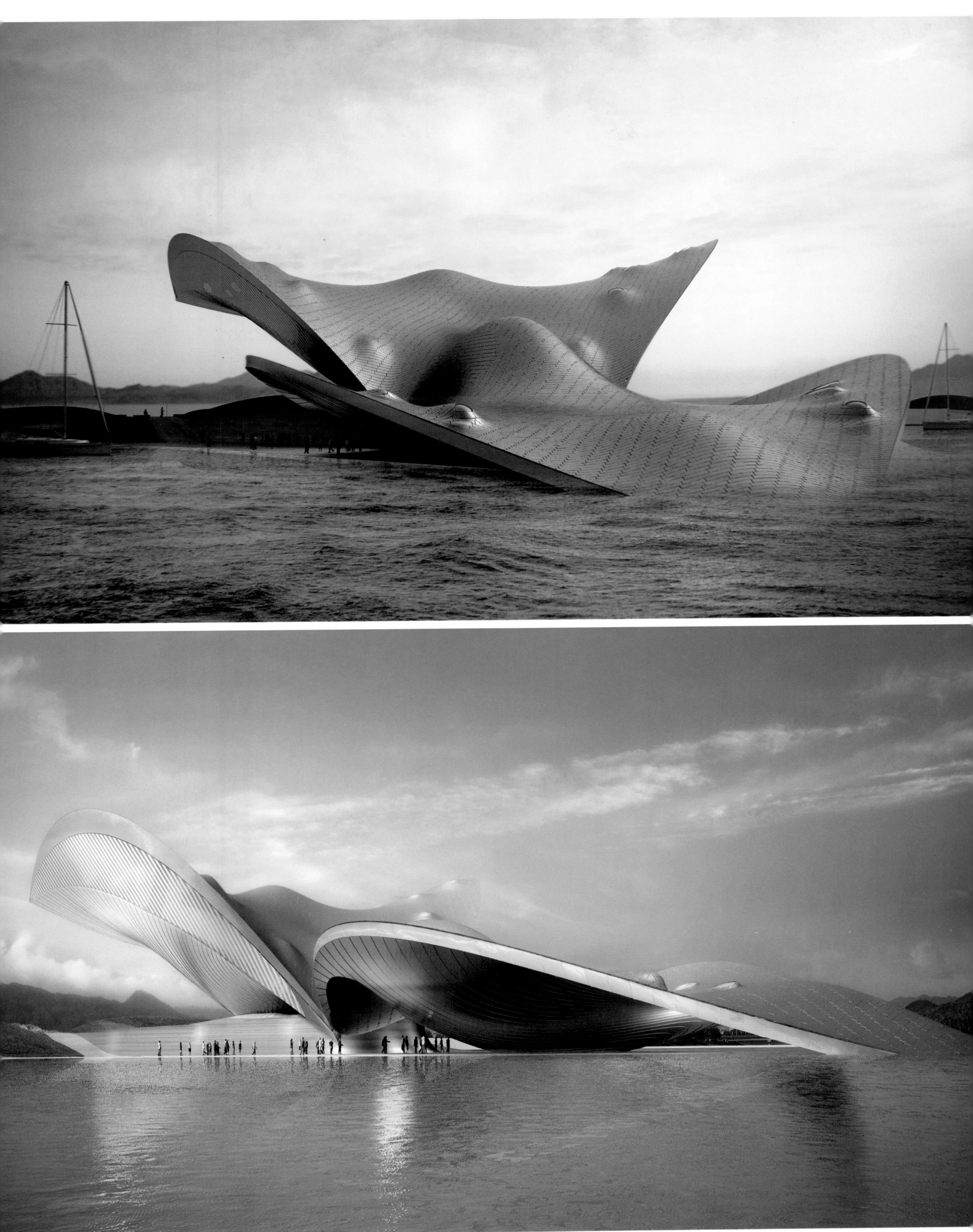

level -1 地下一层平面
Lounge, offices, administration and maintenance, storages, electrical and mechanical room

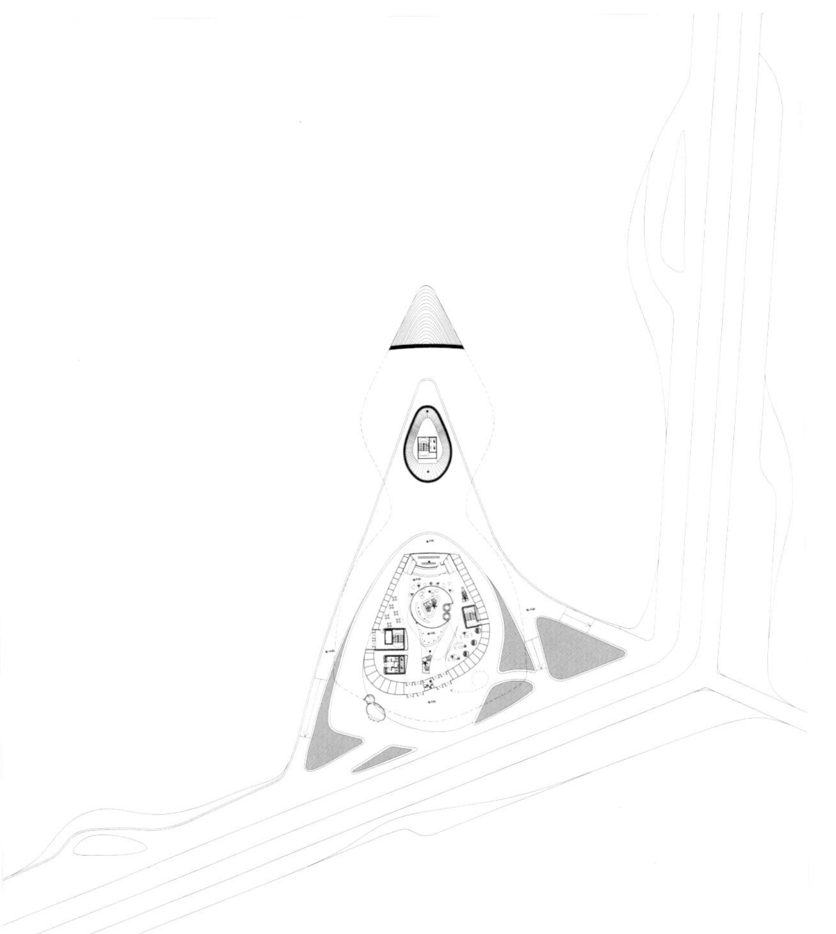

level 0 地面层平面
Lobby, lounge, resting and waiting room

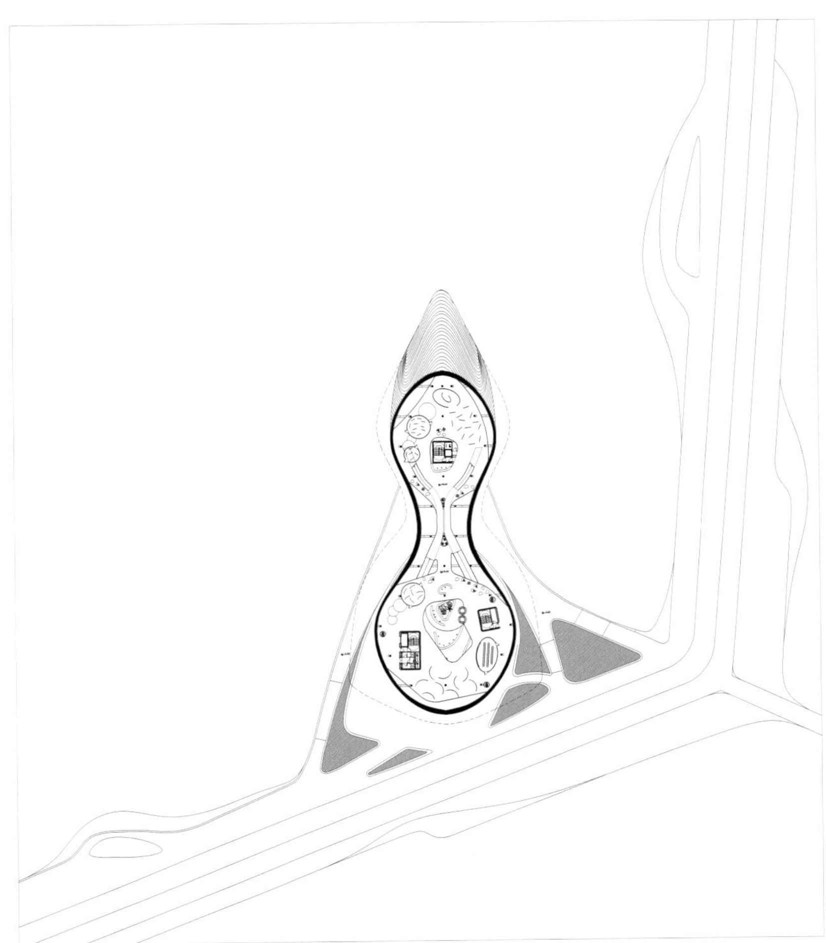

level 1 一层平面
theme exhibition
BPA exhibition

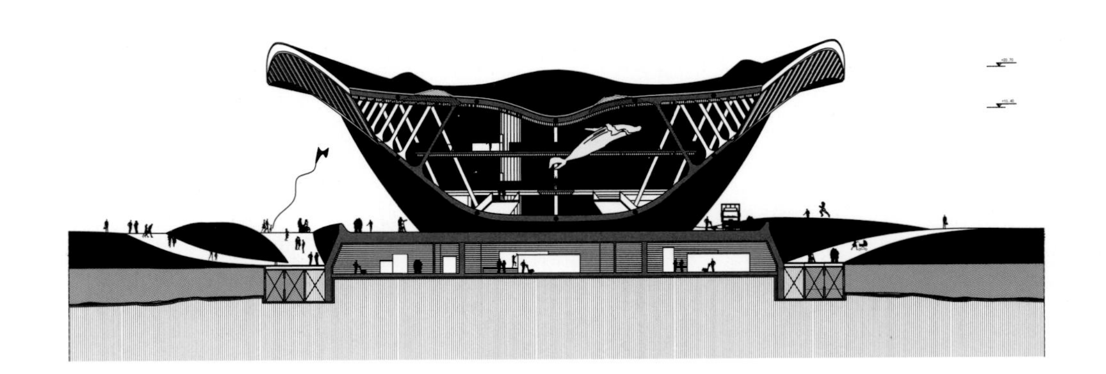

Section 剖面

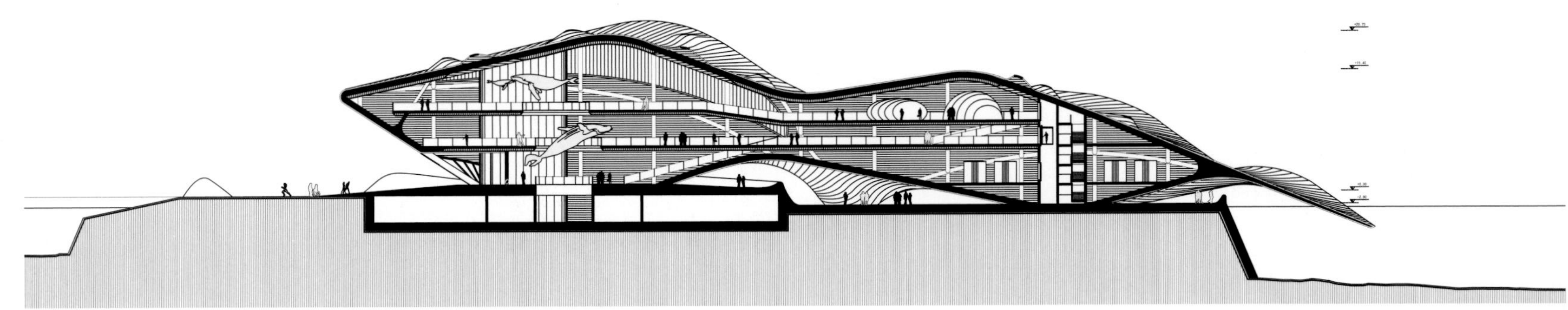

Section 剖面

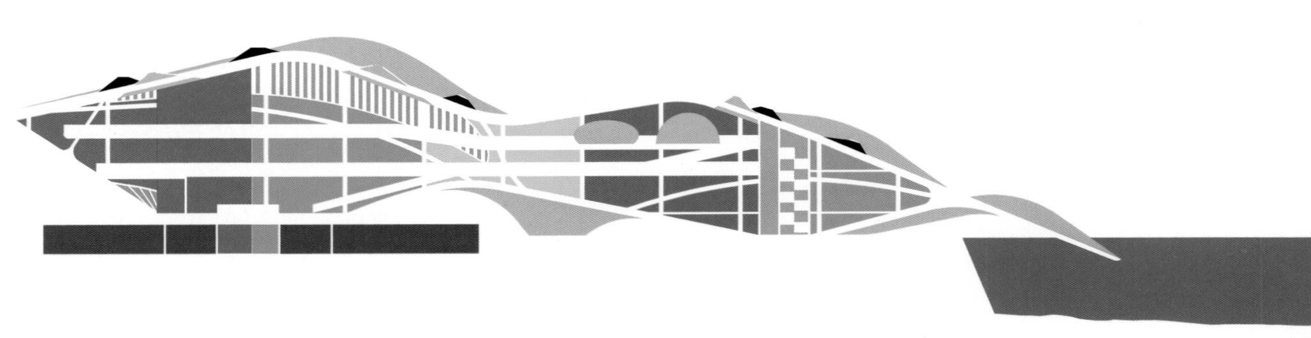

Section 剖面

2012 여수세계박람

Cranium
Melon
Mandible
EXPO 201
ASIA
EUROPE

Library

Bibliotheque Multimedia a Vocation Regionale

卡昂市立公共图书馆

OMA 大都会建筑事务所

Location: Caen, France
Program: Library
Area: 12,700 m^2
Status: Schematic Design
Image Courtesy: OMA

区位：法国，卡昂
功能：图书馆
面积：12 700 m^2
状态：方案设计
图片提供：大都会建筑事务所

The Bibliothèque Multimédia à Vocation Régionale (BMVR) is located at the tip of the peninsula, a focal point of the new development in Caen. The library is designed with two intersecting pedagogic axes which encourage maximum interface between disciplines : human sciences, science and technology, literature, and the arts. With its four protruding planes, the building points to four landmark points in Caen (the Abbaye-aux-Dames in the north, the central train station to the south, the Abbaye-aux-Hommes in the east and the area of new construction in the west), and becomes a symbolic center for the city.
The library consists of two intersecting reading rooms, which encourage maximum interface between the programmed disciplines: human sciences, science and technology, literature, and the arts. In the exterior spaces created by these intersecting reading rooms, the library interacts with its surroundings, opening up to a park, pedestrian pathway and waterfront plaza.
The design of the future BMVR Caen meets the Haute Qualité Environnementale, a standard for sustainable building in France. The sustainable approach responds to local climactic conditions to ensure energy efficiency. Shallow floor plans maximise available natural light, creating the ideal reading environment crucial to a library.

卡昂市立公共图书馆（BMVR）位于半岛的顶端，是卡昂新发展的一个焦点。图书馆设计成两个教学坐标轴的交叉，目的是能最大限度地接触不同科目：人文科学、科学技术、文学和艺术。其指向外界的四个平面指向了卡昂的四个地标（北边是Abbaye-aux-Dames，南边是中央火车站，东边是Abbaye-aux-Hommes，西边是新建筑群），并且成为了城市的象征中心。

图书馆包括两个交叉阅览室，目的是最大化地鼓励不同学科间的接触：人文科学、科学技术、文学和艺术。在这些交叉阅览室的外部空间，图书馆和周围建筑产生互动，并且面向公园、人行道和滨水广场。

未来卡昂BMVR的设计符合法国可持续建筑标准Haute Qualité Environnementale。可持续发展要适应当地的气候条件，以确保能源效益。浅层平面规划最大限度地利用了对图书馆至关重要的自然光，创造了理想的阅读环境。

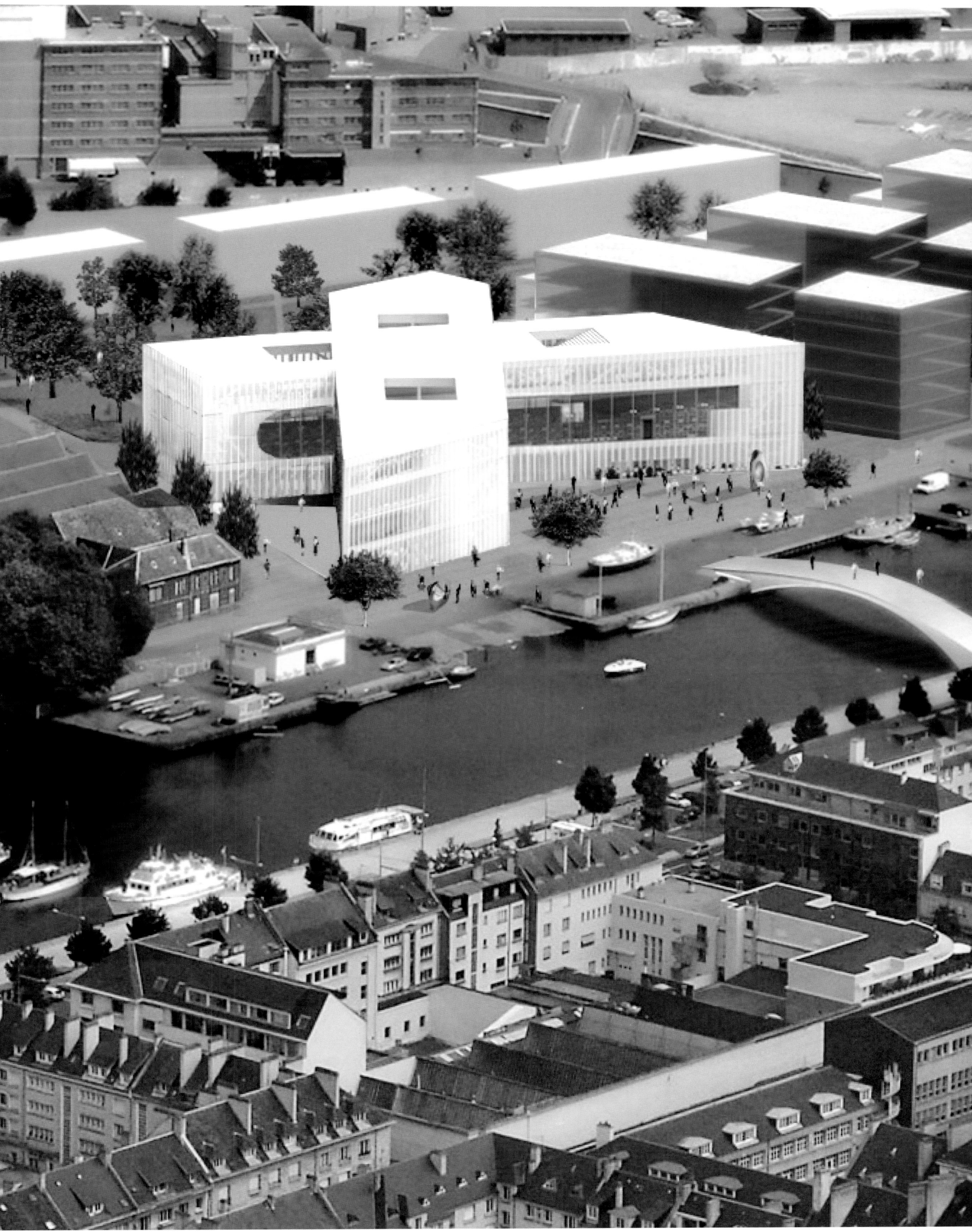

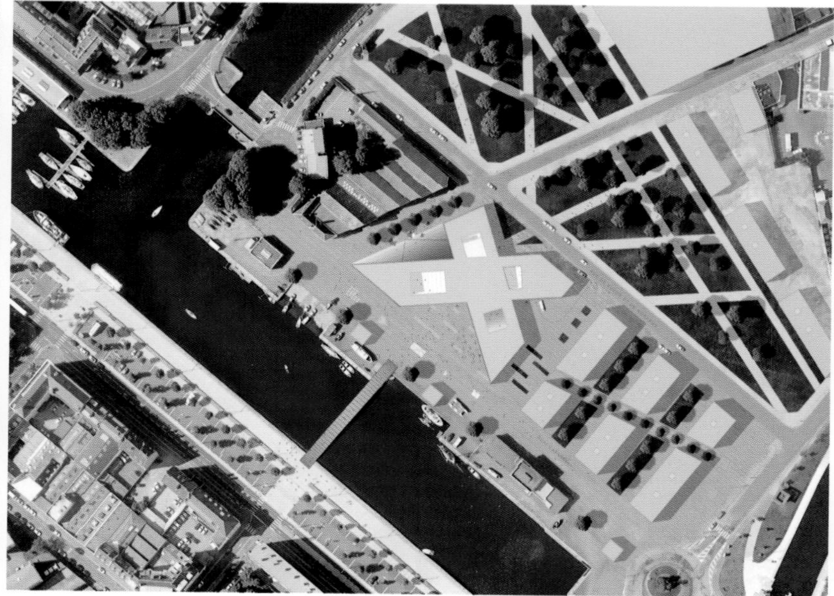

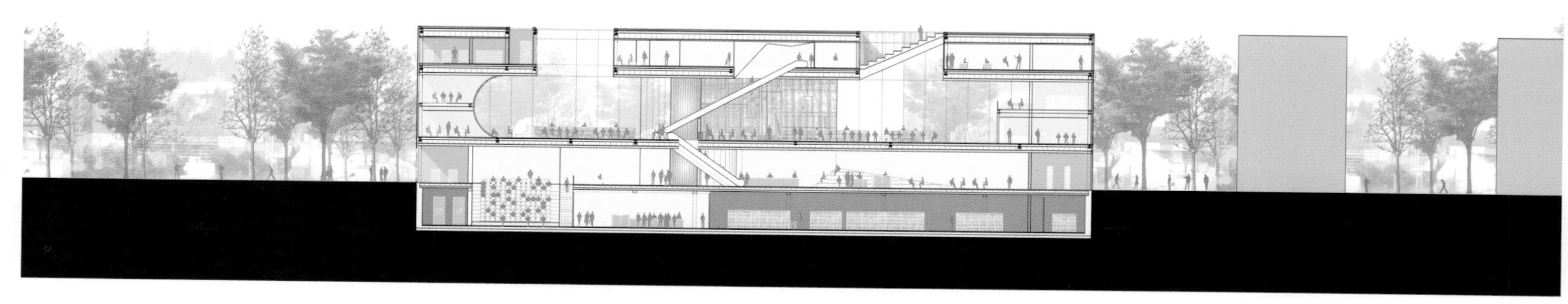

Section 剖面

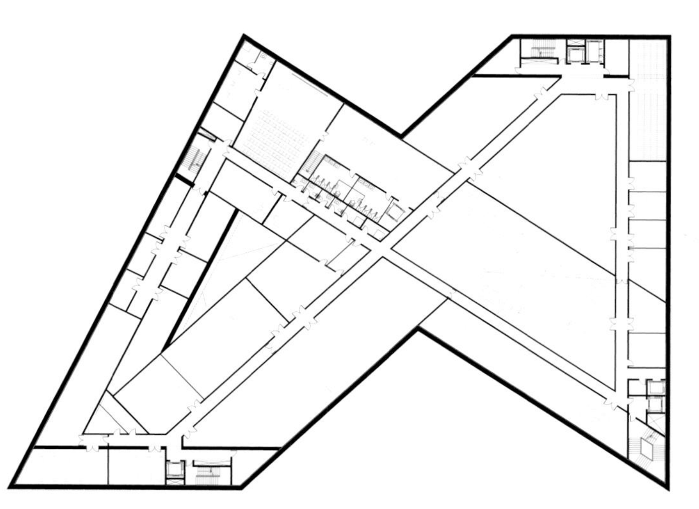

Plan Level -1 地下一层平面

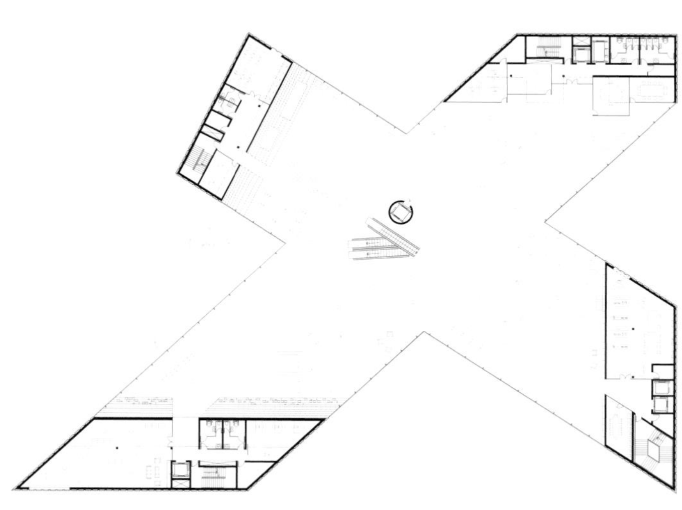

Plan Level 1 一层平面

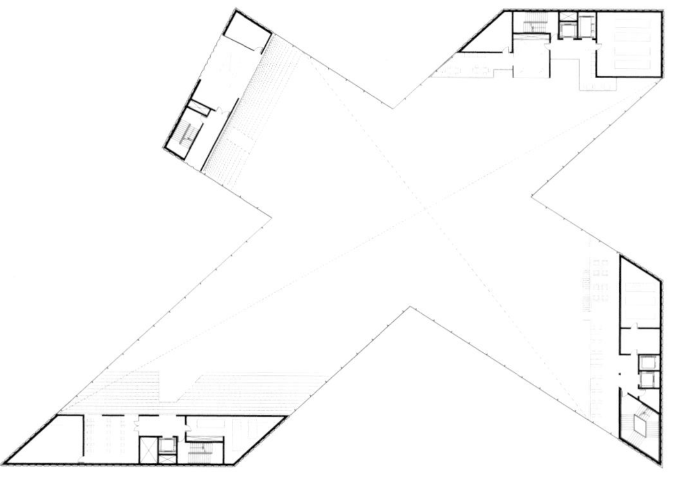

Plan Level 2 二层平面

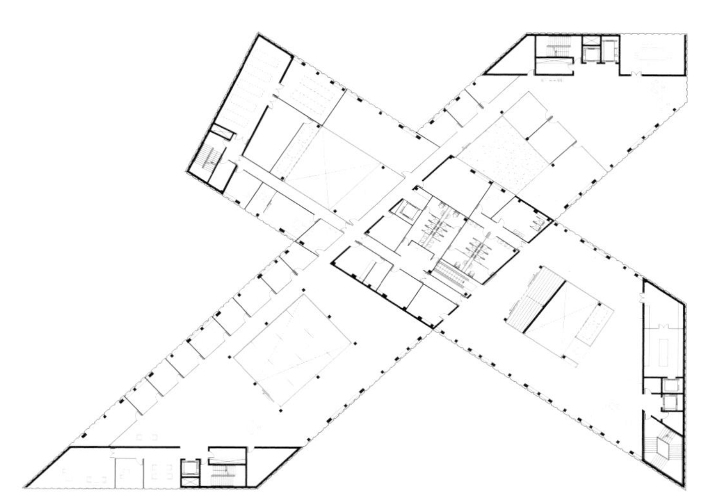

Plan Level 3 三层平面

Tsinghua Library

清华图书馆

Studio arch. Mario Botta　　Mario Botta建筑工作室

Location: Beijing, China
Typology: Library
Area: 20,000 m²
Above ground area: 16,000 m²
Status: Completed in 2011
Image Courtesy: Fu Xing

区位：中国，北京
类型：图书馆
面积：20000 m²
地面面积：16000 m²
状态：2011年竣工
图片提供：傅兴

The project for the library of the Tsinghua University Campus envisages a conic volume housing the central foyer that rises up to the skylight. From this reference space one can see, through a wooden screen, the three levels for the study and the storage of the documents of the many disciplines studied at the Tsinghua University. In the basement and in the support building, other complementary facilities (exhibition spaces, newspaper library, etc...) provide all the services open to the students. The site is on the east side of the campus main road and it is surrounded by the buildings with the classrooms.
The new library lies in a block in which it is envisaged the construction of other buildings and whose south-east corner is characterized by a green area. The library is accessible via different routes that turn around the building. The two main entrances are on the north side where there is also the parking for the several bicycles (the main means of transport within the campus). The insulated masonry work and the stone cover ensure a good energy management during the harsh winters and the hot summers which are typical of the Japanese climate. The sober interior finishes create a pleasant and essential atmosphere that draws attention to the real protagonist, i.e. the book.

清华大学校园图书馆建筑项目设想建造一个锥形的大型建筑，其中心门厅可直通至天窗。透过一道木质屏幕，我们可以看到位于该参照空间内的三层自习室，及清华大学多学科研究文件的储藏室。在地下室和辅楼，诸如展览馆、报纸阅览室等其他配套设施向学生开放，提供各种各样的服务。图书馆位于校园主干道的东边，周围环绕着很多教学建筑。

新图书馆的建址东南角有一片标志性的绿色区域，最初是设想建造其他的建筑。从校园周围的建筑出发，通过不同的路径都能到达图书馆。图书馆的两个主入口设于北侧，那里同样设有自行车停放处（自行车是校园里主要的交通工具）。无论严冬还是酷暑，隔热的石造建筑和石造外观都能确保良好的能源管理。素雅的室内装修在图书馆充斥着图书的必然氛围中增添了一抹愉悦，这种氛围吸引着学生关注真正的主角——书籍。

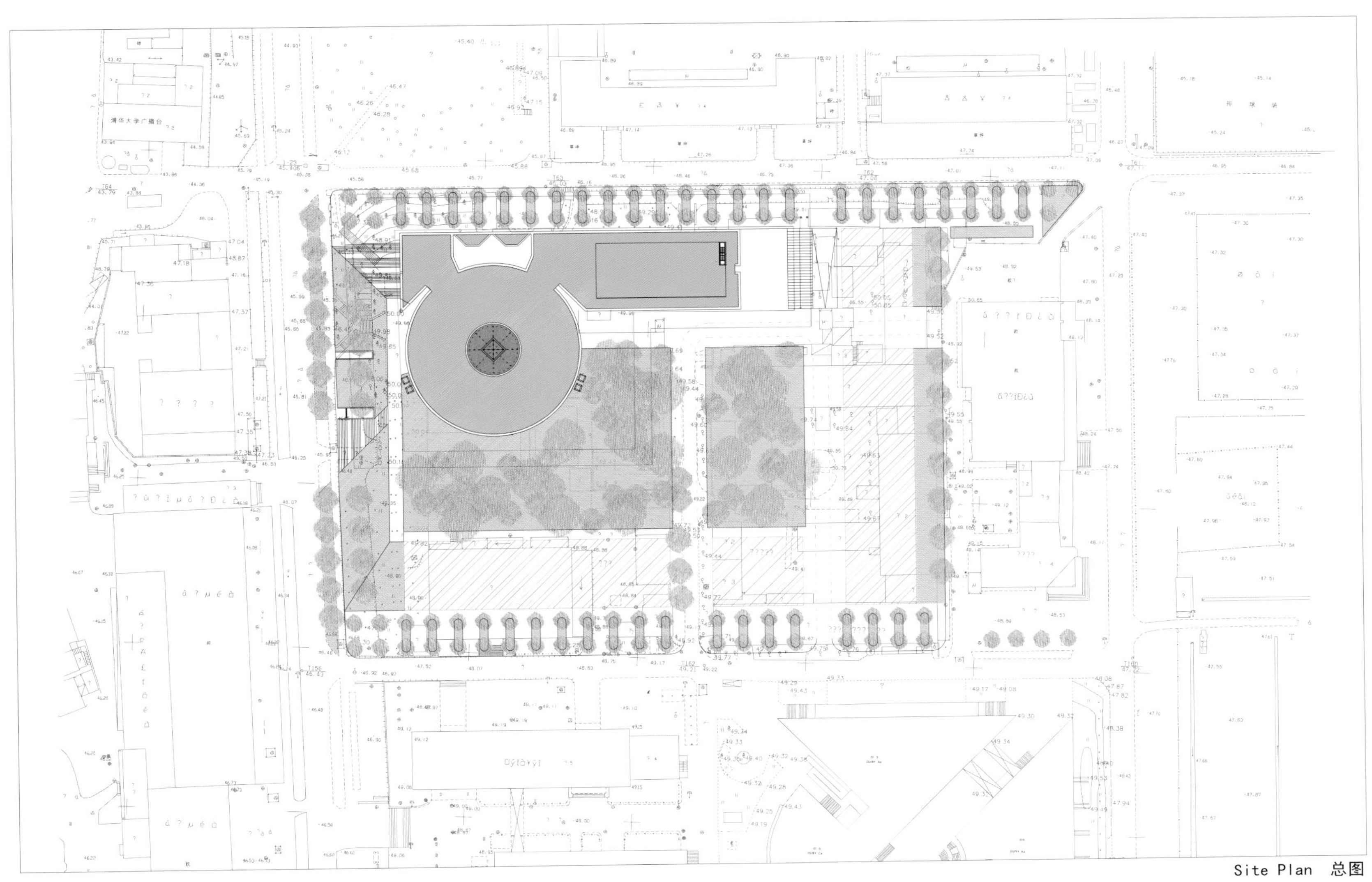

Site Plan 总图

TSINGHUA
M.B. 011

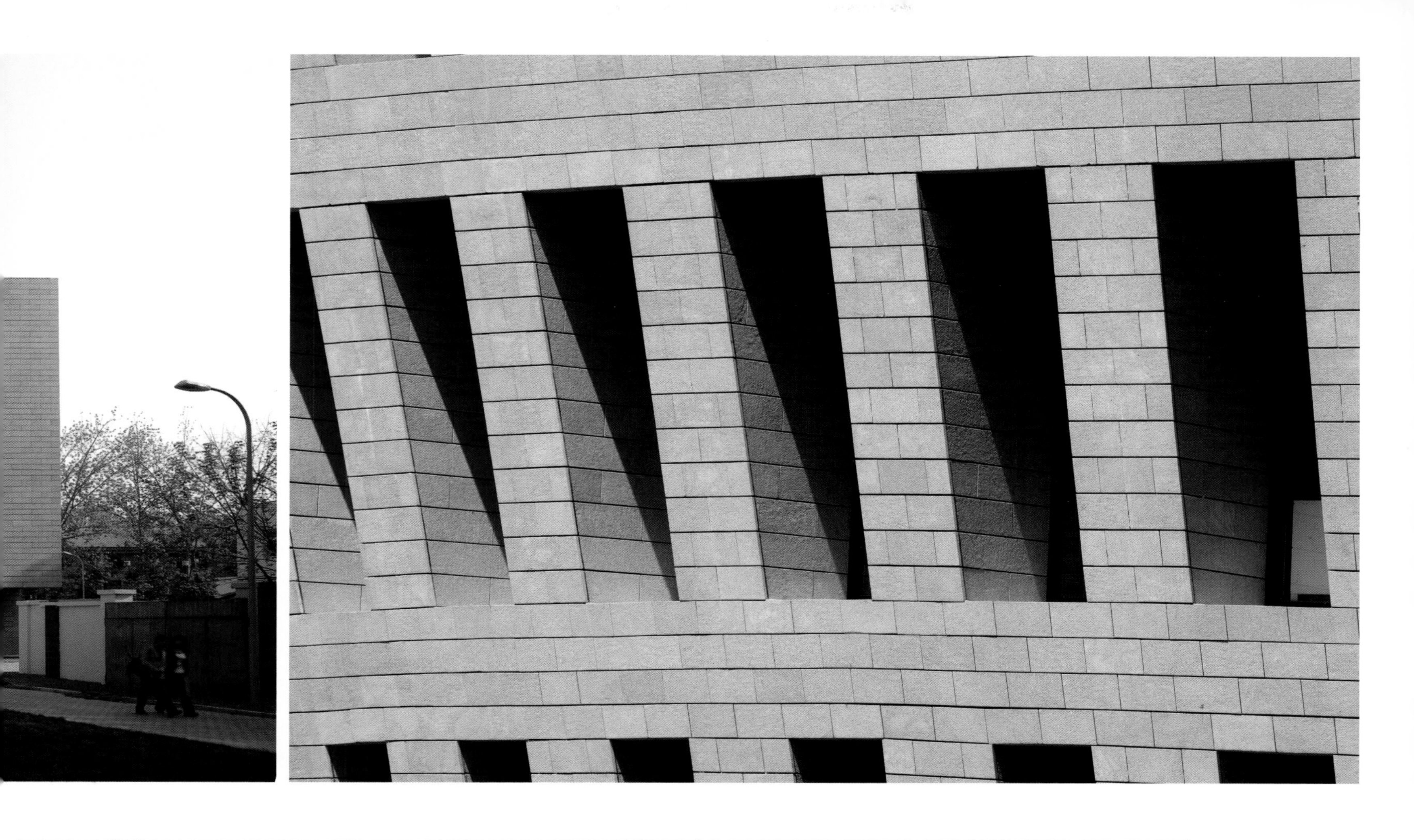

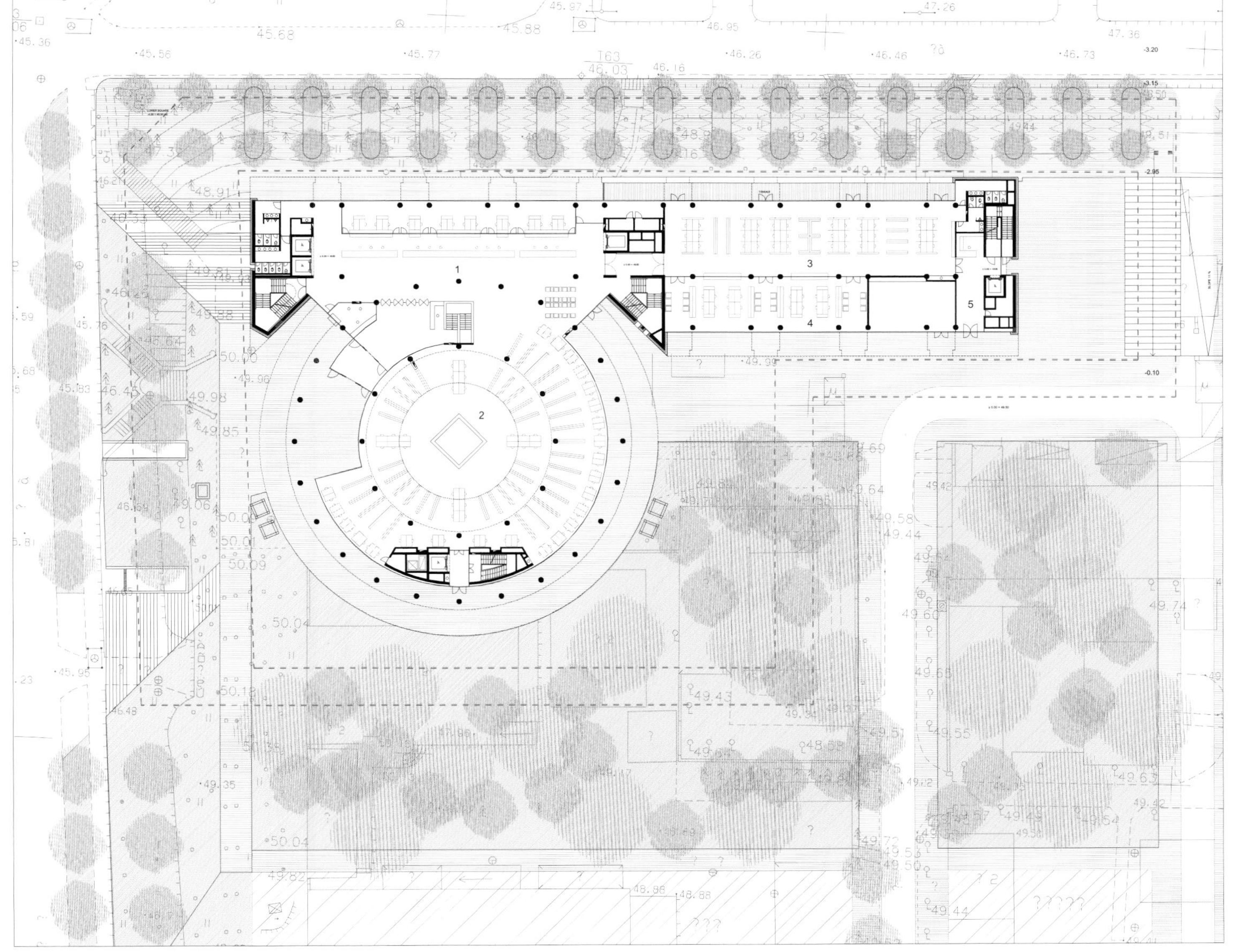

1 ENTRANCE HALL/ RECEPTION DESK

2 OPEN SHELVES AREA

3 INDEX HALL

4 OFFICES

5 LOBBY

Plan Level 0.00 地面层平面

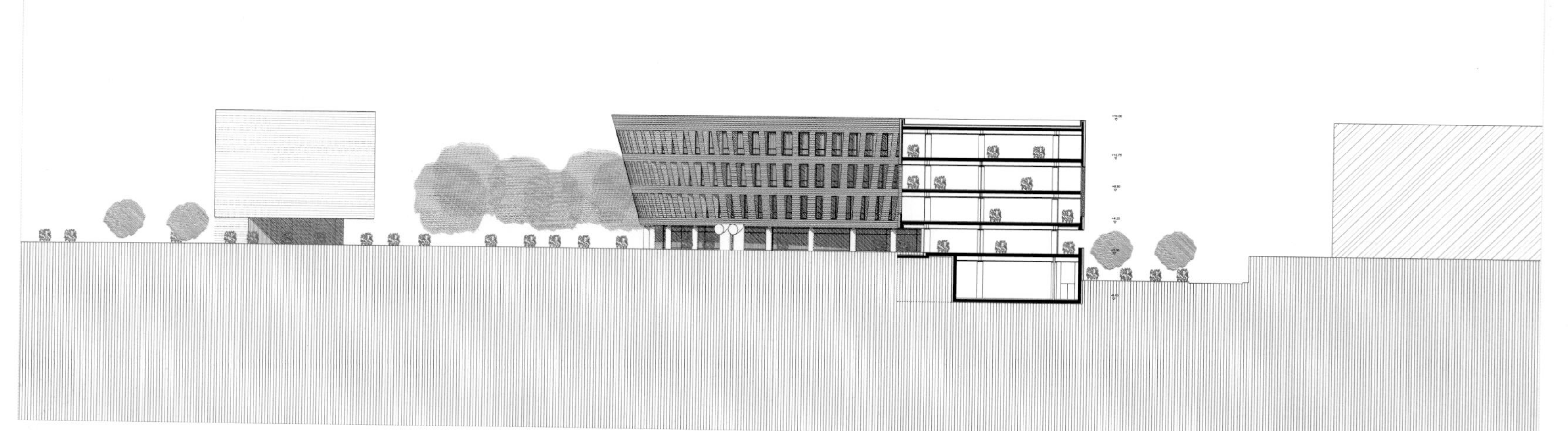

Section 剖面

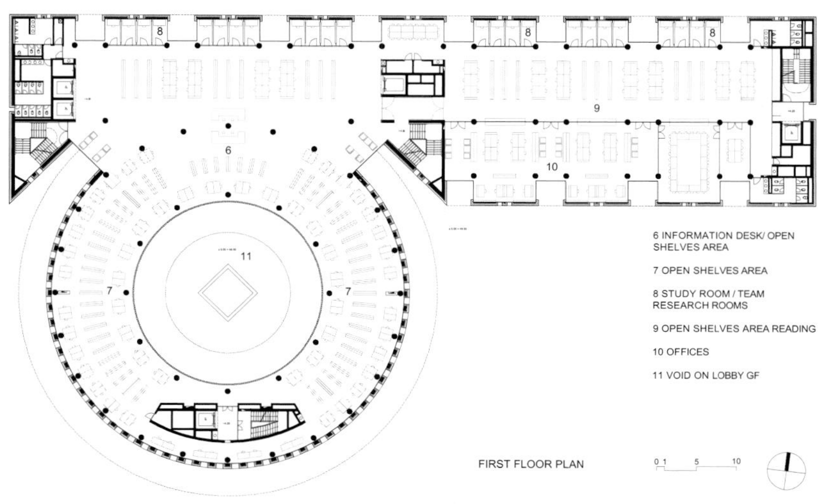

Plan Level 01 一层平面

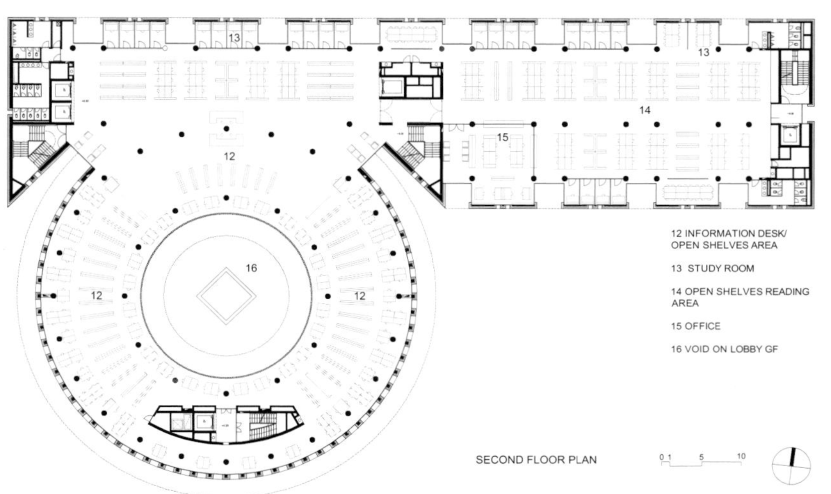

Plan Level 02 二层平面

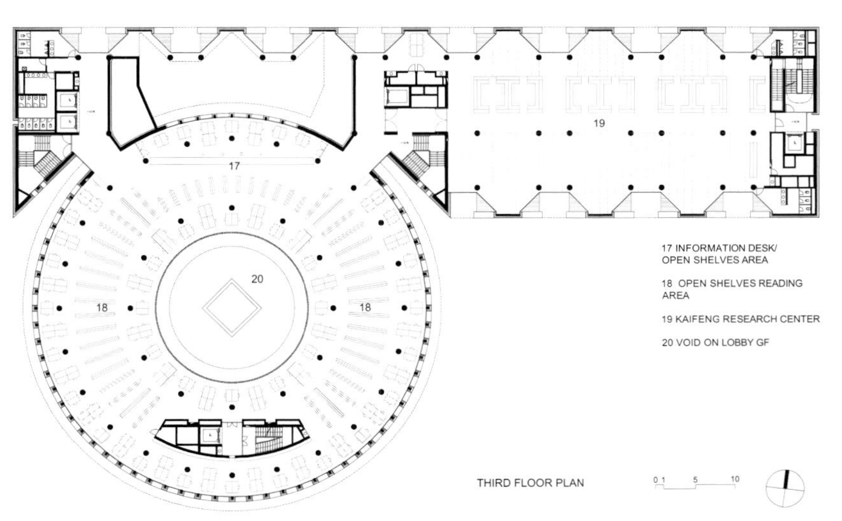

Plan Level 03 三层平面

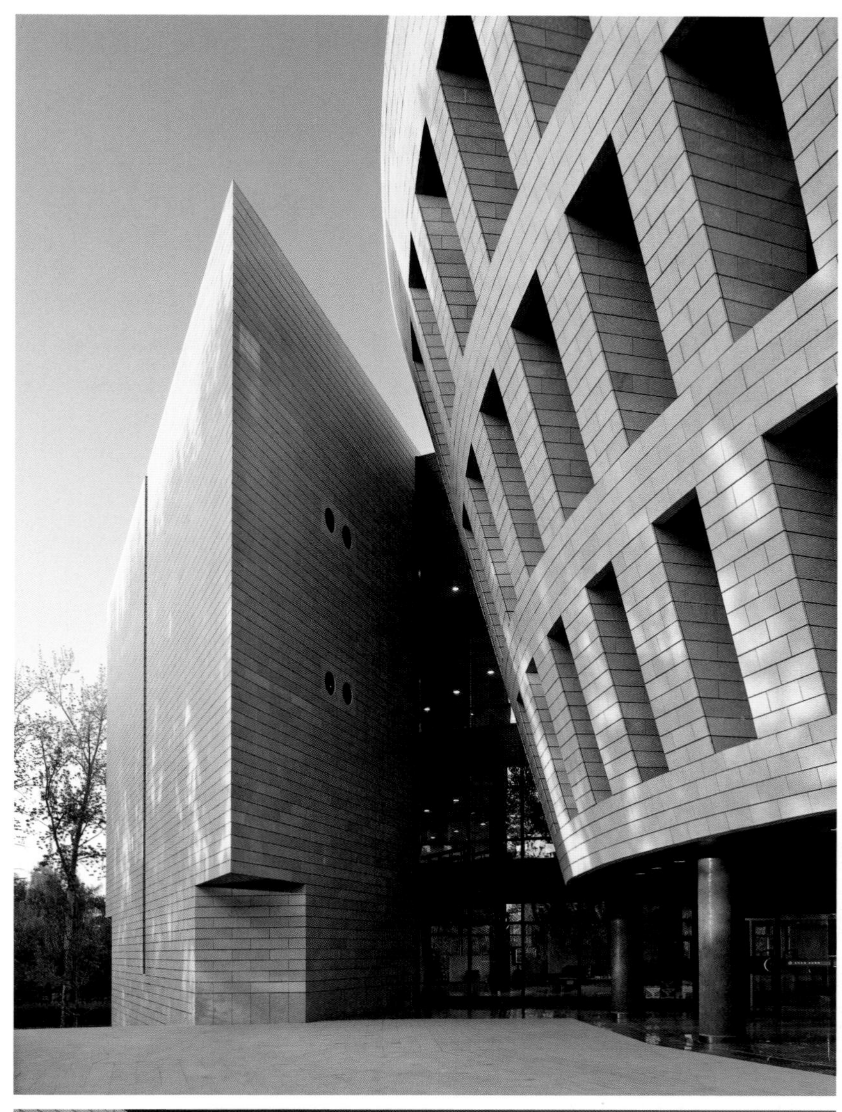

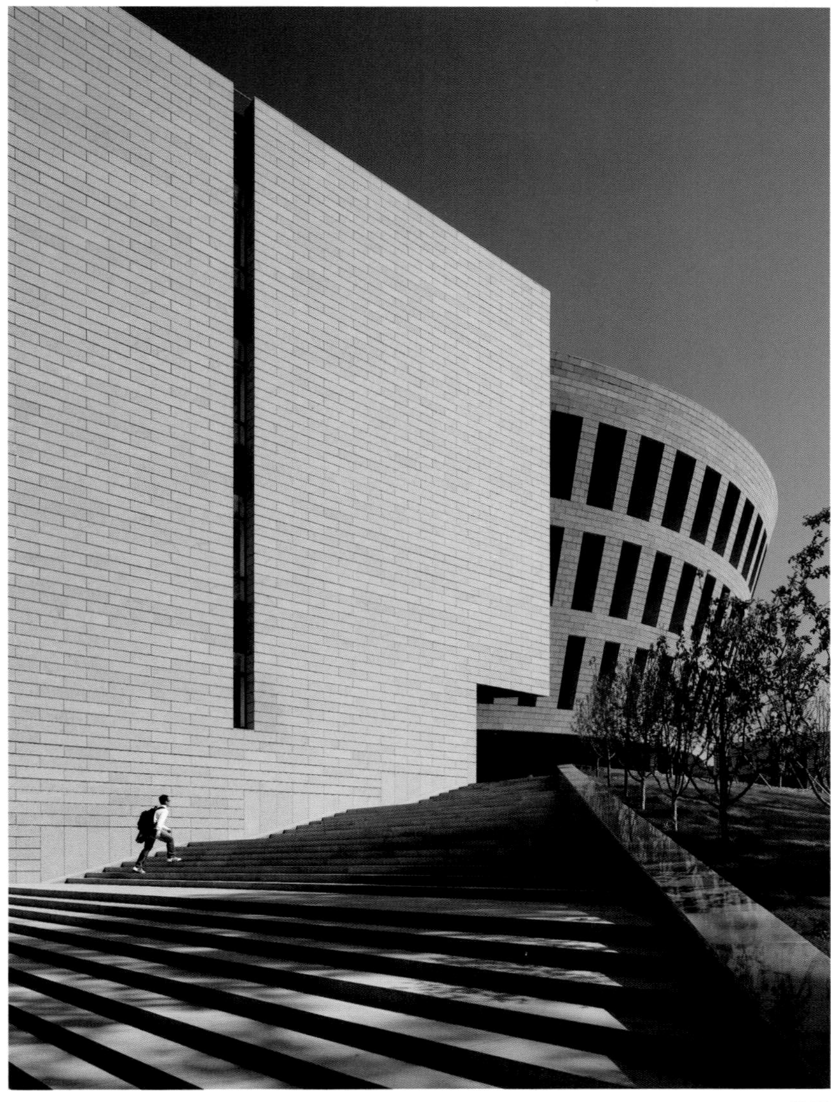

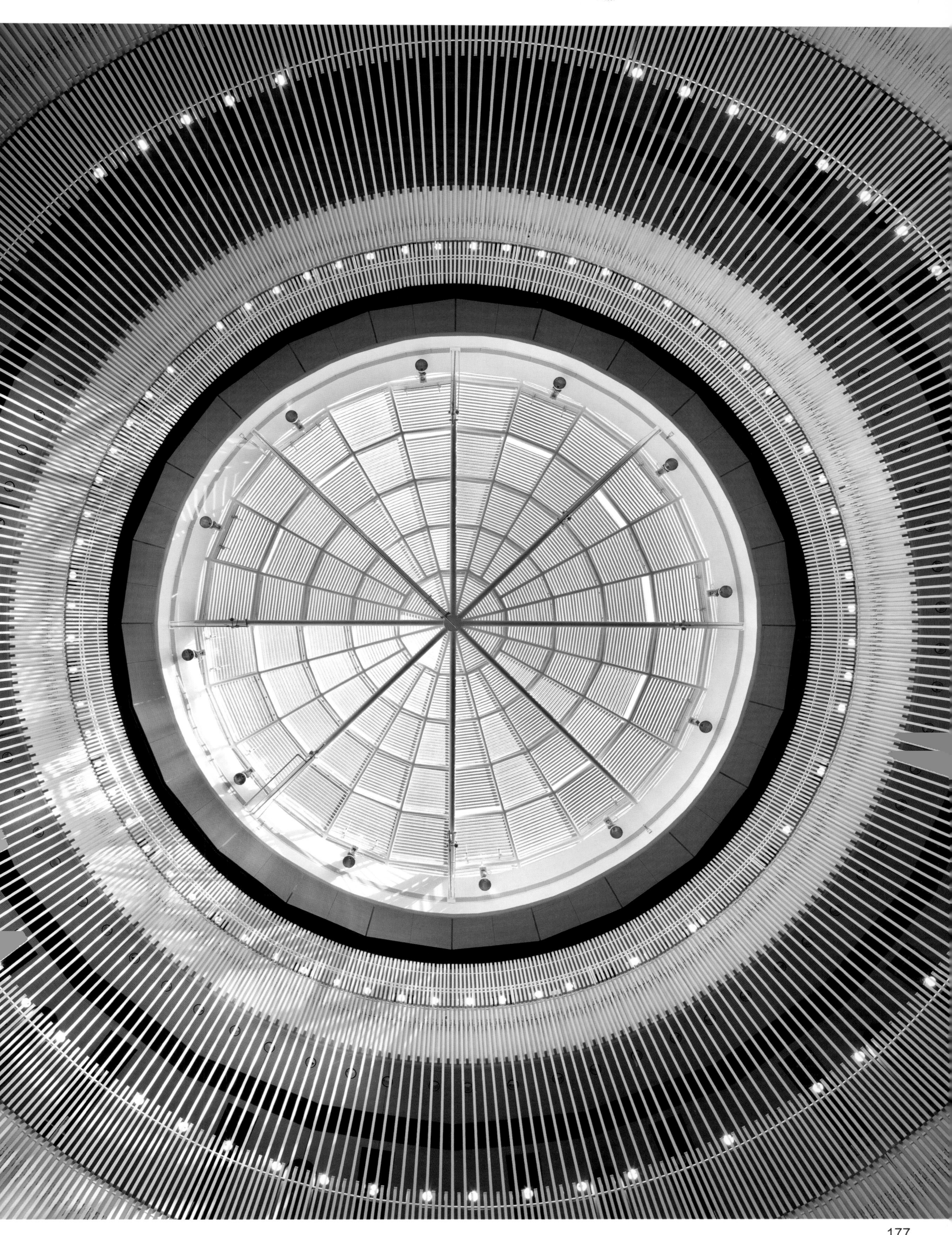

Public

公共

Looptecture F: Tsunami Disaster Preventive Control Centre

Looptecture F: 海啸预防控制中心

Endo Shuhei Architect Institute 远藤秀平建筑研究所

Location: Minami-Awaji city , Hyogo perf. Japan
Typology: Tsunami Disaster Preventive Control Centre
Sit Area: 8,502 m^2
Building Area: 310 m^2
Total Floor Area: 376 m^2
Status: Completed in 2010
Image Courtesy: Endo Shuhei Architect Institute, Yoshiharu Matsumura

区位：日本，兵库县，南淡路市
类型：海啸预防控制中心
基地面积：8502 m^2
占地面积：310 m^2
总建筑面积：376 m^2
状态：2010年竣工
图片提供：远藤秀平建筑研究所，Yoshiharu Matsumura

The function of this architecture is security and controlling all the floodgates located at port of Fukura, enlightening dangerous of the tsunami for tourists, and using as a place of refuge in case of the tsunami warning. For these reasons, it should ensure to keep the spaces of necessary and viewpoint for watching all over the port, also rational shape and structure to against of tsunami and the drift coming after disaster are necessary.
Main floor is placed higher level than assumed height of tsunami and opening the ground level floor allows waves to pass through. It can the outside wall curved as efficient form to disperse stress. This form consists of 7.3 m width belt (curved wall); continuous arcs are constructed and crossing by 6 different centres of circle. Consequently, these arcs are closing at the same point both start and finish.

该建筑的功能即确保安全，控制福库拉港口所有水闸，警示游客海啸的危险，且作为海啸预警时的避难所。出于对这些原因的考虑，该建筑的设计必须保证必要的集散空间并设有港口全景观测点，为抵御海啸及灾难过后的漂流物而设计的合理构造也是必需的。

主楼层高于测算的海啸浪高，地平面楼层开放以便让海浪通过。它的整体结构形状以圆形为主，这样可以分散海啸所带来的巨大压力。该设计由7.3 m宽的带状墙体（即弧形墙体）构成，六个不同的圆心穿过连续的拱形设计。最终，这些拱形结构都在海啸伊始直至终止关闭。

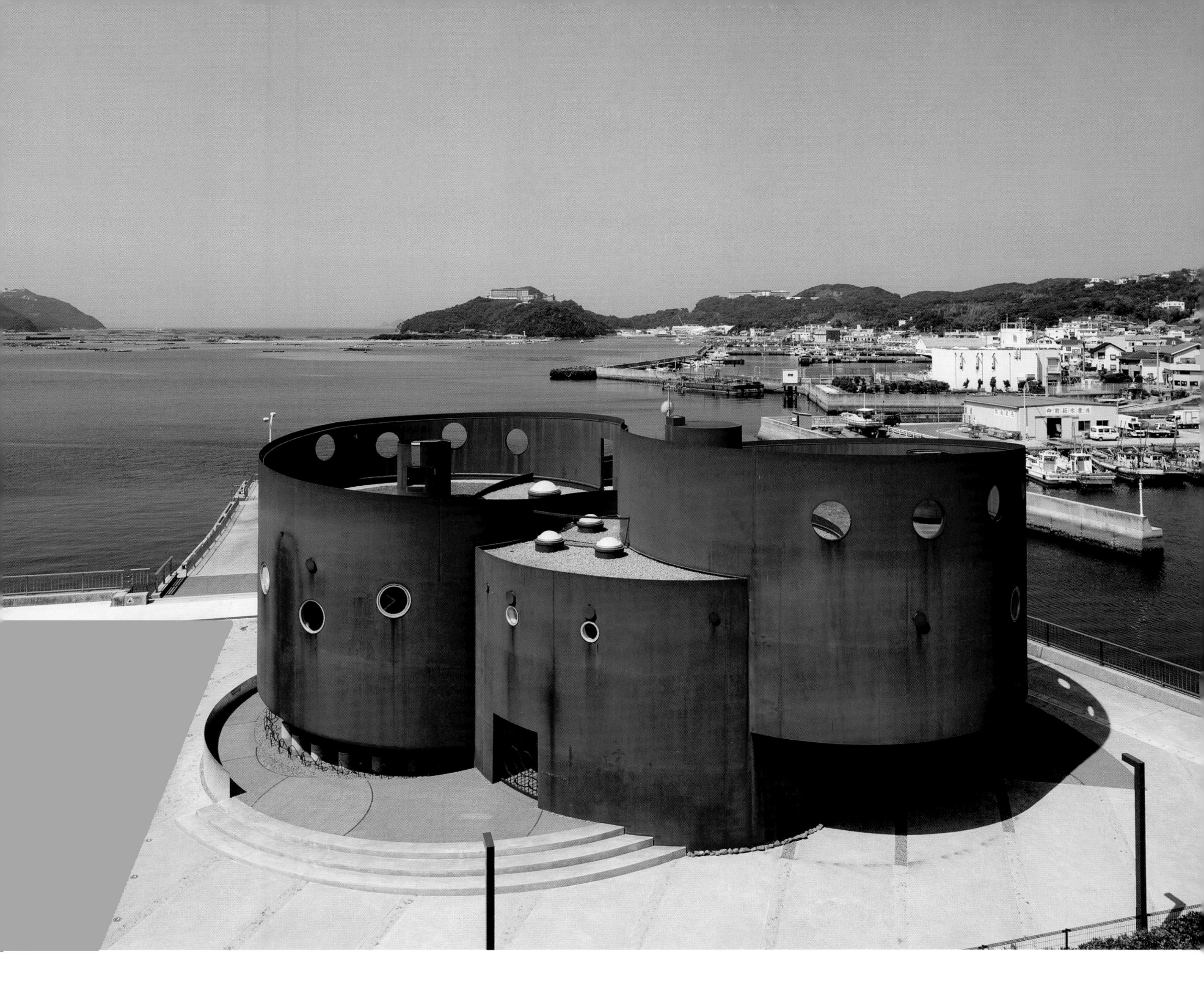

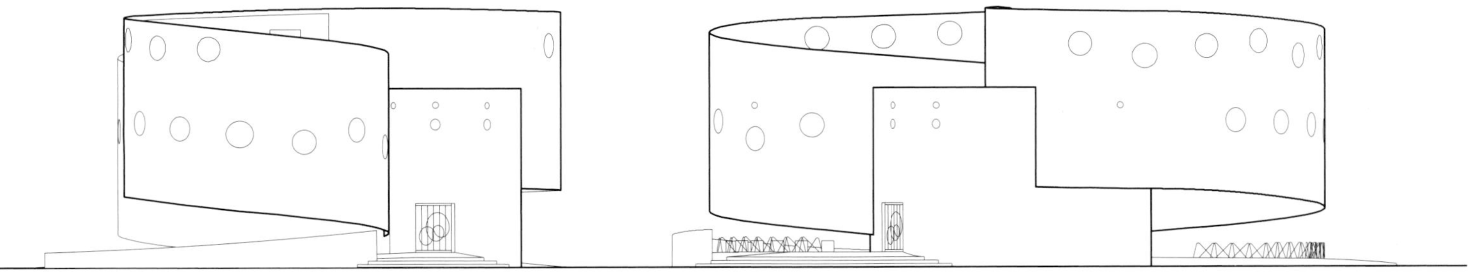

East Elevation S=1/400　　North Elevation S=1/400

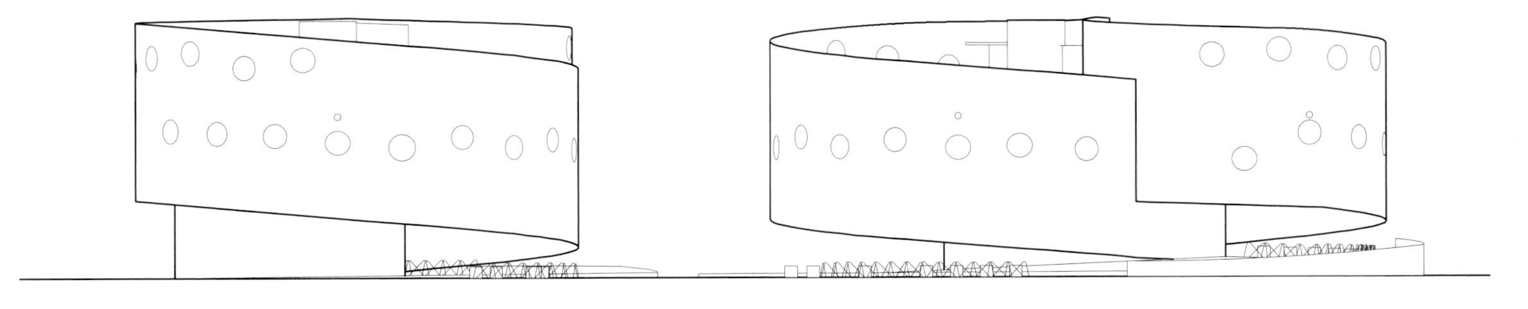

West Elevation S=1/400　　South Elevation S=1/400

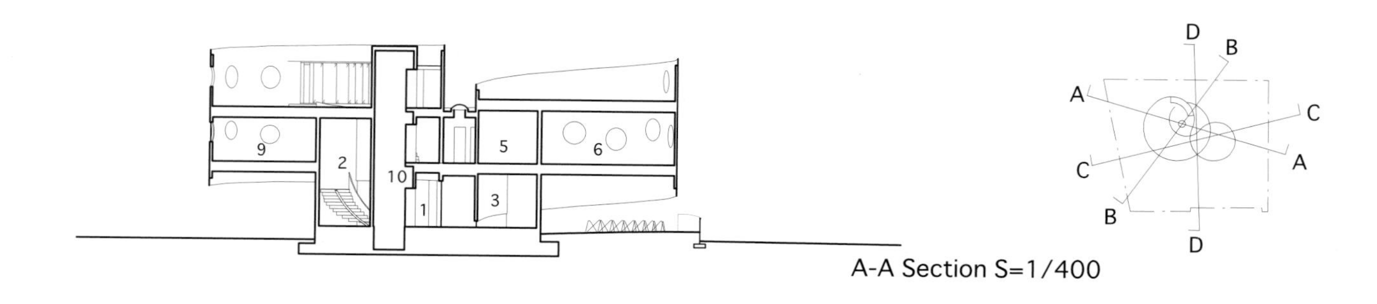

A-A Section S=1/400

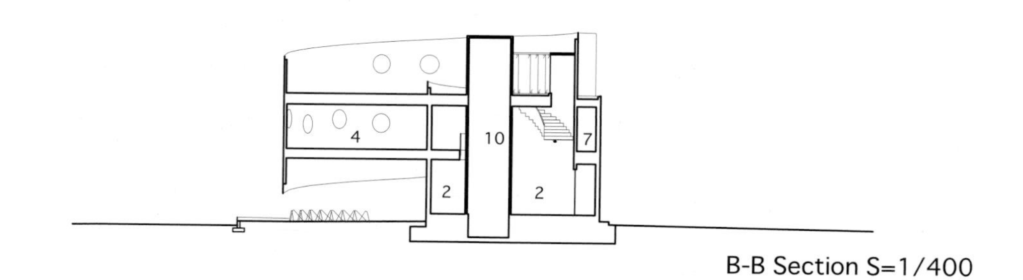

B-B Section S=1/400

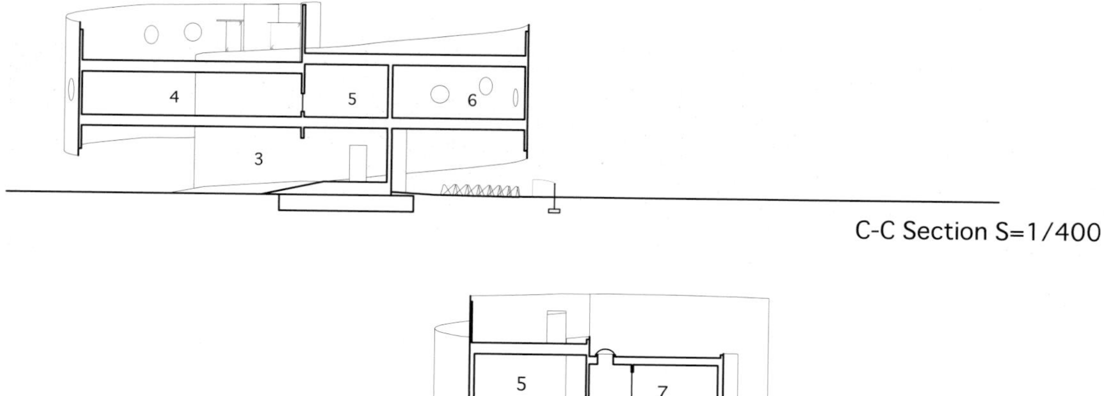

C-C Section S=1/400

D-D Section S=1/400

1. Entrance
2. Hall
3. Piloti
4. Centre control room
5. Exhibition room
6. Disaster learning room
7. Men's lavatory
8. Ladies' Lavatory
9. Machinery room
10. EV

Section 剖面

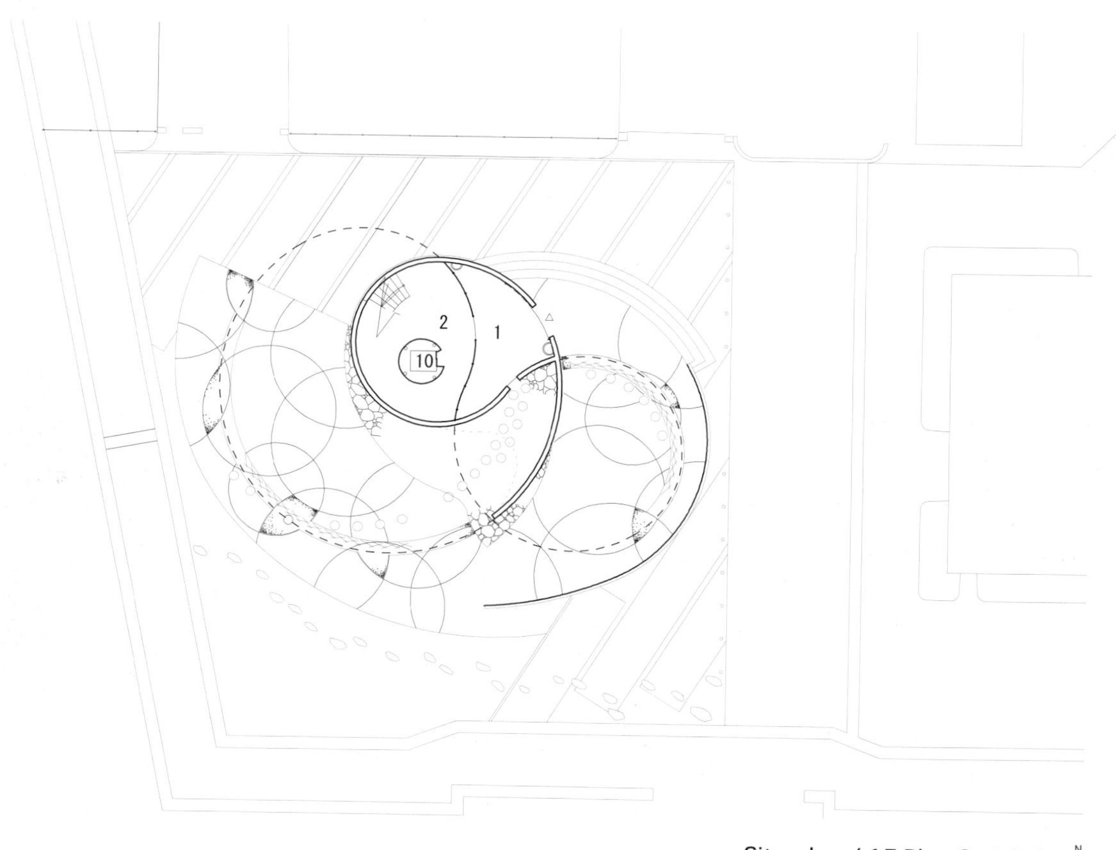

Site plan / 1F Plan S=1/400

1. Entrance
2. Hall
3. Piloti
4. Centre control room
5. Exhibition room
6. Disaster learning room
7. Men's lavatory
8. Ladies' Lavatory
9. Machinery room
10. EV

Site Plan 一层平面

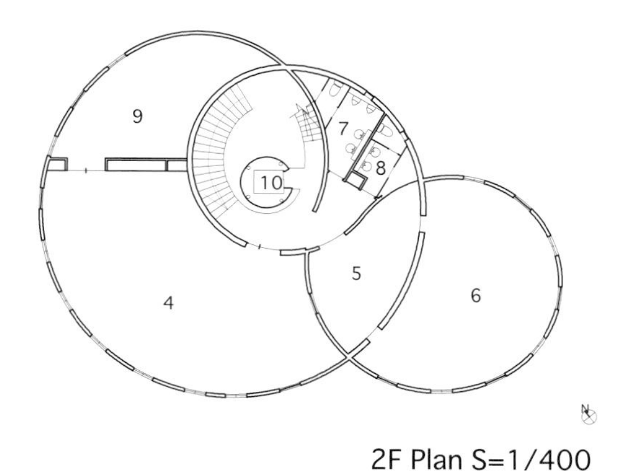

2F Plan S=1/400

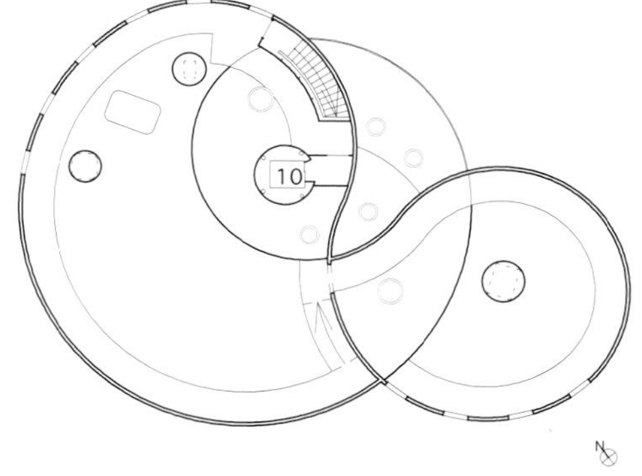

RF Plan S=1/400

1. Entrance
2. Hall
3. Piloti
4. Centre control room
5. Exhibition room
6. Disaster learning room
7. Men's lavatory
8. Ladies' Lavatory
9. Machinery room
10. EV

2F/RF Plan 二三层平面

K.A.I.S.T. Memorial Building

K.A.I.S.T.纪念楼

J. MAYER H. Architects　　J. MAYER H. 建筑事务所

Location: Deajeon, Republic of Korea
Typology: Culture
Status: Invited competition in 2009
Image Courtesy: J. MAYER H. Architects

区位：韩国，大田市
类型：文化
状态：2009年邀请赛
图片提供：J MAYER H. 建筑事务所

The K.A.I.S.T Memorial Building is open to all directions and integrates the landscape inside the ground floor. The central atrium is open all the way to the roof terrace with a panoramic view onto the entire campus and beyond. A central spiral circulation connects all public areas inside the entire building. This spacious atrium is an interactive space that allows visitors to take an elevator up to the terrace and exhibition spaces, and then move down until they arrive again in the lobby. The ground floor and top floor are open public areas. The floors in between are flexible floorplans that house laboratories, meeting rooms and office spaces.

The building constantly changes its appearance. Water curtains underneath the cantilevers animate the building during the day while at night, media screens and projections activate the building beyond official opening hours. It is a new prototype for an academic building that is a pulsating heart of the campus all day and night.

韩国科学技术院纪念楼在各个方向都设有入口，并将当地景观整合到了一楼大厅内部。可以俯瞰学院全景及周边美景的楼顶露台与中央中庭之间四通八达。中央的螺旋式环路将整栋建筑物内的所有公共区域联系在一起。宽敞的中庭作为一个交互空间，使游客可以乘坐电梯上至露台和展览区，下至大厅。建筑一层和顶层均为开放的公共区域。中间各层为设有图书馆、会议室及办公空间的灵活布局楼层。

该纪念楼的外部视觉效果处于不断地变化之中。白天，悬臂之下的水幕赋予该建筑以生机和活力；夜晚，媒体银幕和投影设备则让整栋大楼在非办公开放时间万分灵动。该建筑设计为学院注入了生命力，全天候的脉动跳跃使之成为了学院建筑新典范。

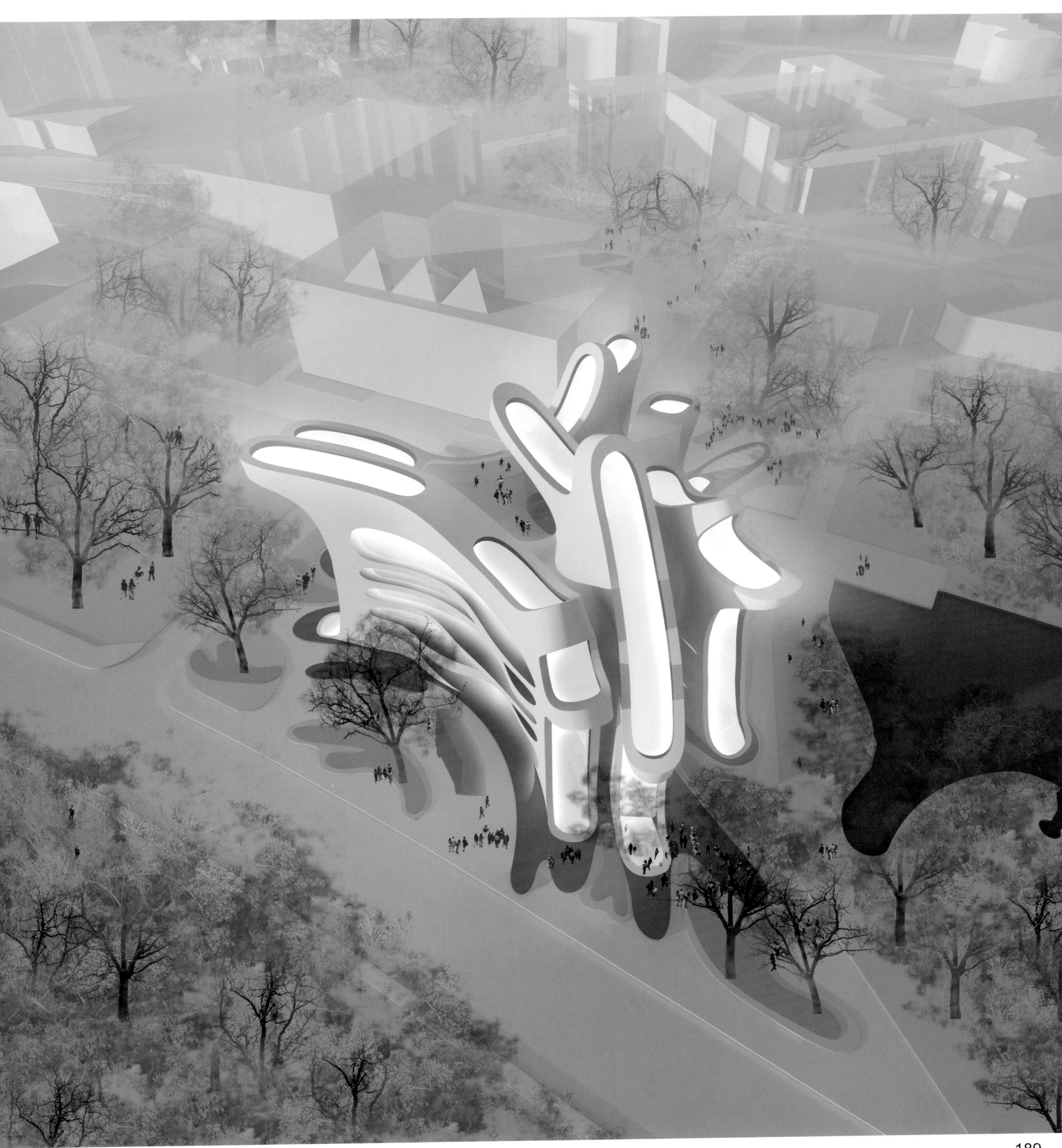

PANORAMA TERRACE
EXHIBITION
AUDIO/VISU
EXHIBITION
MULTI PURPOSE
RESEARCH CENTER
LABS
ANIMAL LABS
RESEARCH CENTER
PUBLIC
RESEARCH CENTER
RESEARCH CENTER
RESTRICTED
EXHIBITION
SHOP
CAFE
MEMORIAL BUILDING

Diagram 分析图

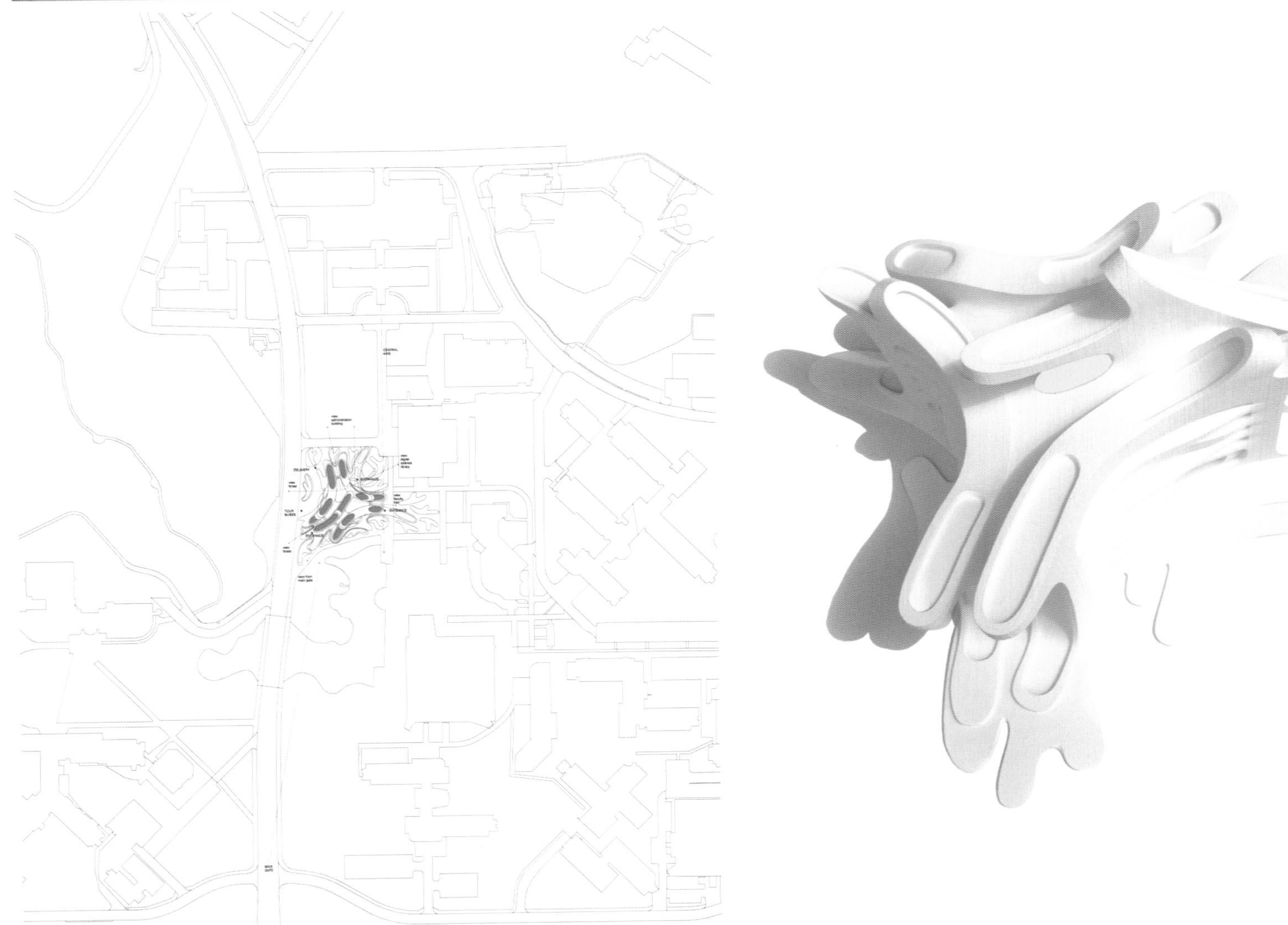

Location 区位

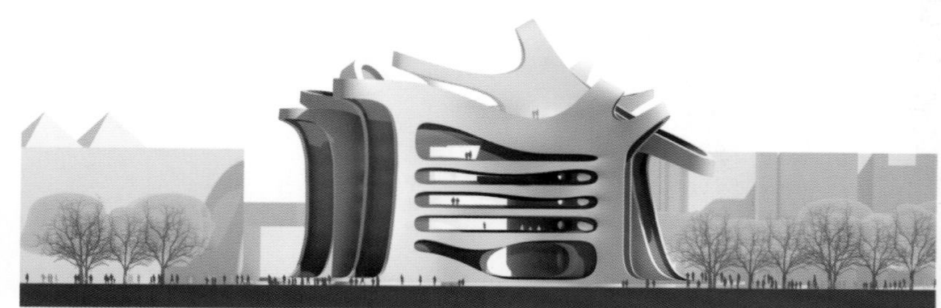

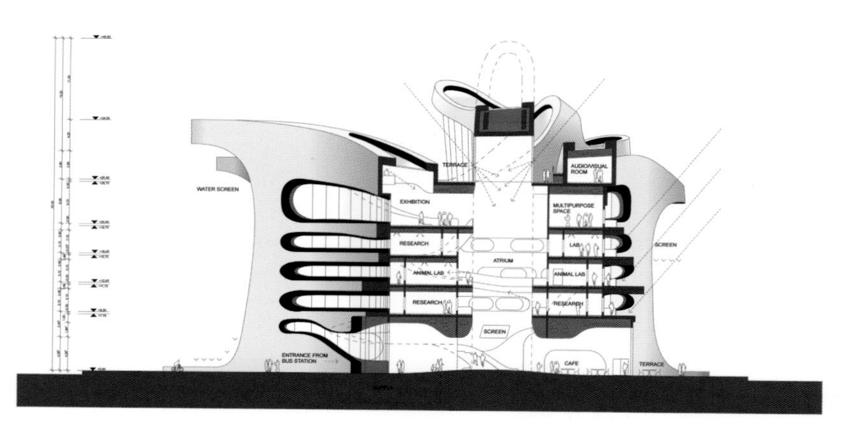

Section 剖面

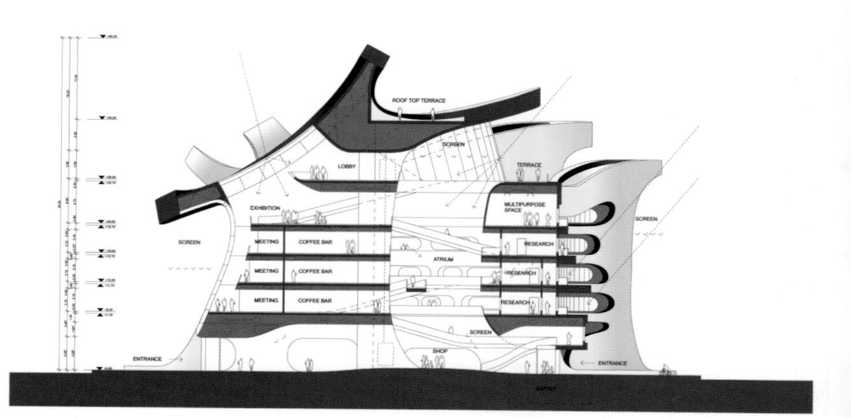

Section 剖面

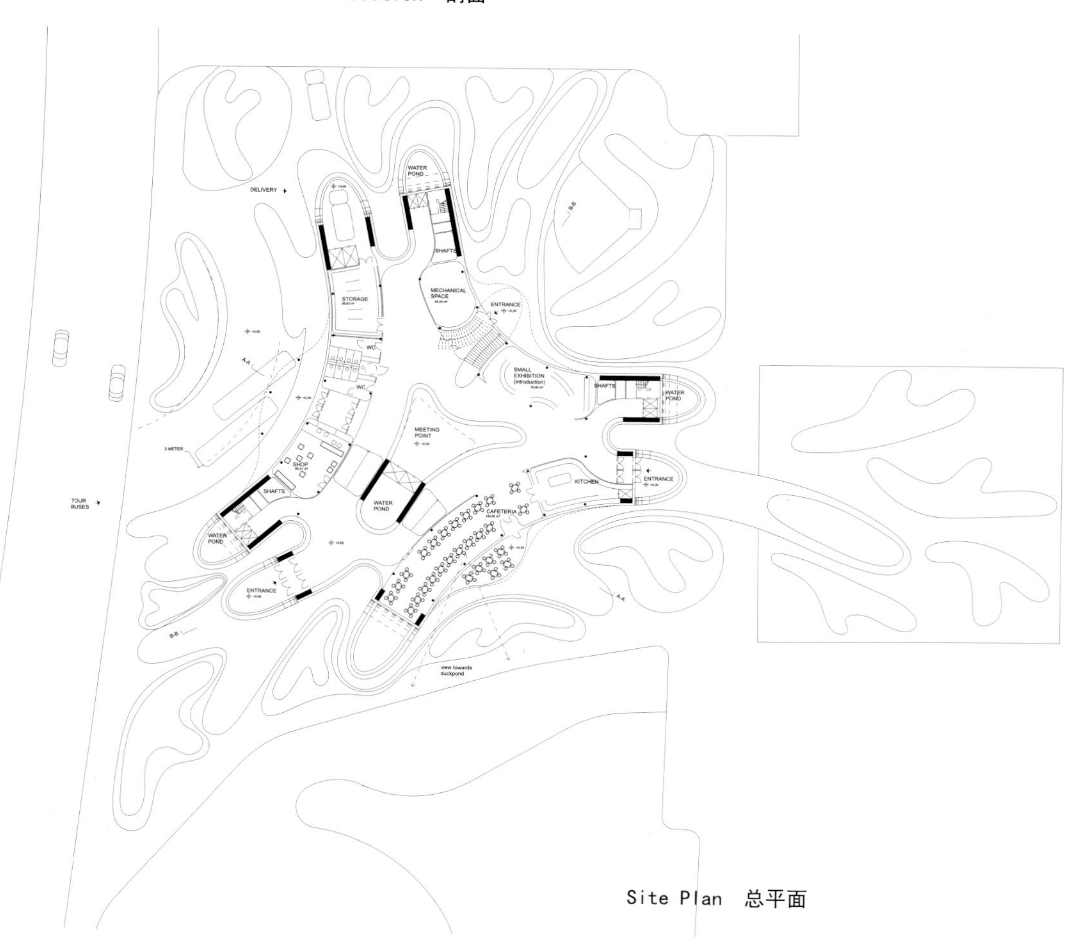

Site Plan 总平面

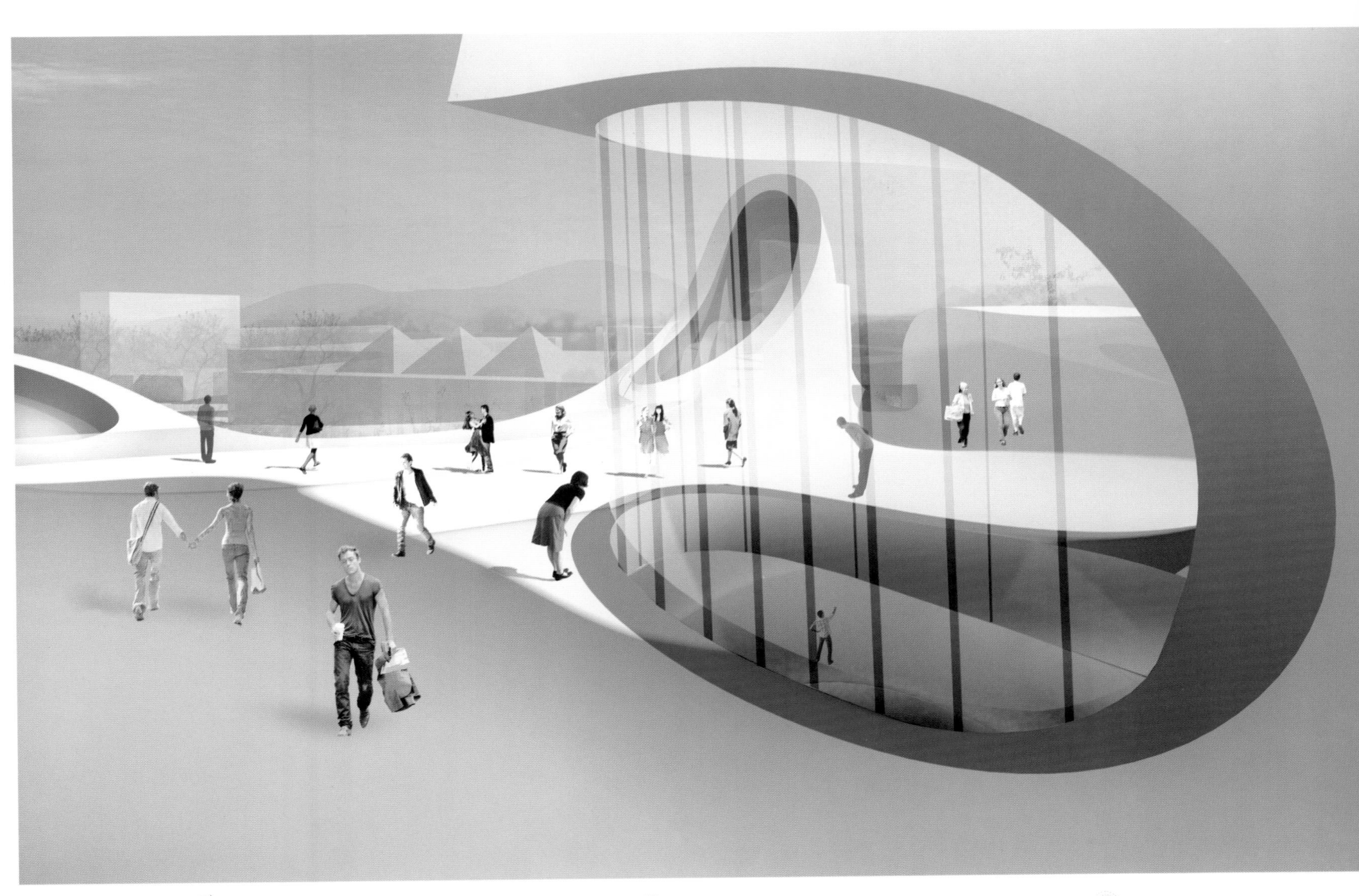

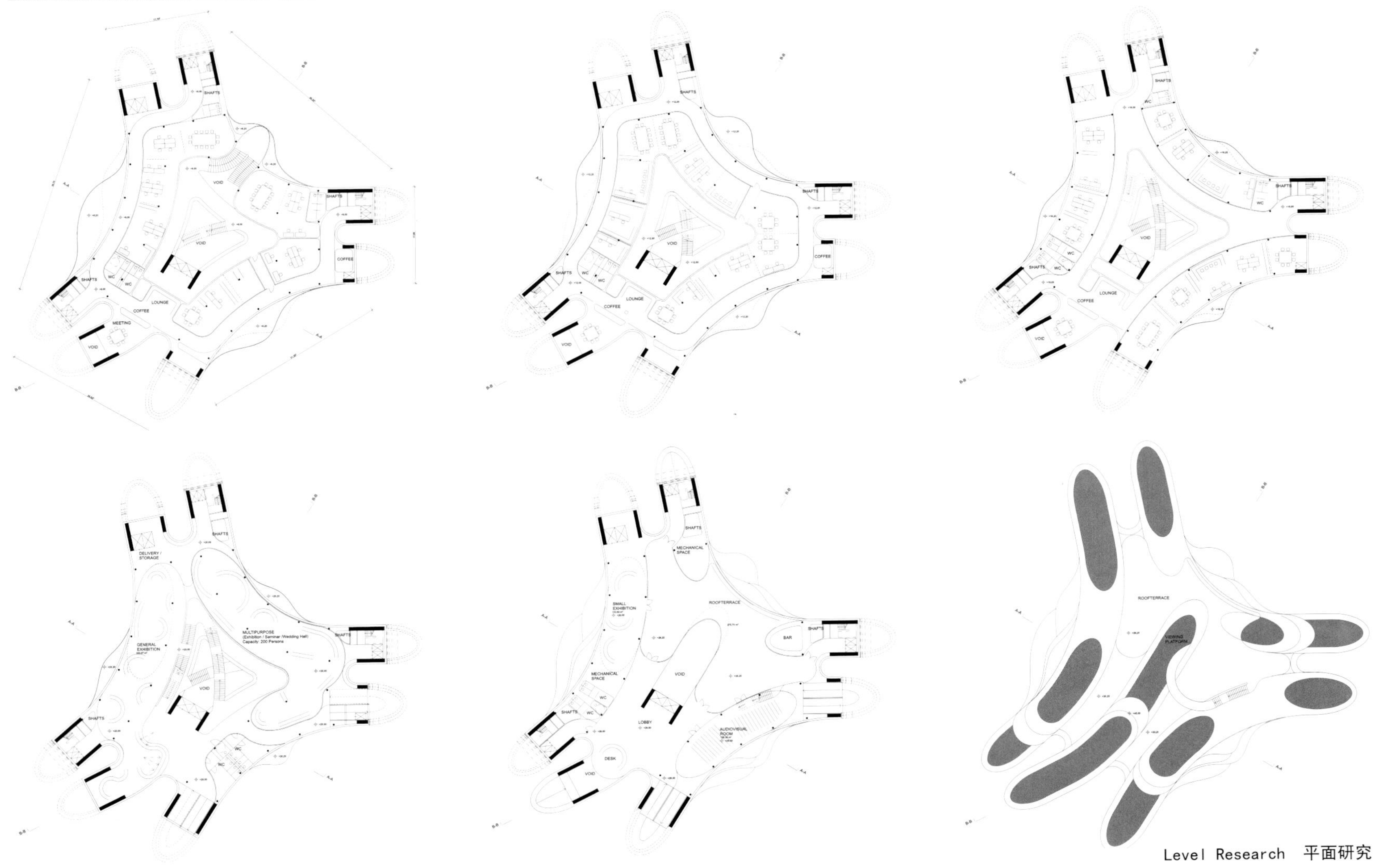

Level Research 平面研究

Redevelopment of Plaza de la Encarnacion

la Encarnacion广场新开发方案

J. MAYER H. Architects　　J. MAYER H. 建筑事务所

Location: Seville, Spain
Typology: Public
Status: Completed in 2011
Image Courtesy: David Franck , Fernando Alda

区位：西班牙，塞维利亚
类型：公共
状态：2011年竣工
图片提供：David Franck , Fernando Alda

"Metropol Parasol", the Redevelopment of the Plaza de la Encarnacíon in Seville, designed by J. MAYER H. architects, becomes the new icon for Seville, a place of identification and to articulate Seville's role as one of the world's most fascinating cultural destinations. "Metropol Parasol" explores the potential of the Plaza de la Encarnacion to become the new contemporary urban centre. Its role as a unique urban space within the dense fabric of the medieval inner city of Seville allows for a great variety of activities such as memory, leisure and commerce. A highly developed infrastructure helps to activate the square, making it an attractive destination for tourists and locals alike.

The "Metropol Parasol" scheme with its impressive timber structures offers an archaeological museum, a farmers market, an elevated plaza, multiple bars and restaurants underneath and inside the parasols, as well as a panorama terrace on the very top of the parasols. Realized as one of the largest and most innovative bonded timber-constructions with a polyurethane coating, the parasols grow out of the archaeological excavation site into a contemporary landmark, defining a unique relationship between the historical and the contemporary city. "Metropol Parasols" mix-used character initiates a dynamic development for culture and commerce in the heart of Seville and beyond.

位于恩卡纳西翁广场的塞维利亚新地标——“城市景观遮阳伞”，是由J.MAYER.H建筑事务所设计建造的广场重建项目。这是一个标志性的地方，向世界清晰呈现了塞维利亚这个世界上最迷人的文化目的地。“城市景观遮阳伞”探索着恩卡纳西翁广场的潜力，并将这一区域打造成为新的市中心所在地。坐落于塞维利亚星罗棋布的中世纪市中心建筑物之中，“城市景观遮阳伞”扮演了一处独特的都市空间，人们可以在其中开展诸如纪念、休闲、商业的各类活动。高度发达的基础设施更是激活了整个广场，使其成为了吸引游客和当地居民的旅游目的地。

“城市景观遮阳伞”建筑设计中的木质结构让人印象深刻，遮阳伞内的考古博物馆、农贸市场、高层广场、各式酒吧、地下餐厅以及伞顶的全景露台一应俱全。作为经聚氨酯涂料粉刷过的最大且最具创造力的木质结构之一，对“城市景观遮阳伞”的定位已然超出考古发掘选址的层面，而提升为现代地标了，它定义了老城区与现代城区间的独特联系。塞维利亚“城市景观遮阳伞”多元化的特点，开启了其心脏地带和周边地区文化及商业动态发展的新篇章。

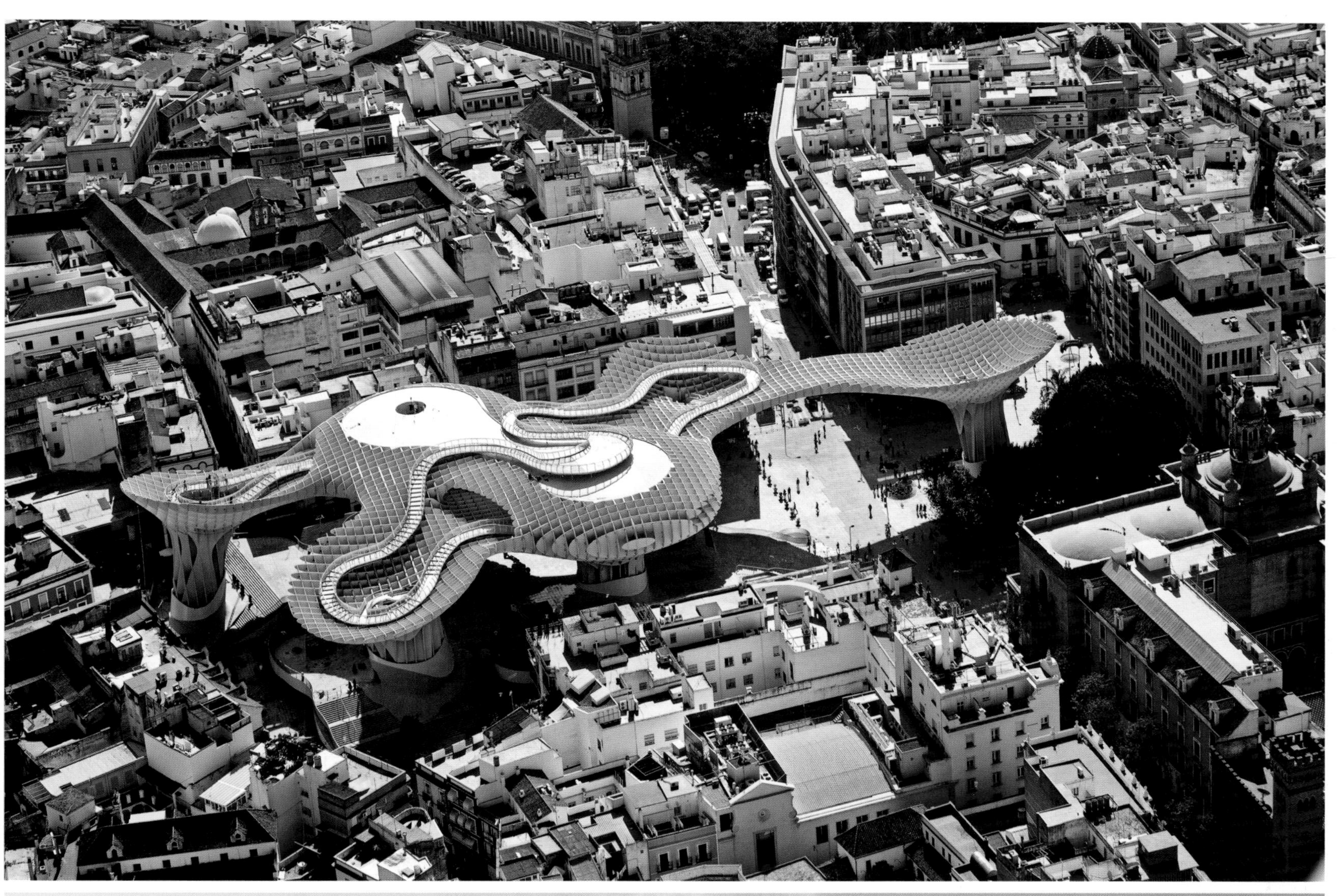

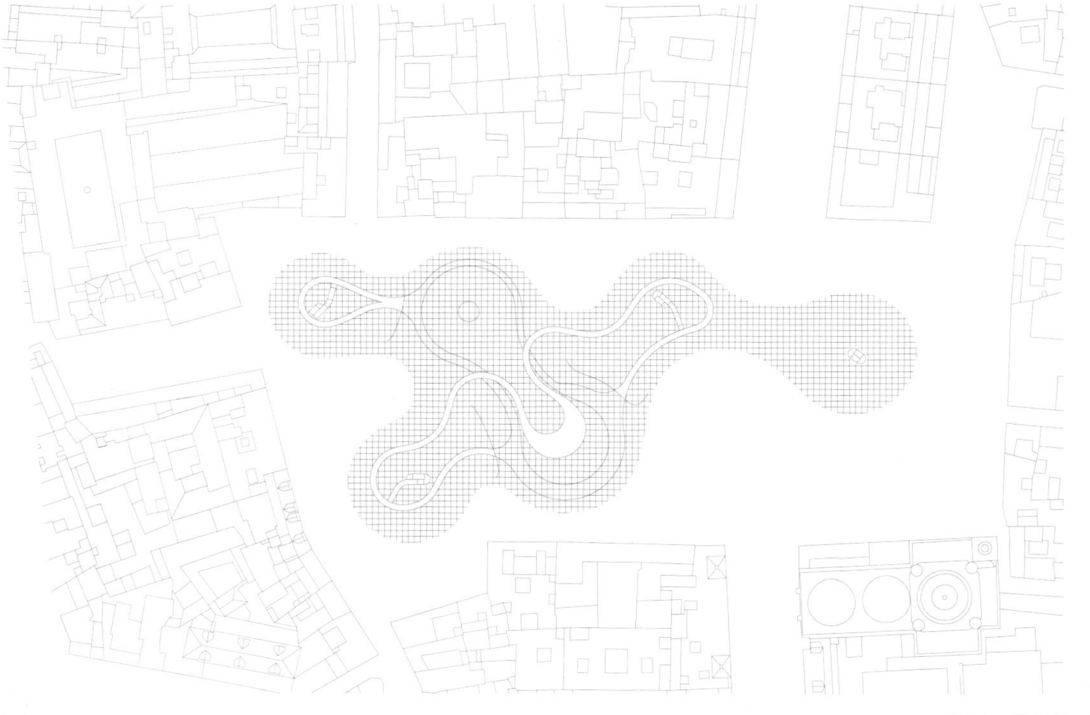

Sky Walk　人行天桥流线

medias carmen

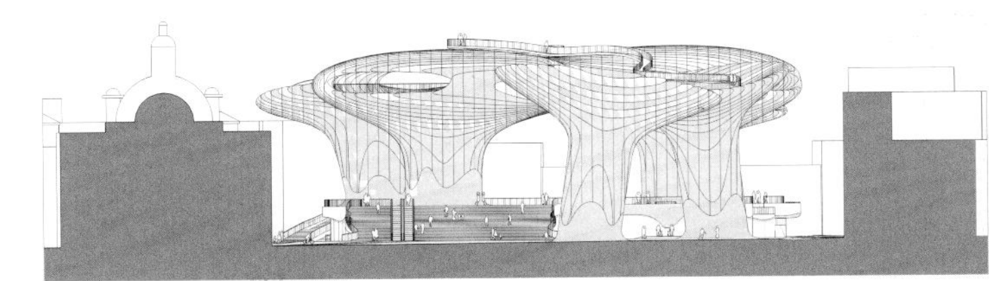

Elevation 立面

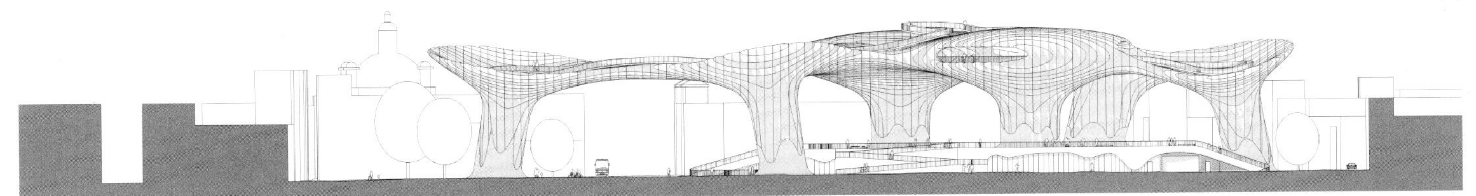

Elevation 立面

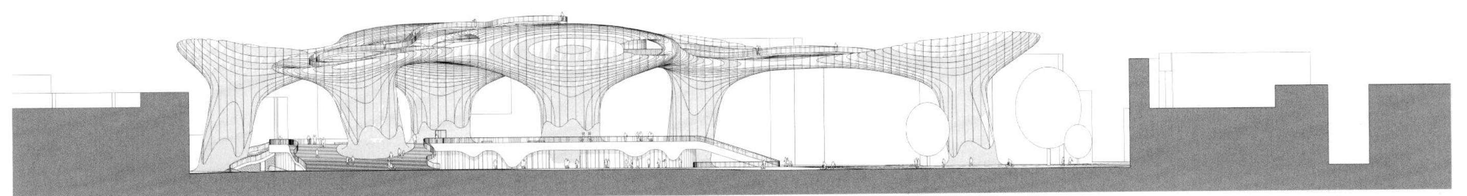

Elevation 立面

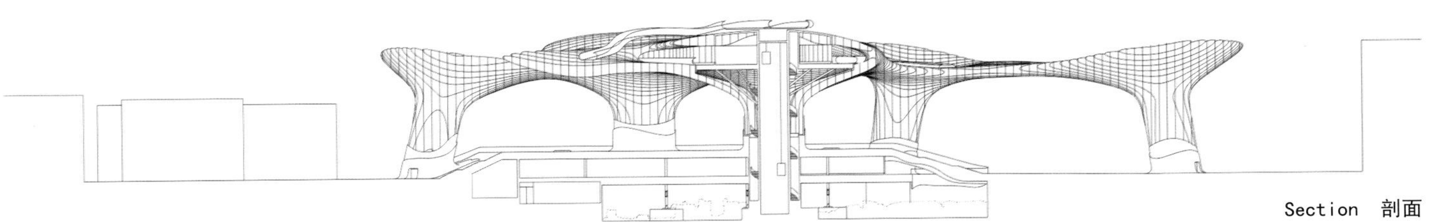

Section 剖面

Una nueva forma de ver Sevilla

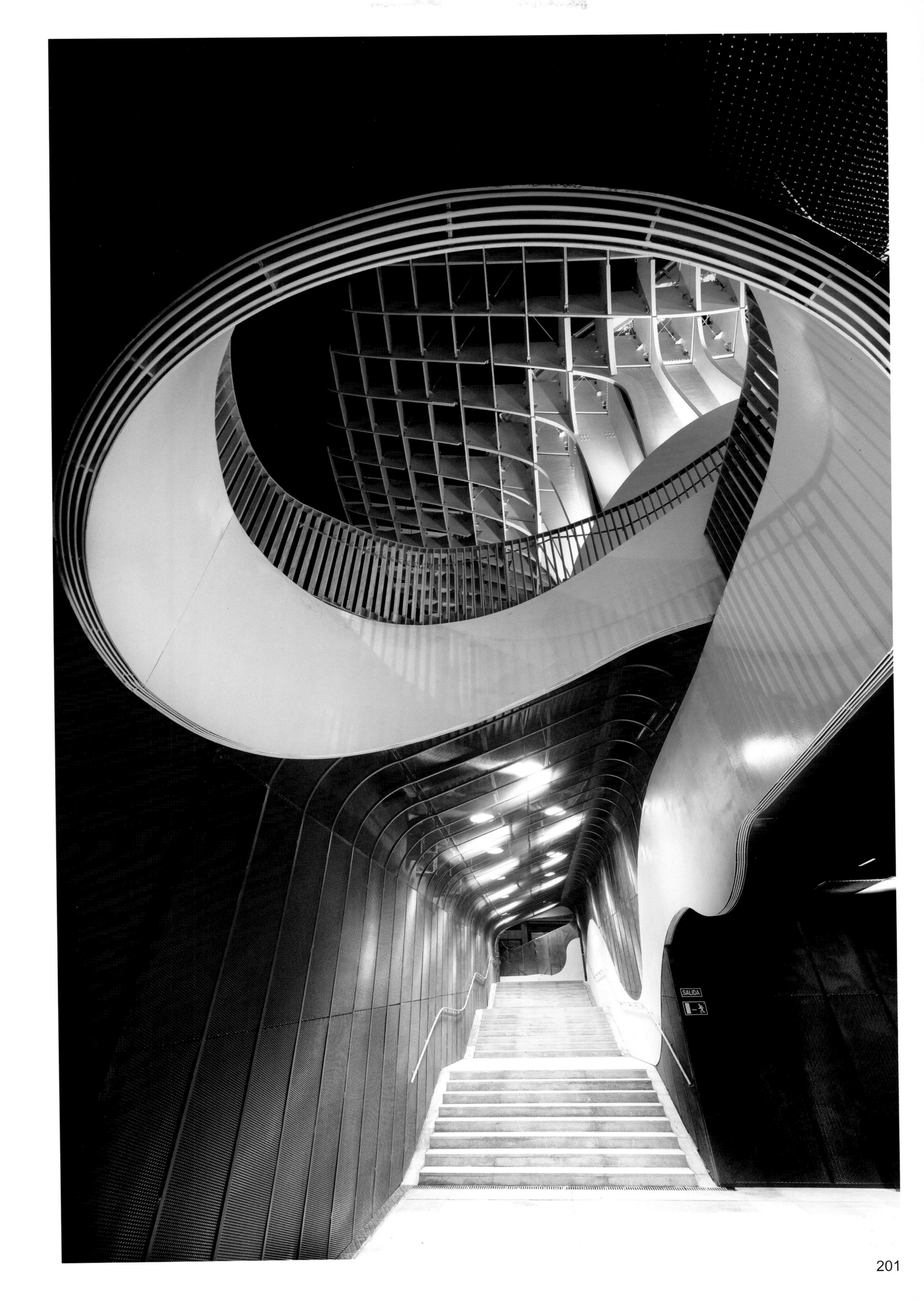
SALIDA

Plan 平面

Bubbletecture H

泡泡筑H

Endo Shuhei Architect Institute　　远藤秀平建筑研究所

Location: Sayo-cho, Hyogo Pref.
Typology: Tourist Centre
Site Area: 5,000 m²
Building Area: 968 m²
Gross Floor Area: 995 m²
Status: Completed in 2008
Image Courtesy: Endo Shuhei Architect Institute, Yoshiharu Matsumura

区位：兵库县，Sayo-cho
类型：游客中心
基地面积：5000 m²
占地面积：968 m²
总建筑面积：995 m²
状态：2008年竣工
图片提供：远藤秀平建筑研究所，Yoshiharu Matsumura

The owner requested that all people who will visit this place including the inhabitants of Hyogo prefecture, improves the interest for global environmental concerns and will be able to experience various approaches as the place of environmental study.

We thought about creating the new environment architectural space that could share the point of contact with nature and environment providing a keyword called "the circulation" in relation with nature for the request.

The site is on the steep slope of the north side in the forest. After integrating all functions that the client requested at the design term into three, required area and volume were set according to the function and structure. The two functions were arranged in parallel on flat land (old town road was there) that had remained at a high level in the site, and another was arranged having floated from the slope at the same level of other two. This is because it is assumed very important to make use of the limited flat land, to keep natural landform as large as possible, and to minimize the influence of construction against peripheral natural environment. The form of the building shows the rational shape that connects these three functions for plane and section.

Japanese cypress thinning wood log that is used for superstructure is provided with the intention of realizing the light structure as well as standardizing and improving the work efficiency of the timber using, and fixing the amount of CO_2 emission. The weather resistance steel board of 1.2 mm thickness is used for outside roof and wall finish. This steel board shows the characteristic that doesn't rust any more after stabilizing process of the initial rust.

Moreover, we tried to green a roof and wall partially applying moss that grows by moisture on an atmospheric inside and turf that grows on the soil containing high retentivity of water.

The material was chosen not only for the characters of the maintenance free or the low maintenance but also of the quality that acquires the form the oneness with a natural spectacle obtaining the expression of architecture that cooperates with the natural environment that changes and grows up.

业主要求所有来此参观的游客，包括兵库县的居民都必须提升自身对全球环境问题的关注，且能够以各种途径作为环境研究的场所。

我们考虑创建一个能够在某一点上与大自然形成关联的新环境建筑空间，也在思索着一个能按要求与大自然建立联系并提出“循环”这一关键词的环境。

基地位于森林北侧的一个陡坡上。在按照客户要求将所有功能归结为三大设计术语之后，必要区域及空间体积就依据功能和结构设计好了。两大功能在基地高处的平地（老城马路所在地）上并行排列，而另一功能则悬浮在与另两个功能相同标高的斜坡上。从而可以利用有限的平地资源，最大限度地保留自然地貌，并将建筑对外部自然环境的影响最小化。这一建筑模式展示了连接这三大功能的平面与剖面的合理形式。

用于上层建筑的日本柏树细原木也显示出该建筑追求轻结构的意图以及对木材使用效率的提升和标准化及调整二氧化碳排放量的考虑。1.2mm厚的耐候钢板用于外部屋顶和墙面。这种钢板在对最初生锈处进行稳定化处理后体现出了不再生锈的特点。

不仅如此，我们还试图在屋顶和墙面局部利用在室内大气湿度下生长的苔藓，和于高水分保持力的土层表面生长的草皮实现绿化。

建筑选材不仅要考虑其不需维修或只需进行少量维修的特点，同时也要考虑材料的品质是否可以形成与自然景观相融的建筑物，使得建筑与不断变化与生长的自然环境相和谐。

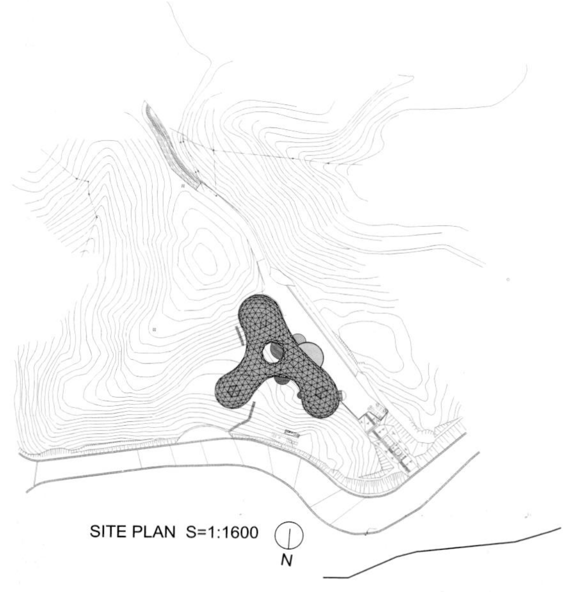

Site Plan 基地平面

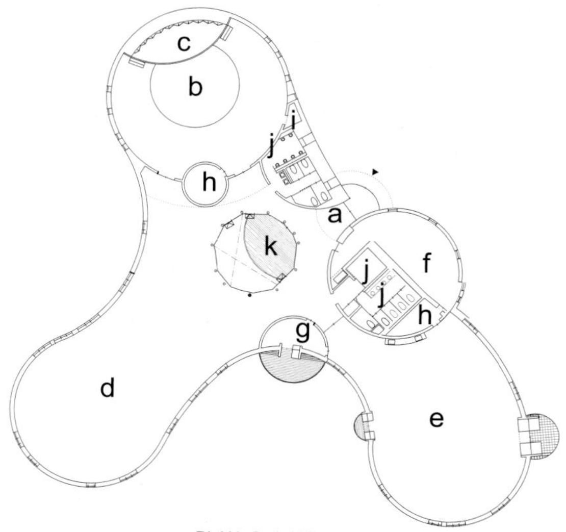

PLAN S=1:400

a:Entrance
b:Theater room
c:Stage
d:Library and information area
e:Workshop room
f:Office
g:Waiting room
h:Storage
i:Operating space for theater room
j: Lavatory
k: Courtyard

Plan 平面

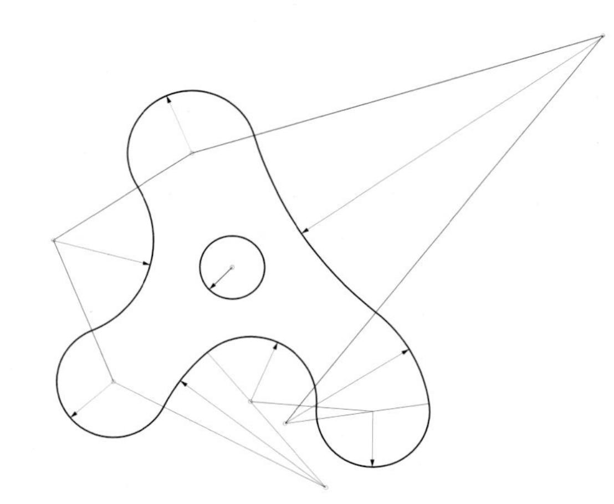

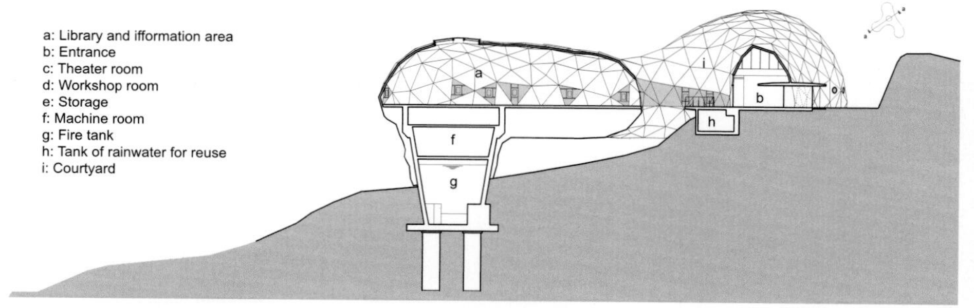

a-a SECTION S=1:400

a-a Section a-a剖面

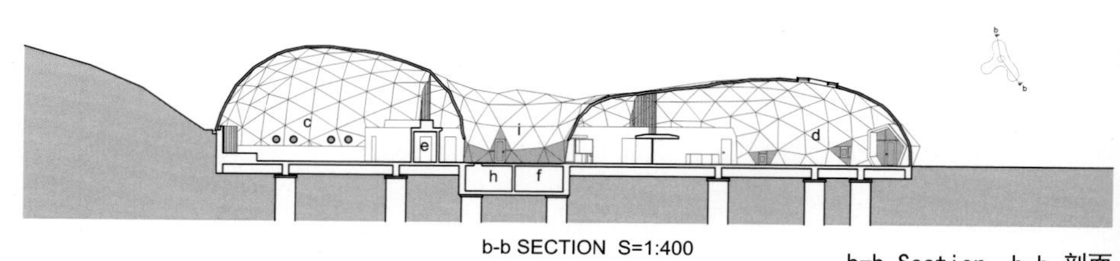

b-b SECTION S=1:400

b-b Section b-b 剖面

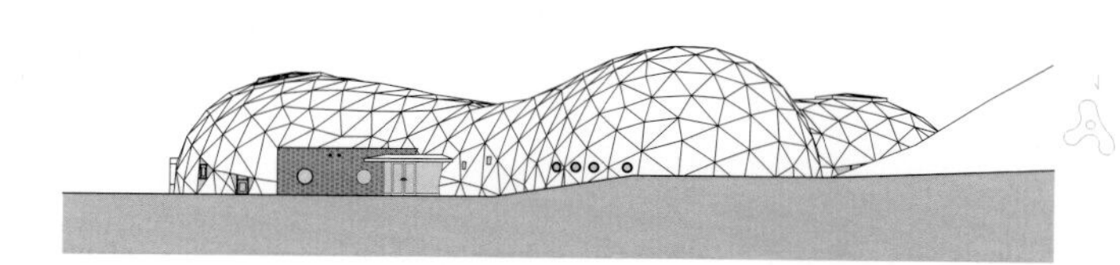

SOUTH ELEVATION S=1:400

South Elevation 南立面

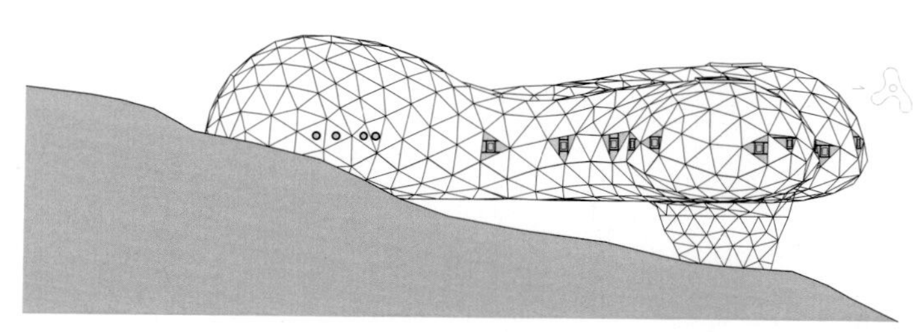

EAST ELEVATION S=1:400

East Elevation 东立面

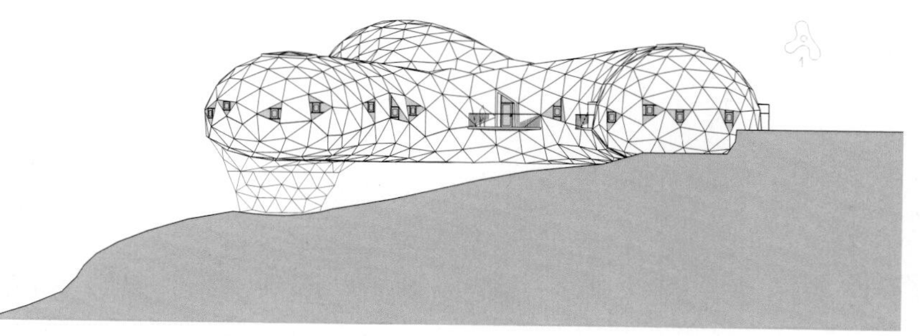

NORTH ELEVATION S=1:400

North Elevation 东立面

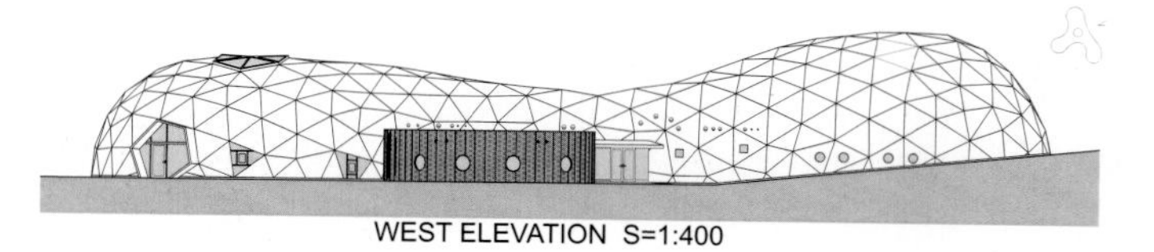

WEST ELEVATION S=1:400

West Elevation 西立面

Information Booth / Restroom at Henningsvær, Norway

挪威亨宁斯维尔询问处及休息室

Jarmund / Vigsnæs AS Architects MNAL　Jarmund / Vigsnæs AS 建筑事务所

Location: Henningsvær, Lofoten Islands, Norway
Typology: Public
Gross Floor Area: 80 m²
Image Courtesy: John Stenersen

区位：挪威，罗浮敦群岛，亨宁斯维尔
类型：公共
总建筑面积：80 m²
图片提供：John Stenersen

On a large horizontal stretch of parking, the information booth welcomes the visitors with open doors for the 2 summer months. For the rest of the year the building has a different, closed expression.

挪威的亨宁斯维尔拥有一望无际的停车场，夏季两个月全面开放的询问处欢迎来自世界各地的游客。而一年内其余的时间该建筑则呈现出一派截然不同的封闭式景致。

FOTENS HUS

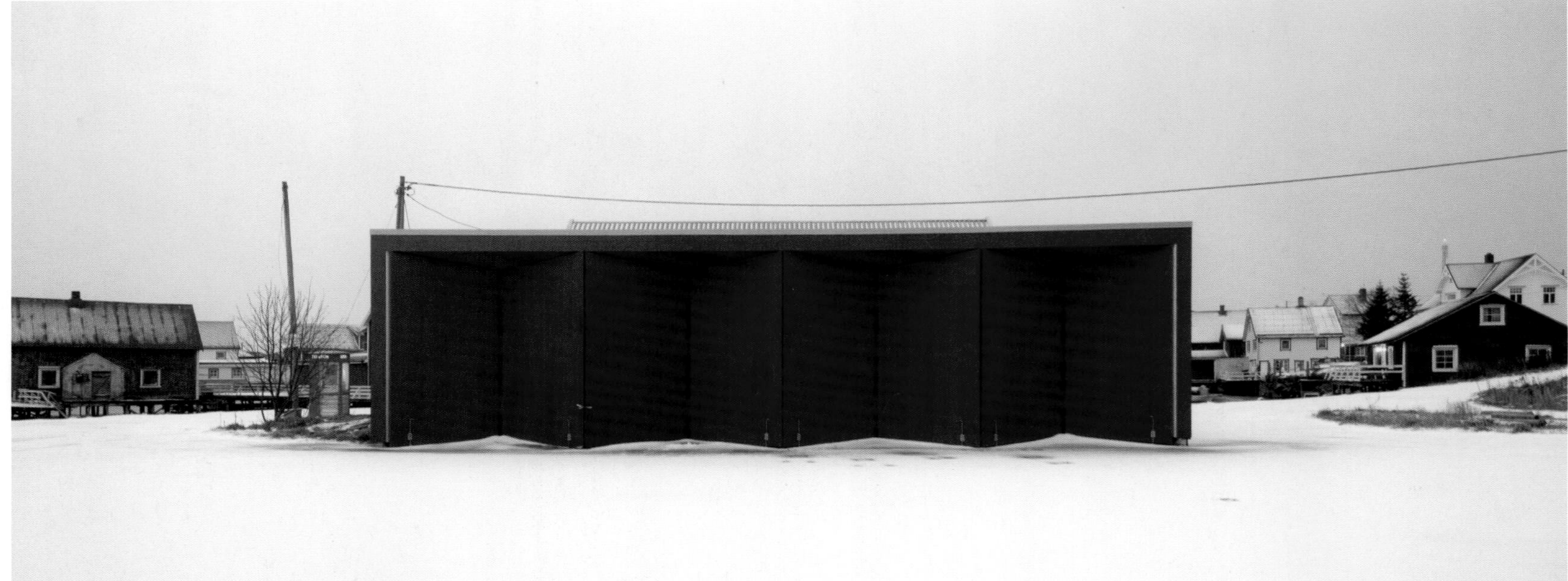

S HUS

Solberg Tower & Rest Area

索尔伯格塔及休息区域

SAUNDERS ARCHITECTURE AS　　SAUNDERS建筑事务所

Location: Sarpsborg, stfold, Norway
Typology: Public
Building Area: 2,000 m^2
Status: Completed in 2010
Image Courtesy: Bent Rene Synnevag

区位：挪威，萨尔普斯堡，东福尔
类型：公共
占地面积：2 000 m^2
状态：2010年竣工
图片提供：Bent Rene Synnevag

Sarpsborg is a green, flat and calm piece of South Norway and a traditional stopover for travellers on the route to and from Sweden.

As Sarpsborg is one of the first tastes of Norway the travellers from Sweden experience, it was important for the client that they would be able to slow down and spend time discovering the surrounding nature. The local forest and coastline form a beautiful, yet largely unknown part of the country. The neighbouring highway's speed and noise only enhance the traveller's need for a break and re-connection with nature, so a green resting space was on the top of the list. A low walled ramp spirals around the rest area, defining the 2,000 m^2 area's limits, while spring-flowering fruit trees adorn the courtyard. Within it, Saunders designed seven small pavilions working with graphic designer Camilla Holcroft, showcasing information on the local rock carvings from the Bronze Age, an exhibition, which continues on the ramp's walls.

The structures also offer the option for temporary artist exhibitions. The flatness of the landscape meant that the beauty of the surrounding nature could only be enjoyed from a certain height, so the creation of a tower quickly became a main part of the brief.

Finally, the design style and aesthetic aspect were developed in relation to the environs existing architecture, minimal and geometrical contemporary shapes were chosen, contrasting with the local farming villages' more traditional forms. The main materials used were beautifully-ageing CorTen steel for the exterior walls and warm oiled hard wood for the courtyard's design elements and information points. Local slate and fine gravel pave the ground level.

索尔伯格是挪威南部的一片平坦地域，草木葱茏，一片祥和，也是往返挪威瑞典旅游线路的游客的必经之处。

由于索尔伯格是到瑞典的游客最初体验挪威风情的旅游胜地之一，那么对旅行者而言，放慢脚步并花上些时日探索身边的大自然就变得至关重要了。当地的森林及海岸线描绘出了一幅美丽的画卷，这也正是挪威所不为人知的广阔疆域之所在。在邻近高速公路上的急速行驶和随之而来的噪声使游人迫切渴望能够途中小憩，重返大自然的怀抱，因此，绿色环保的休息场所便成为了游人所关注的重中之重。休息场所低矮的螺旋式弧形墙界定了2 000 m^{2}的有限区域，而春季枝头挂满花朵的果树点缀了整个庭院。在休息庭院内，桑德斯与美术设计师卡蜜拉·霍尔克罗夫特合作设计的七栋楼阁，成为了展示当地青铜器时代的石刻艺术品所传达信息的展览厅，该展览也包括弧形斜墙本身。

该建筑设计也为临时的艺术展览提供了场馆。平坦的地质景观意味着人们欣赏周边自然美景的角度被局限在特定的高度，因此索尔伯格塔的拔地而起很快便成为了吸引人眼球的主要部分。

最后，索尔伯格塔的设计风格和美学视角也发展成与市郊的现存建筑相映成趣的风格模式，同样选取微型几何体当代建筑造型，与当地农庄的更为传统的建筑风格形成鲜明对比。外部墙体建造选用经过完美年代化处理的耐候钢为建材，而对于庭院设计元素和信息点的设计则采用了经暖油处理过的硬质木材。地面由当地的板岩和细砾石铺设而成。

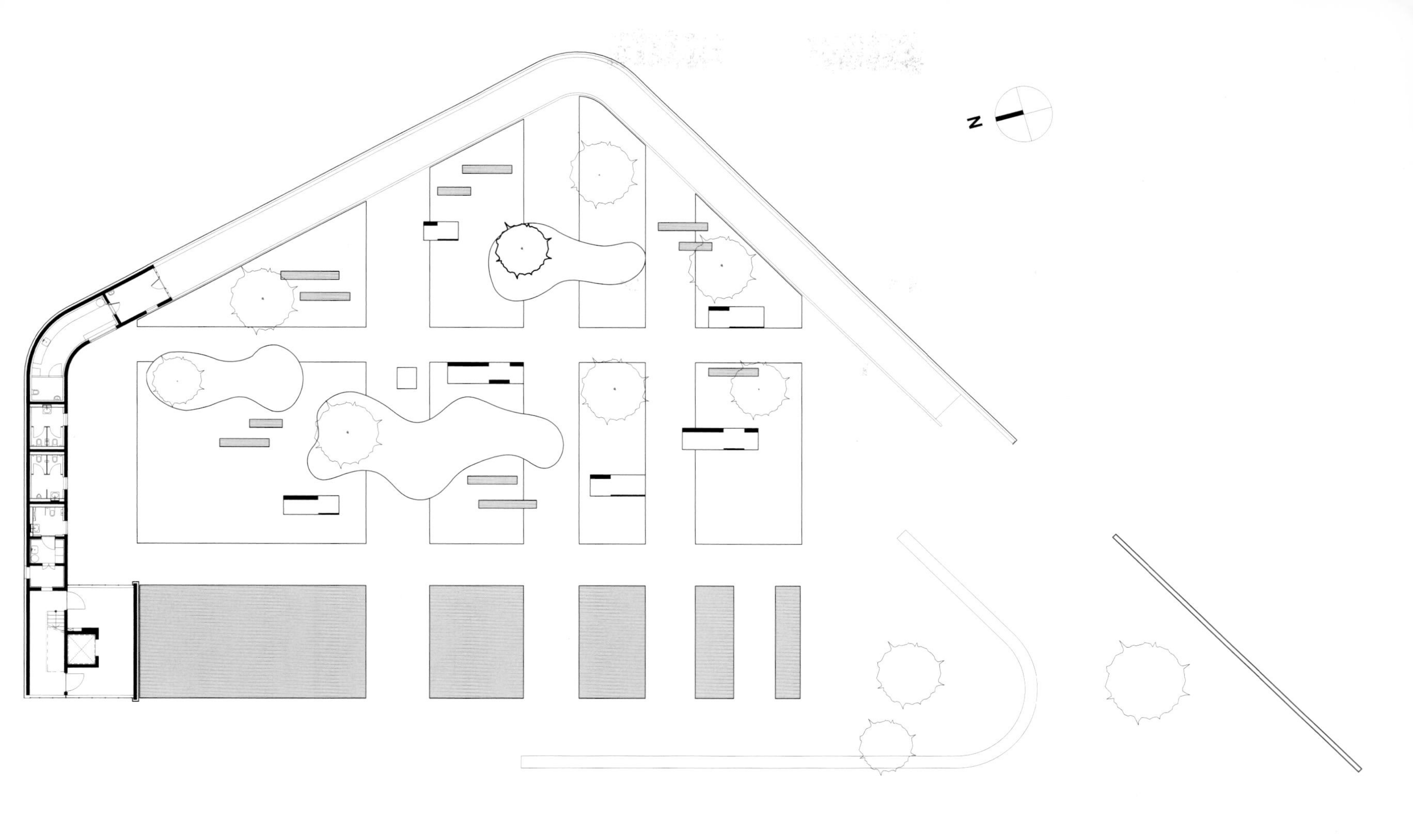
N

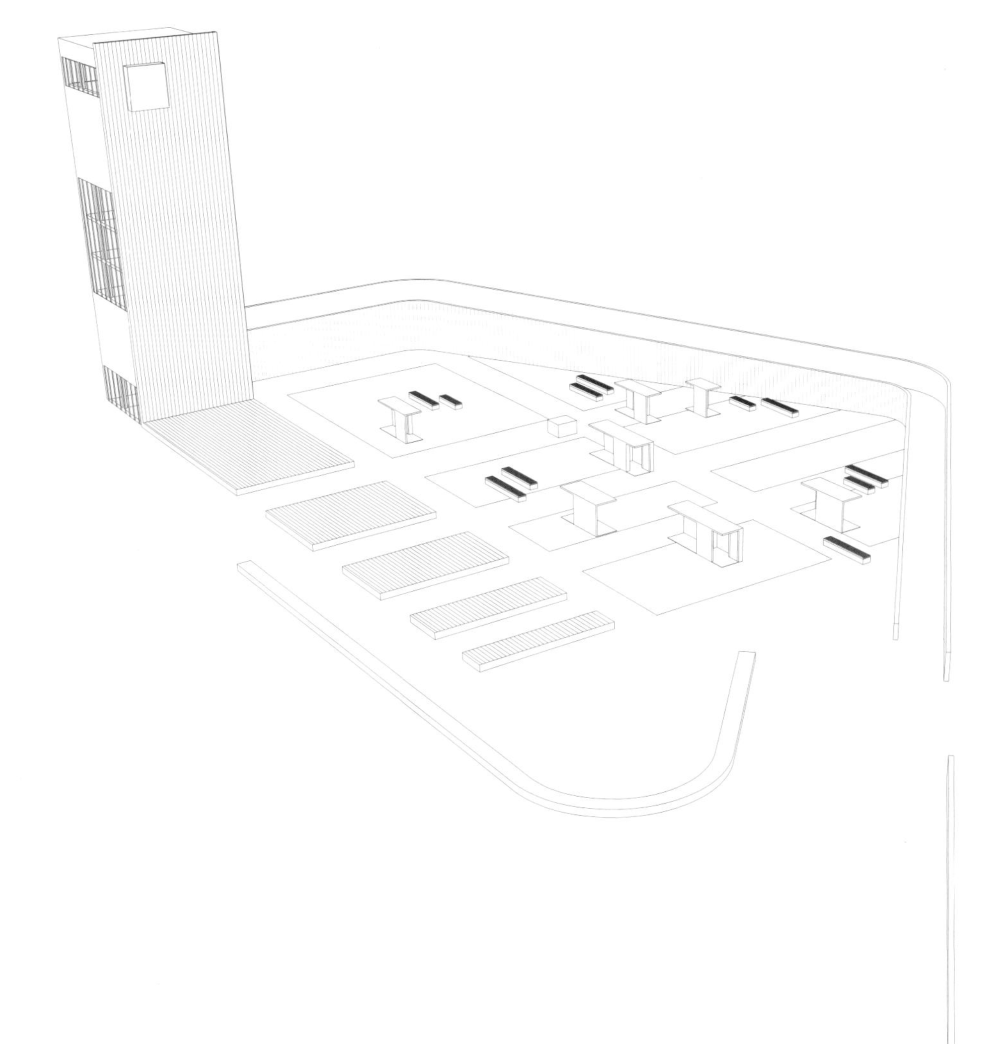

ET LANDSKAP
I ENDRING

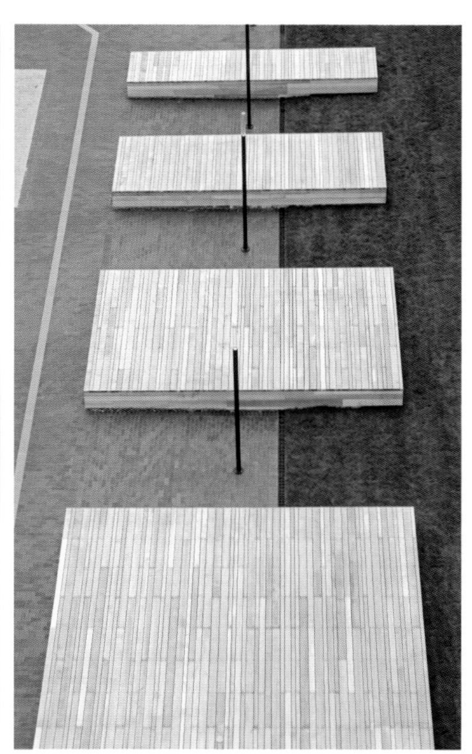

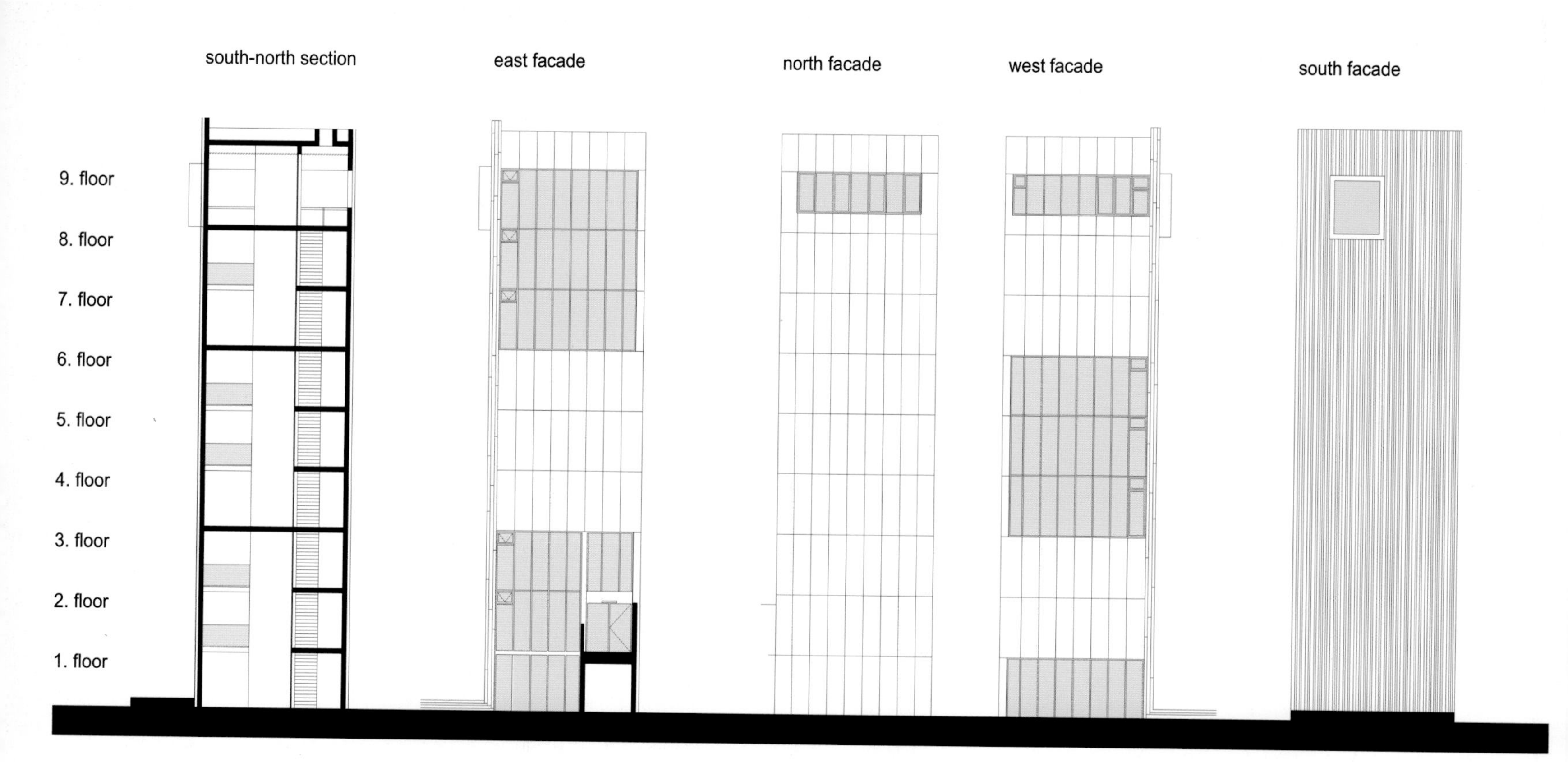
south-north section
east facade
north facade
west facade
south facade
9. floor
8. floor
7. floor
6. floor
5. floor
4. floor
3. floor
2. floor
1. floor

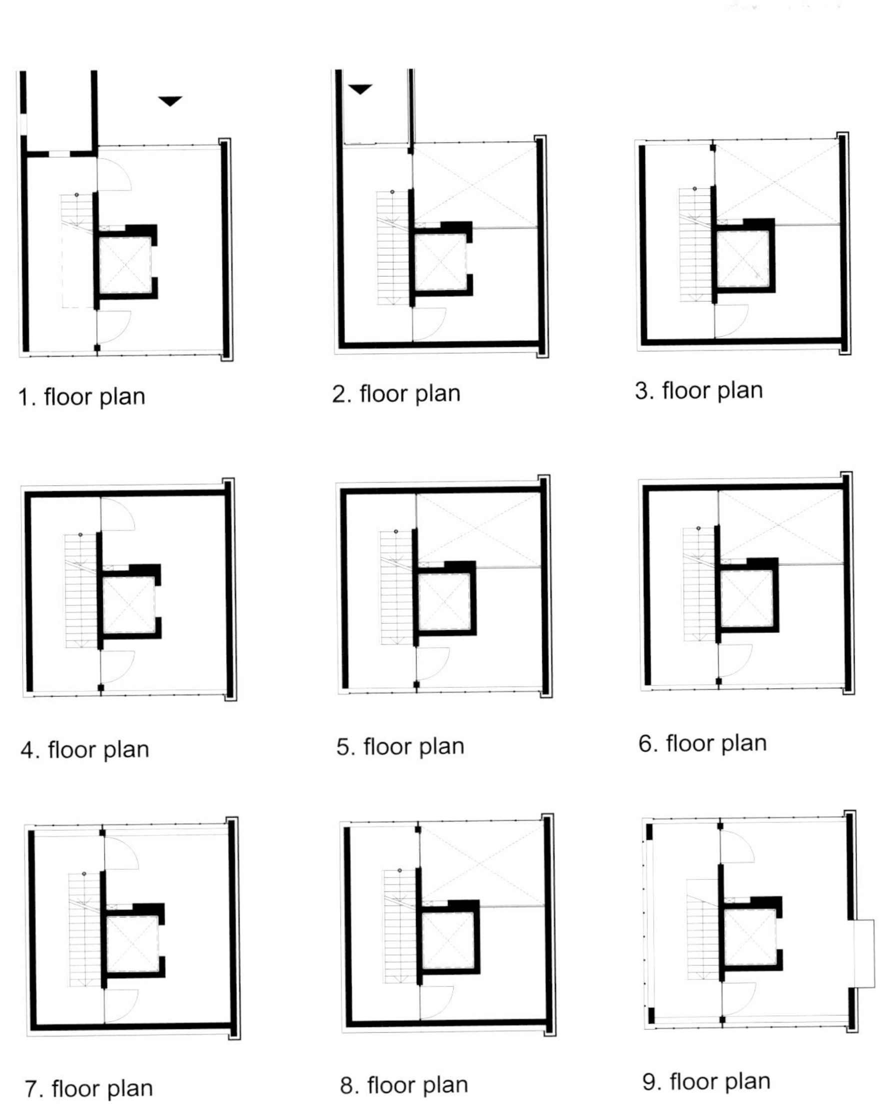

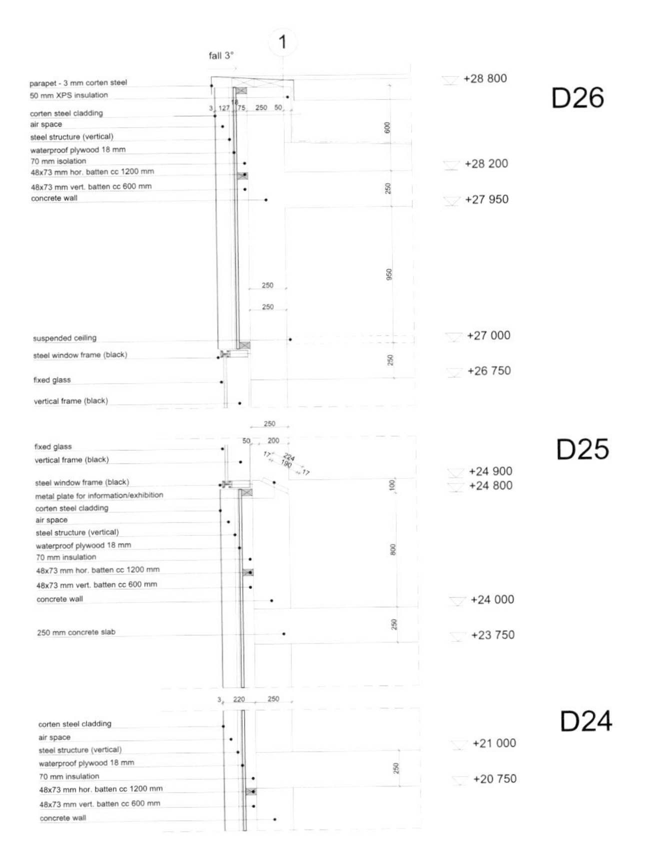

east-west section
from south

D26
D25
D24

Bogotá Chamber of Commerce – Chapinero

波哥大商会——Chapinero

DANIEL BONILLA ARQUITECTOS　　DANIEL BONILLA建筑事务所

Location: Bogotá, Colombia
Typology: Public
Area: 11,436.35 m^2
Status: Completed in 2009
Image Courtesy: Sergio Gómez y Rodrigo Dávila

区位：哥伦比亚，波哥大
类型：公共
面积：11 436.35 m^2
状态：2009年竣工
图片提供：Sergio Gómez y Rodrigo Dávila

The urban reaction of the scheme celebrates the collective value of the program to the city, opening the ground floor space to the public, with a diagonal pass through the building.
The tilted first floor plan generates an interesting/additional dynamic to that of the traditional platform space, while allowing a generous auditorium space below. A contribution to the scheme, in terms of floor area gained, and more profoundly, is in generating a symbolic space, cum utility, particular for its locality within the city.
The building's program is separated into two volumes, one for the general public activities and the other for the support spaces. An interstitial space, as an operational crack or "court", encourages the relationship between the two volumes, while providing the entry of natural light and ventilation.
The volume's undulating plywood envelope is inspired on the kinetic art distortion effects. It is held under the Bogotá Chamber of Commerce institutional frame.

该建筑设计在都市范围内的强烈反响是其对于城市而言的集体价值的庆祝，该项目的一层空间面向大众开放，一条斜线贯通了整栋建筑。

倾斜设计的二层平面图赋予传统平台空间以有趣/额外的动态之感，同时也创造了位于其下方的宽敞无比的礼堂空间。就增加的建筑面积而言，其对该规划的一大贡献在于它进一步创造了一个象征性的空间和附属价值，对于坐落于城内这一特殊建址而言尤是如此。

建筑项目分成两大建筑体量，其一用于公众活动，另一体量则是辅助功能空间。作为操作台的一个间隙空间，促进了两大体量间的良好关系，与此同时也实现了自然光照明和良好通风。

建筑体量内是波状起伏的胶合板外壳，其设计灵感源自活动艺术的失真效应。活动是在波哥大商会的制度框架下举办的。

CAMARA

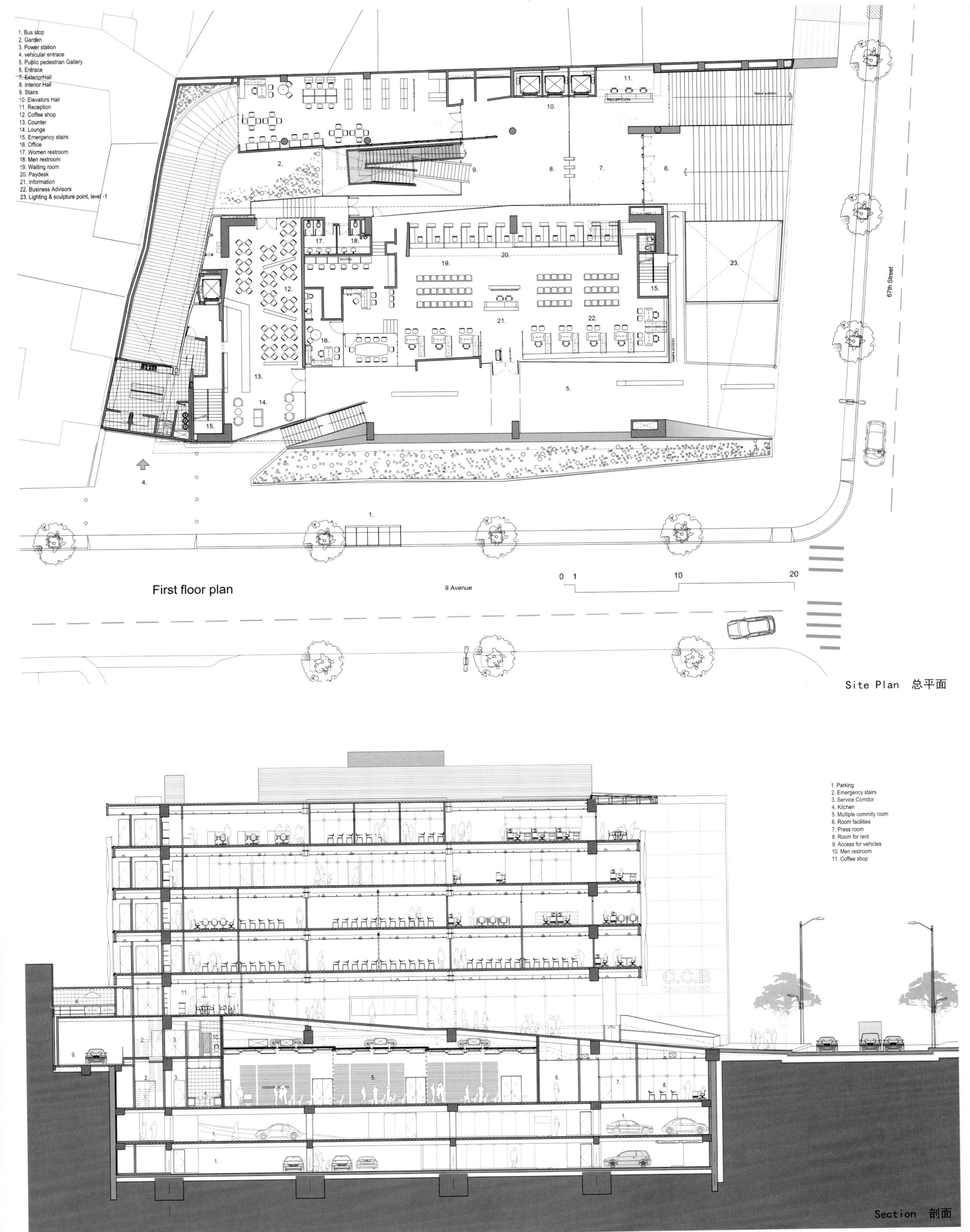

First floor plan

Site Plan 总平面

Section 剖面

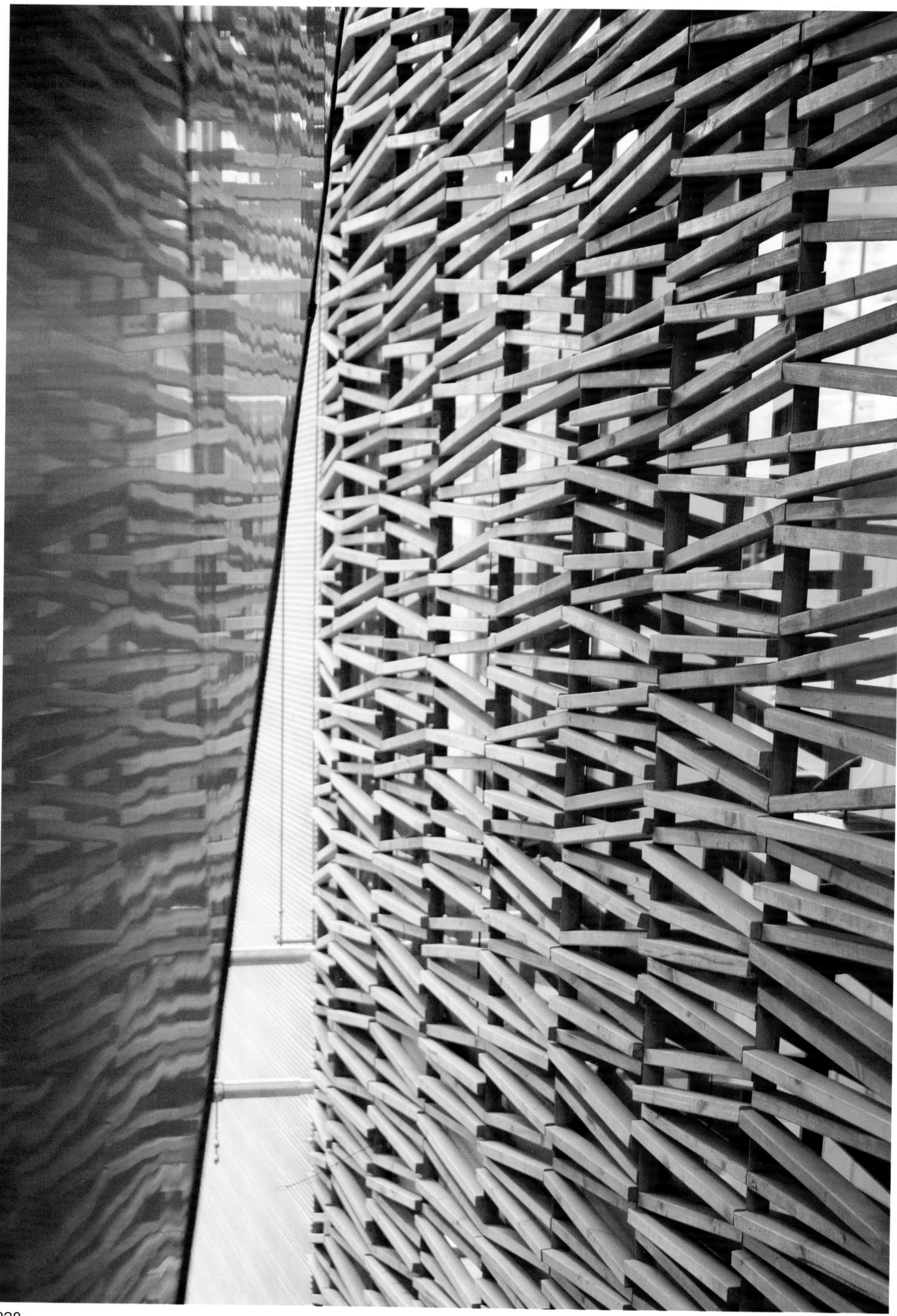

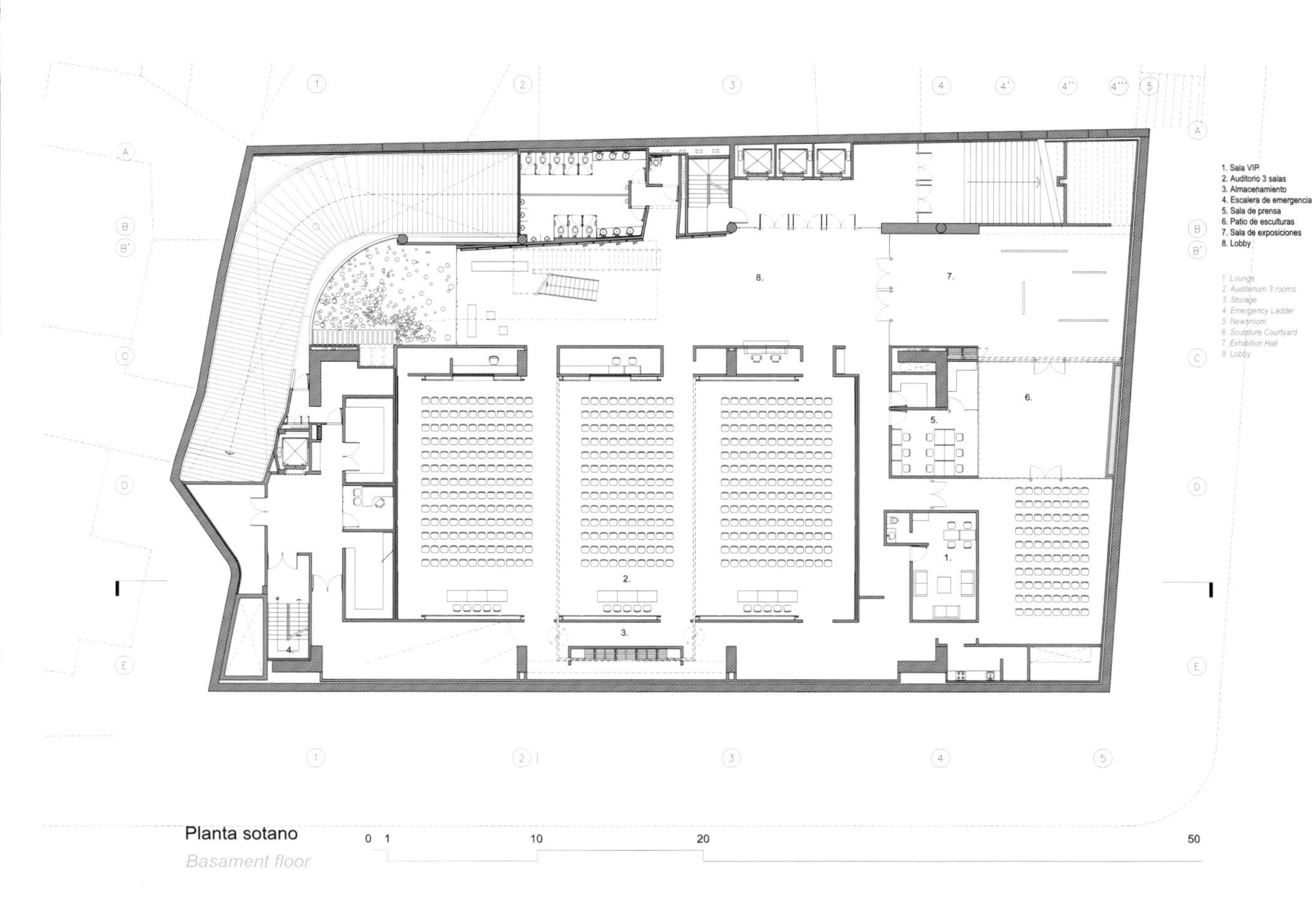
1. Sala VIP
2. Auditorio 3 salas
3. Almacenamiento
4. Escalera de emergencia
5. Sala de prensa
6. Patio de esculturas
7. Sala de exposiciones
8. Lobby
1. Lounge
2. Auditorium 3 rooms
3. Storage
4. Emergency Ladder
5. Newsroom
6. Sculpture Courtyard
7. Exhibition Hall
8. Lobby
Planta sotano
Basament floor
0 1 10 20 50

Plan 平面

Mopti Komoguel Centre for Earth Architecture

莫普提·克莫格尔泥土建筑游客中心

Diébédo Francis Kéré • Architect　　Diébédo Francis Kéré建筑事务所

Location: Mopti, Mali
Program: Tourist Centre
Status: Completed
Image Courtesy: Francis Kere

区位：马里，莫普提
功能：游客中心
状态：已完成
图片提供：Francis Kere

The construction of the Mopti Komoguel Centre for Earth Architecture accomplishes the activities of the AKTC in Mopti after the restoration of the mosque and the construction of a new sewerage system. The construction site has been gained by a backfill at the waterside of an interior lake and so the lakeside has been made accessible for public use. The program itself responds both to the needs of the district management Komoguel as to the visitors of the cultural facilities of the district. The building is clearly structured and its building height responds to the existing fabric without compromising the view on the mosque. The visitor centre, according to its program is divided into three different buildings which are connected by two roof surfaces. All the walls and the barrel vaults are constructed of compressed earth blocks which are very suitable to the climatic conditions for their natural temperature buffer and therefore guarantee comfortable indoor temperature.

莫普提·克莫格尔泥土建筑游客中心项目是在清真寺修复和新下水道系统的建造之后完成的，它的场地是通过填充一个内部的湖区而得到的，这样湖边的空间也能作为公共场所使用，项目本身是对地区发展的回应，同时也为来当地参观文化设施的游客提供了便利。建筑清晰地罗列，高度与周围的环境相呼应，同时它不影响人们对清真寺的视线。这个游客中心被分为三个不同的建筑，它们由两个屋顶表面连接。所有的墙壁和桶形穹窿都由压制土层做成，这种材料非常适合当地气候，能为室内提供舒适的温度和环境。

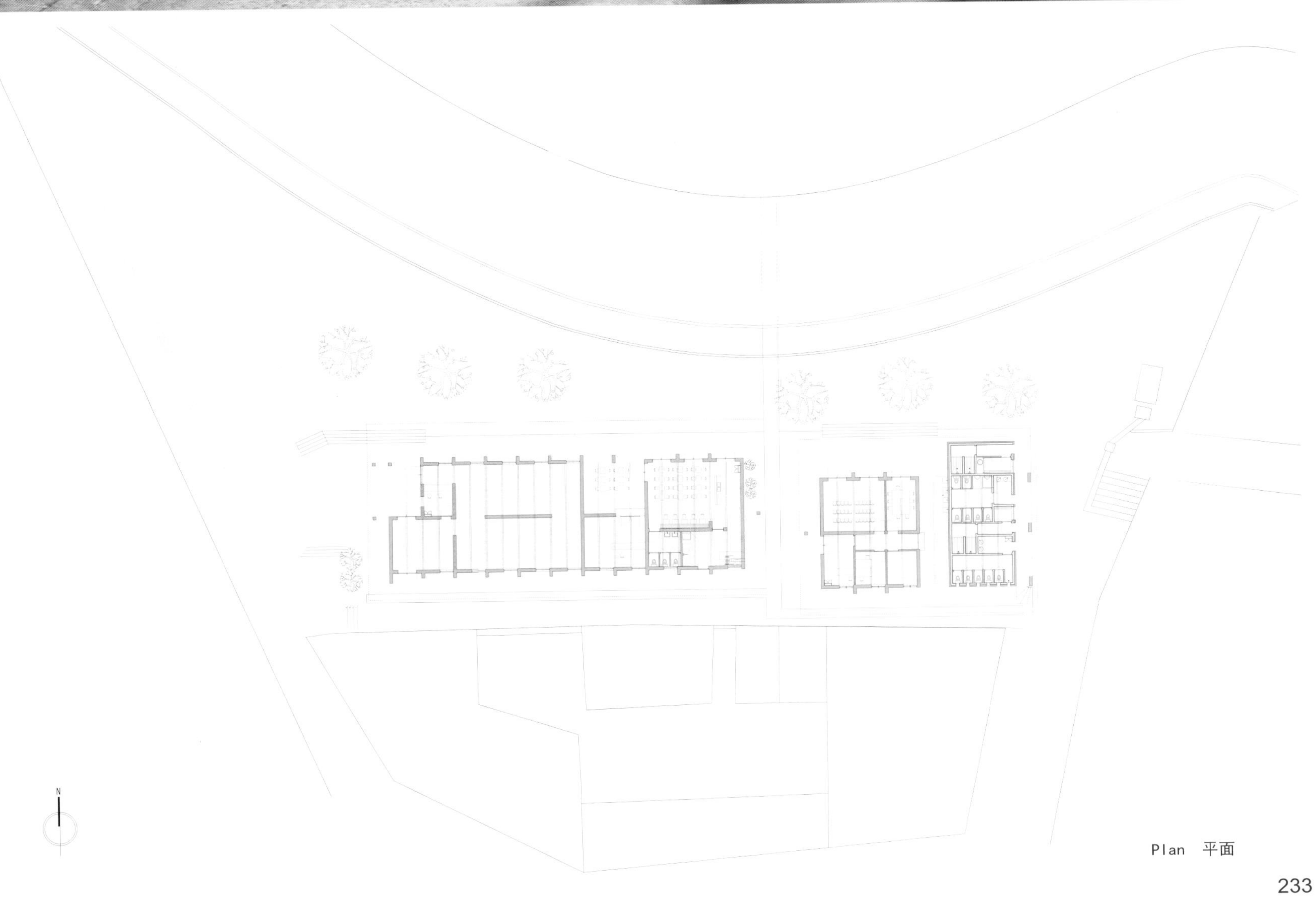

Plan 平面

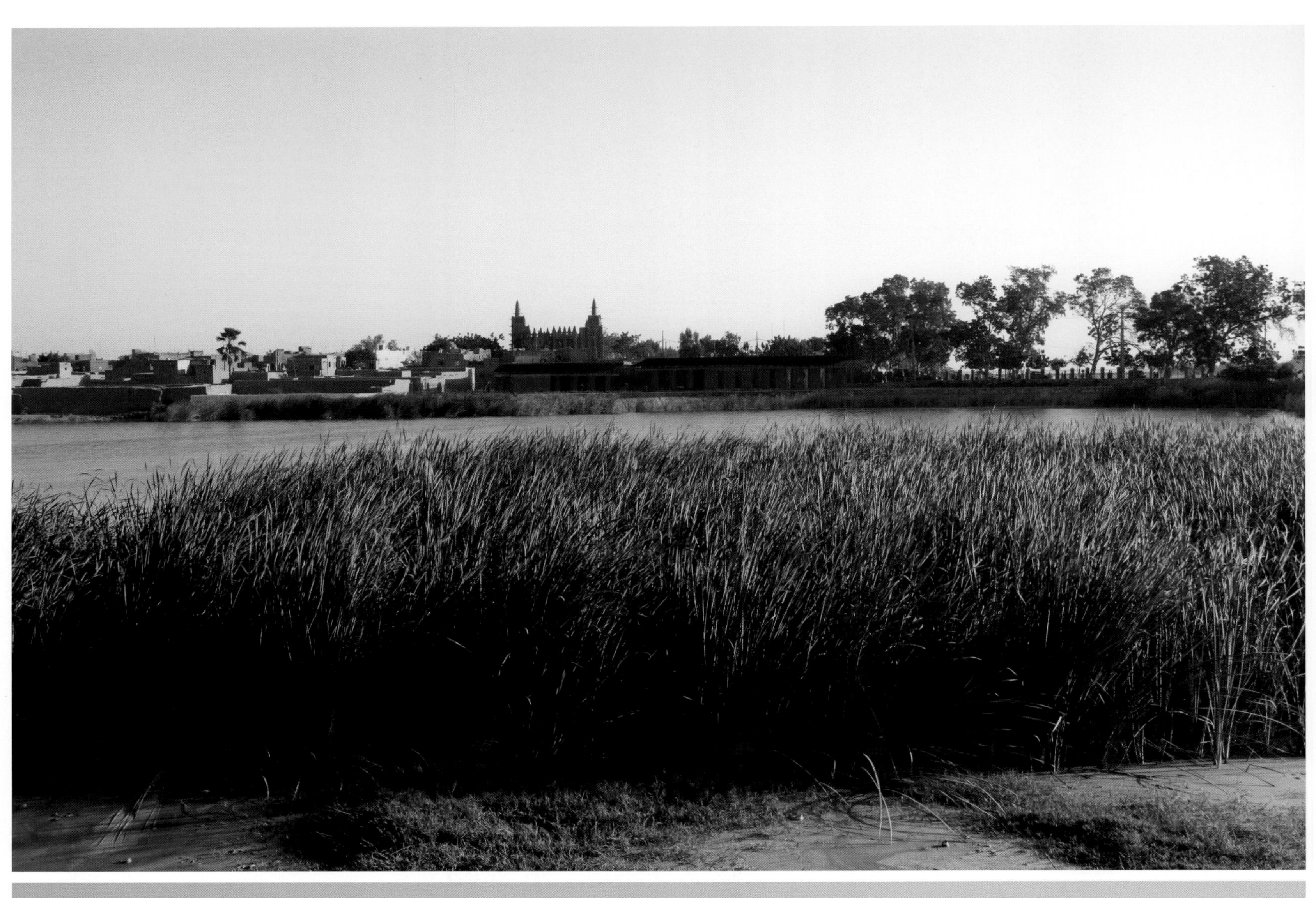

de l'architecture en

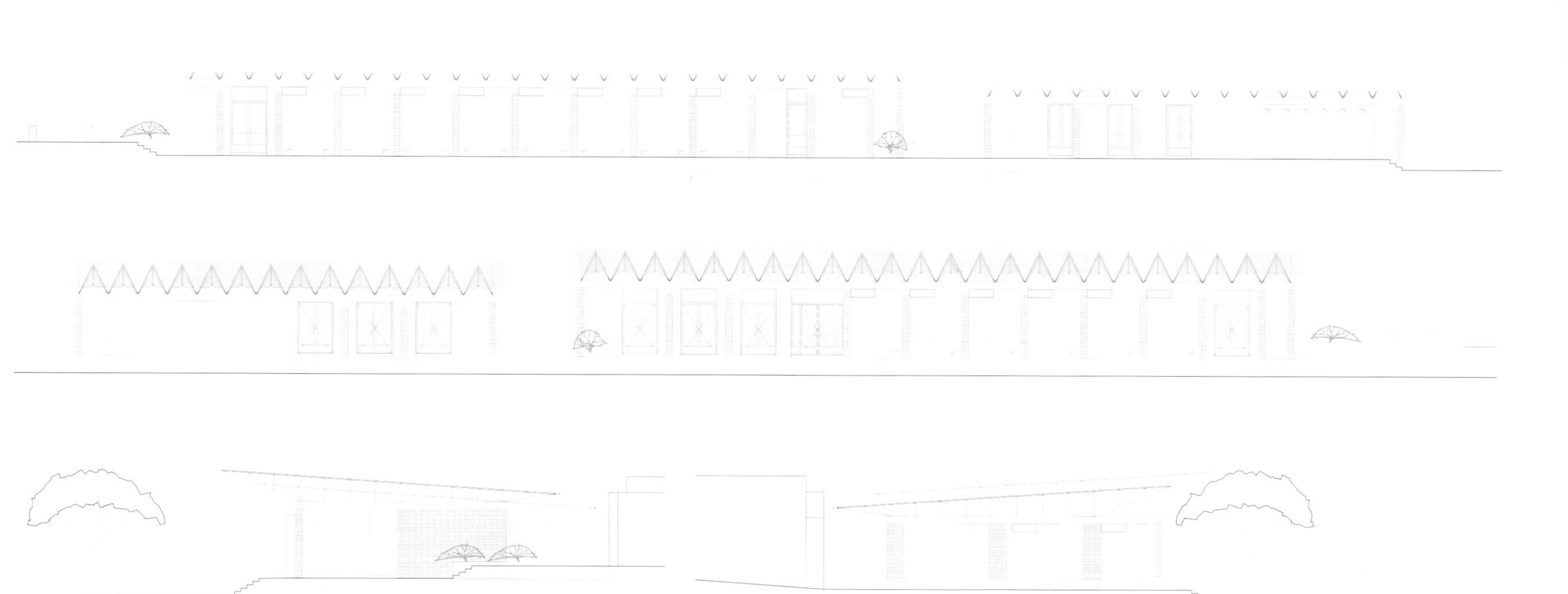

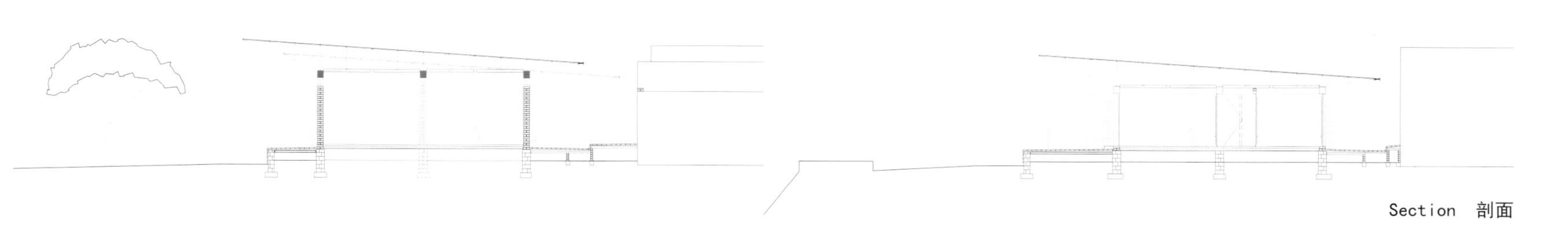

Elevation 立面

Section 剖面

Mopti
La Grande Mosquée
de Mopti

Borgund Stave Church Visitors Centre

博尔贡木构教堂游客中心

Askim /Lantto　　Askim /Lantto建筑事务所

Location: Borgund, Norway
Program: Visitors Centre
Building Area: 580 m^2
Status: Completed
Image Courtesy: Jiri Havran

区位：挪威，博尔贡
功能：游客中心
占地面积：580 m^2
状态：已完成
图片提供：Jiri Havran

Borgund Stave Church has a unique structure, and represents the best of historic Norwegian wooden architecture. The overall goal of the Historical Monuments in the planning of the new visitor centre has been to restore and enhance the experience of what has been the original landscape around Borgund Stave Church. The church is located in a bowl shape in the landscape, with the road and the river in front to the east.

The centre is located with visual contact to the church and the surrounding landscape. The new building volume follows the existing landscape, terrain shape, creating a horizontal movement across the valley.

The parking area is moved from the front of the church behind the visitor centre to the north. There are plans to lower the road in front of the church, which will enhance the experience of the church's original landscape.

The facade of the building has simple detailing, in contrast to the church's more expressive form, with rich ornamentation and vertical appearance. The facades are made of heartwood of pine and should be left untreated, and will be coloured grey by the weather. The building has two distinct openings with large glass openings, one towards the entrance, one to the church and the river. The interior of the building reflects the internal division of internal space and functions are visually and physically separated from the building's exterior walls.

Surfaces are mainly based on the material's own colour and patina, to provide an expression linked to the local architectural expression. The walls are covered with rough wooden fibreboards and birch as the main material of wood. The building's exterior walls are glazed with dark silicate painting on the inside, in contrast to the exterior woods, own colour.

博尔贡木构教堂结构独特，是挪威木质建筑史上的杰出代表。在新游客中心规划中的历史古迹的总体目标是恢复和强化博尔贡木构教堂周边的原始景观体验。教堂坐落于碗状的景观地形之中，教堂前面是一直向东的道路及河流。

该中心位于与教堂及其周边景观有视觉交流的位置。新建筑体量依据现有的景观地貌，创造了一个横跨山谷的水平向的流动。

停车场从位于游客中心后方的教堂前移至北面，我们还规划降低教堂前方路面，以便让游客更好地感受教堂的原始景观的风貌。

教堂更具表现力的样式配以各式装饰和垂直型外观与建筑外观简单的细节设计形成鲜明对比。其外墙的建材选自松木心材原木，会在天气的影响下逐渐变成灰色。该建筑设有两大各不相同的玻璃开口，一个直通入口，一个通向教堂和河流。室内布局反映了建筑的内在分隔，且从视觉上及形态上，建筑的外墙也可反映内部的功能分区。

建筑表皮主要凭借材料本身的颜色和光泽，使得该建筑与当地的建筑表达产生关联。外墙覆有表面粗糙的木质纤维板和桦木板。室内墙面喷涂了表面质感光滑如镜的黑硅酸盐涂料，与外墙木材的原色形成了对比。

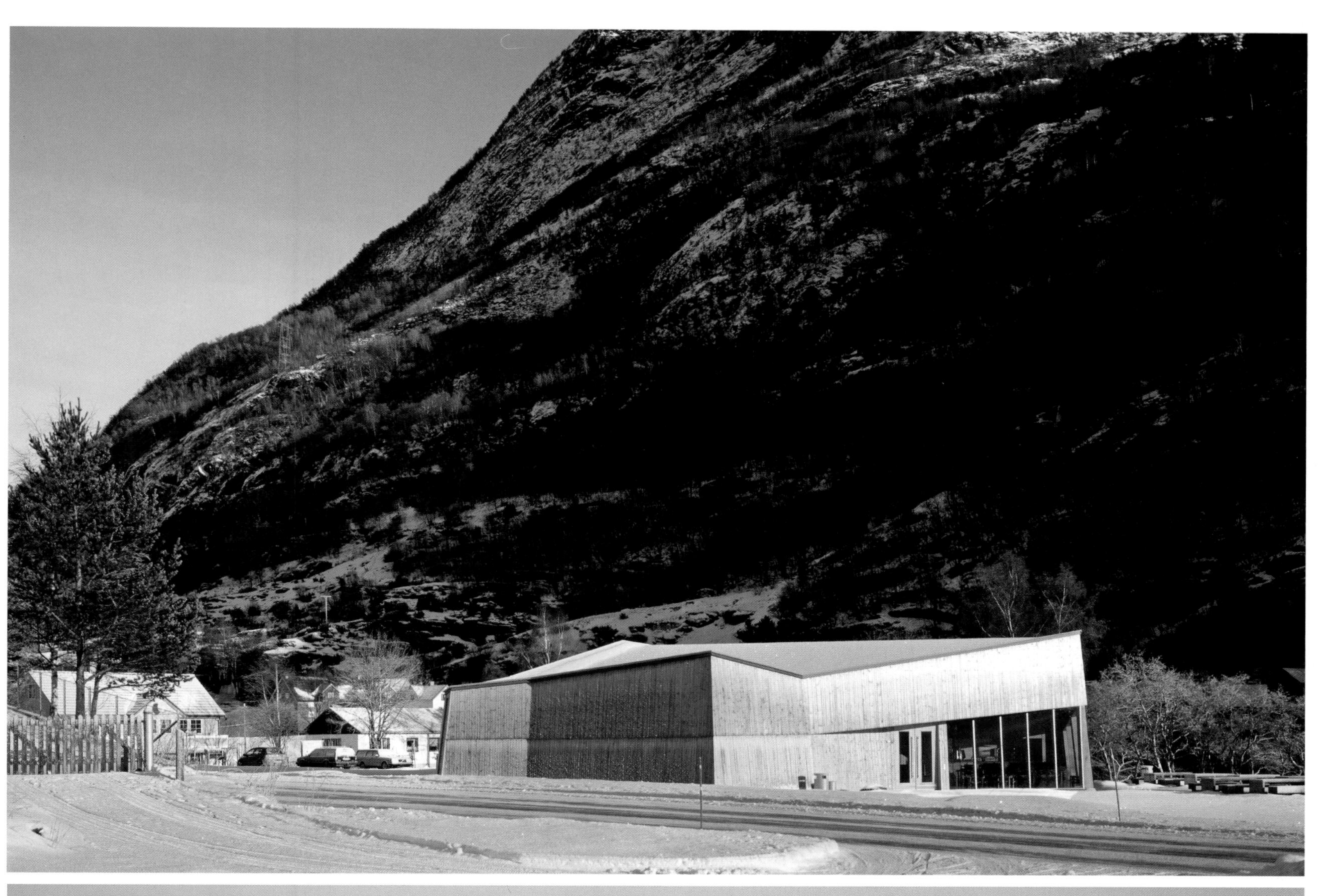

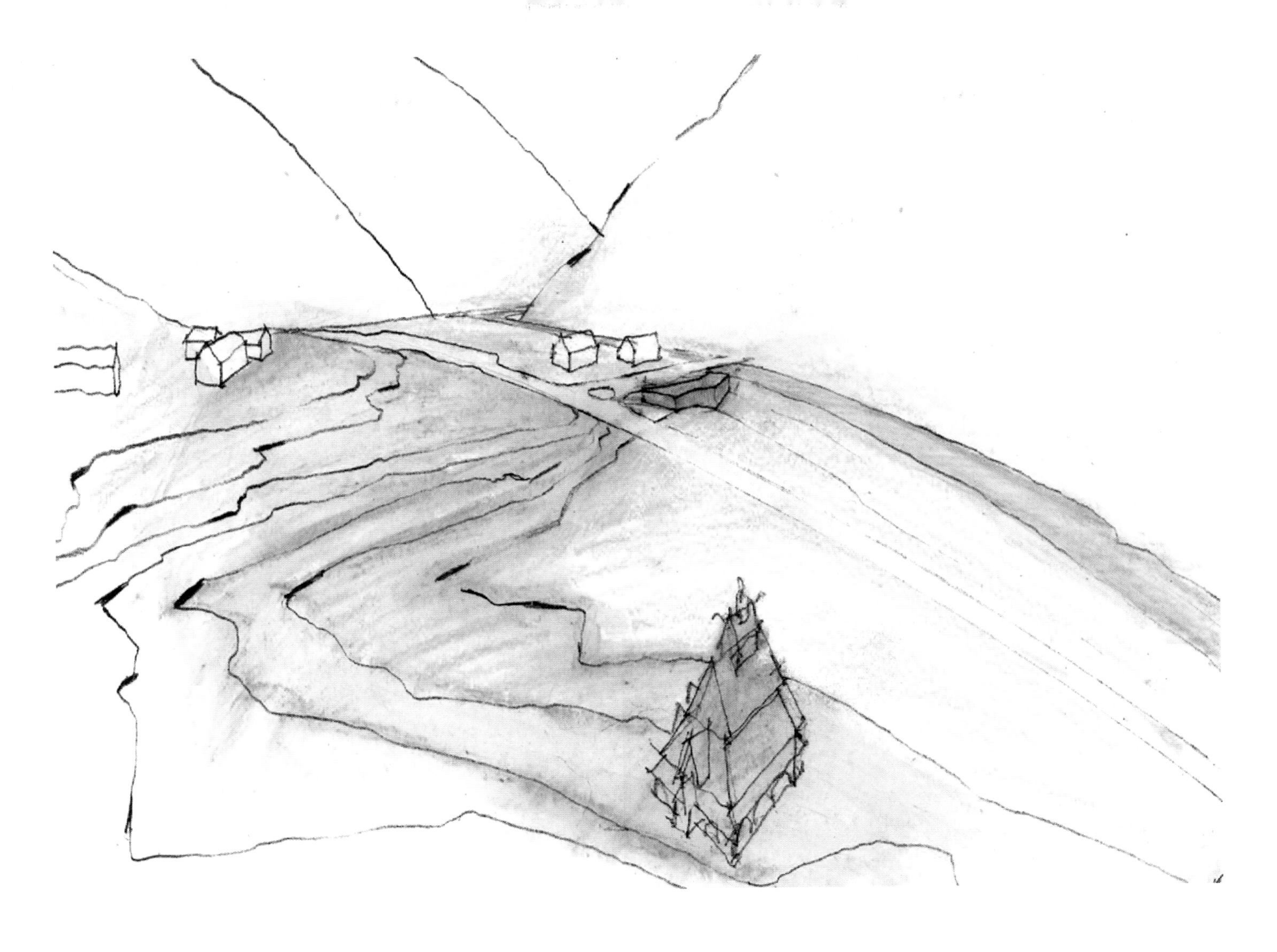

Site Plan 总平面

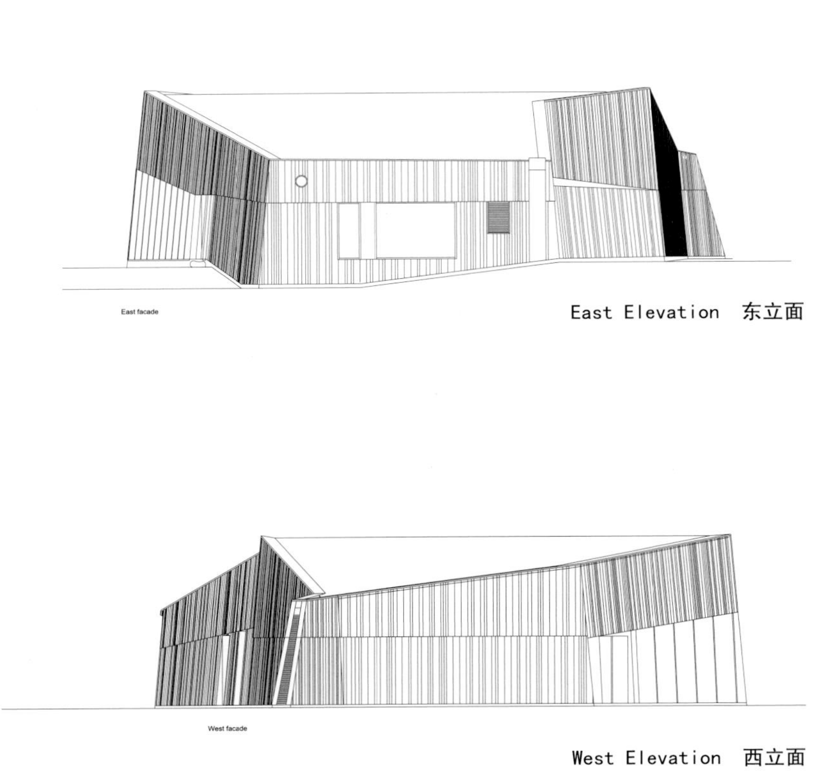

East Elevation 东立面

West Elevation 西立面

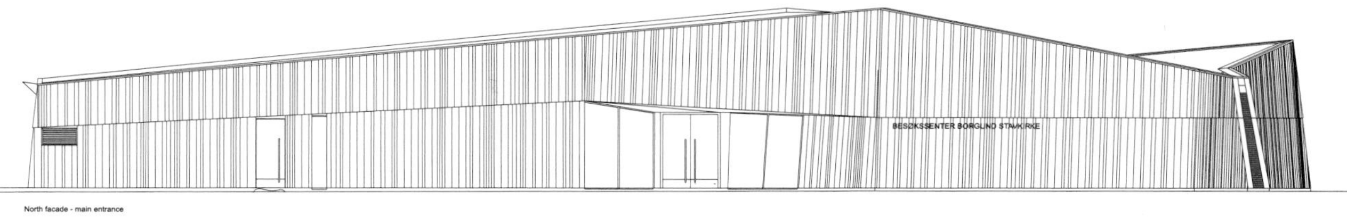

North Elevation 北立面

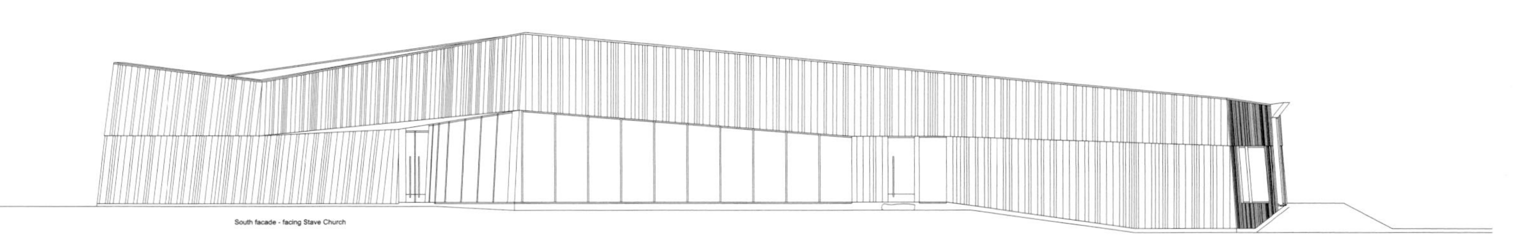

South Elevation 南立面

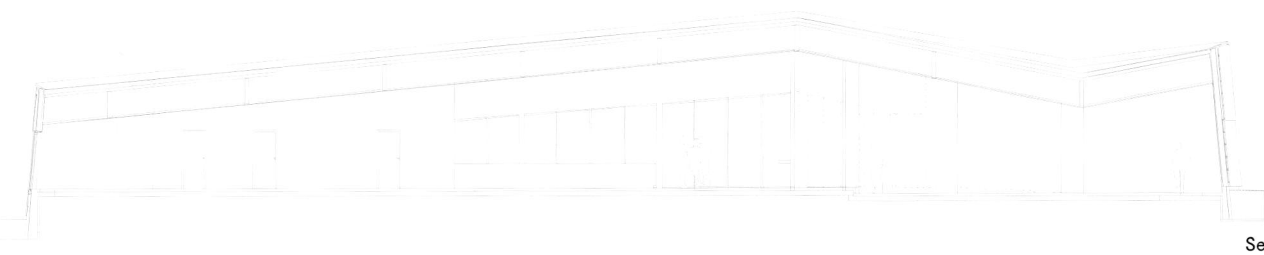

Section 剖面

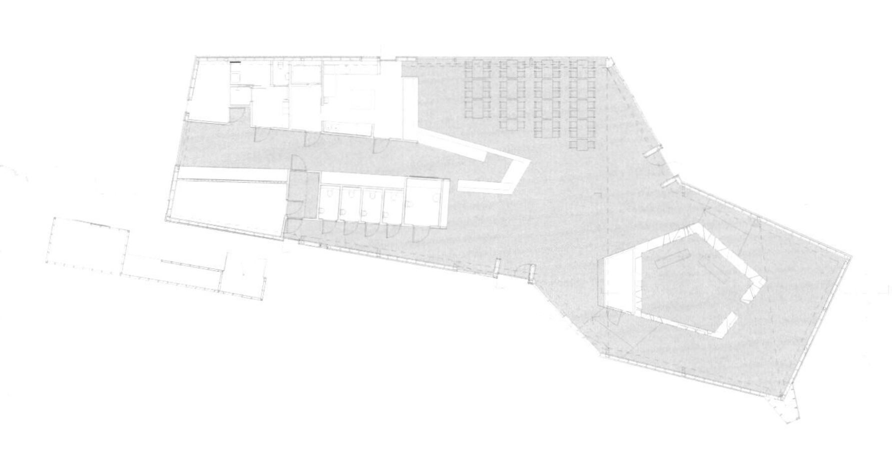

Plan 平面

Gurisenteret - Outdoor Stage and Visitor Centre

Gurisenteret——户外舞台和游客中心

Askim/Lantto Askim/Lantto建筑事务所

Location: The Edøy Island
Program: Visitor Centre
Area: 620 m^2
Status: Completed in 2009
Image Courtesy: Bjarne Ytrøy

区位：Edøy岛
功能：游客中心
面积：620 m^2
状态：2009年竣工
图片提供：Bjarne Ytrøy

This centre was developed for the historical play "Gurispelet", situated on medieval Edøy. It also gives room for different cultural events such as concerts, plays and movies.
The centre is cut into a sandy hill, on the edge of an agricultural landscape including an eight-hundred-year old church. The new amphitheatre shields the stage from the quayside.
To maintain possible movement in the landscape adjacent to centre, a roof terrace has been formed over the visitor centre as an extension to the surrounding farmland.
The rooftop terrace creates a good viewing platform for surrounding landscape.
The entire structure is done in timber, partly as solid wood construction.

该中心是为了中世纪Edøy的历史剧"Gurispelet"的表演而建造的，它也为不同的文化活动，如音乐会、戏剧和电影提供了场所。

这个中心切入一座沙丘，位于农业景观的边缘，其中还包括一个具有800年历史的老教堂。新圆形剧场的巨型顶棚保护舞台免受码头区域的影响。

为保持该中心周围景观的可能动态，已于游客中心上空架起了一座屋顶平台，作为向周边农田延伸的拓展地带。

楼顶平台是一个良好的观赏平台，能够俯瞰周围的景观。

整个结构是由木材建造的，其中一部分是由实木建造的。

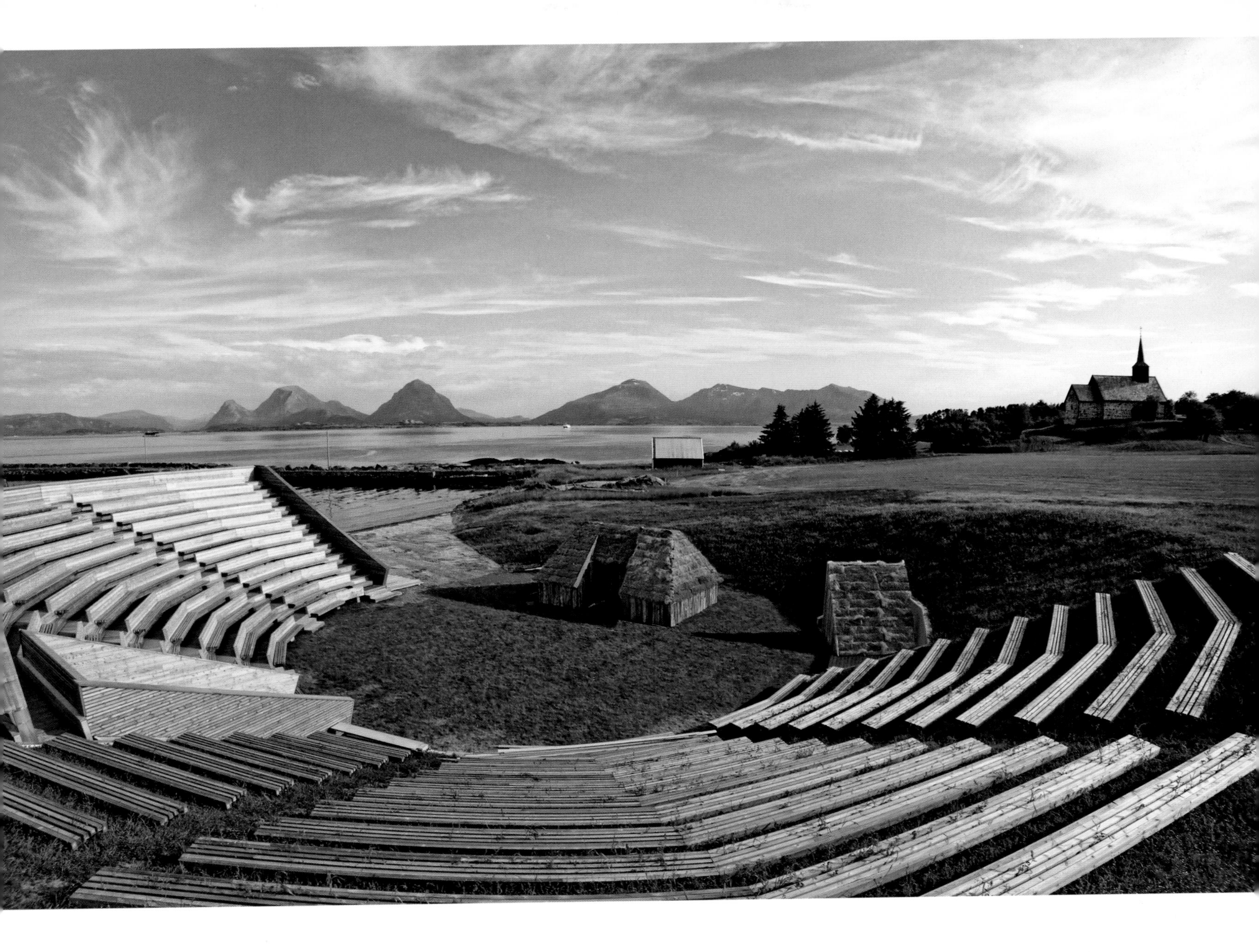

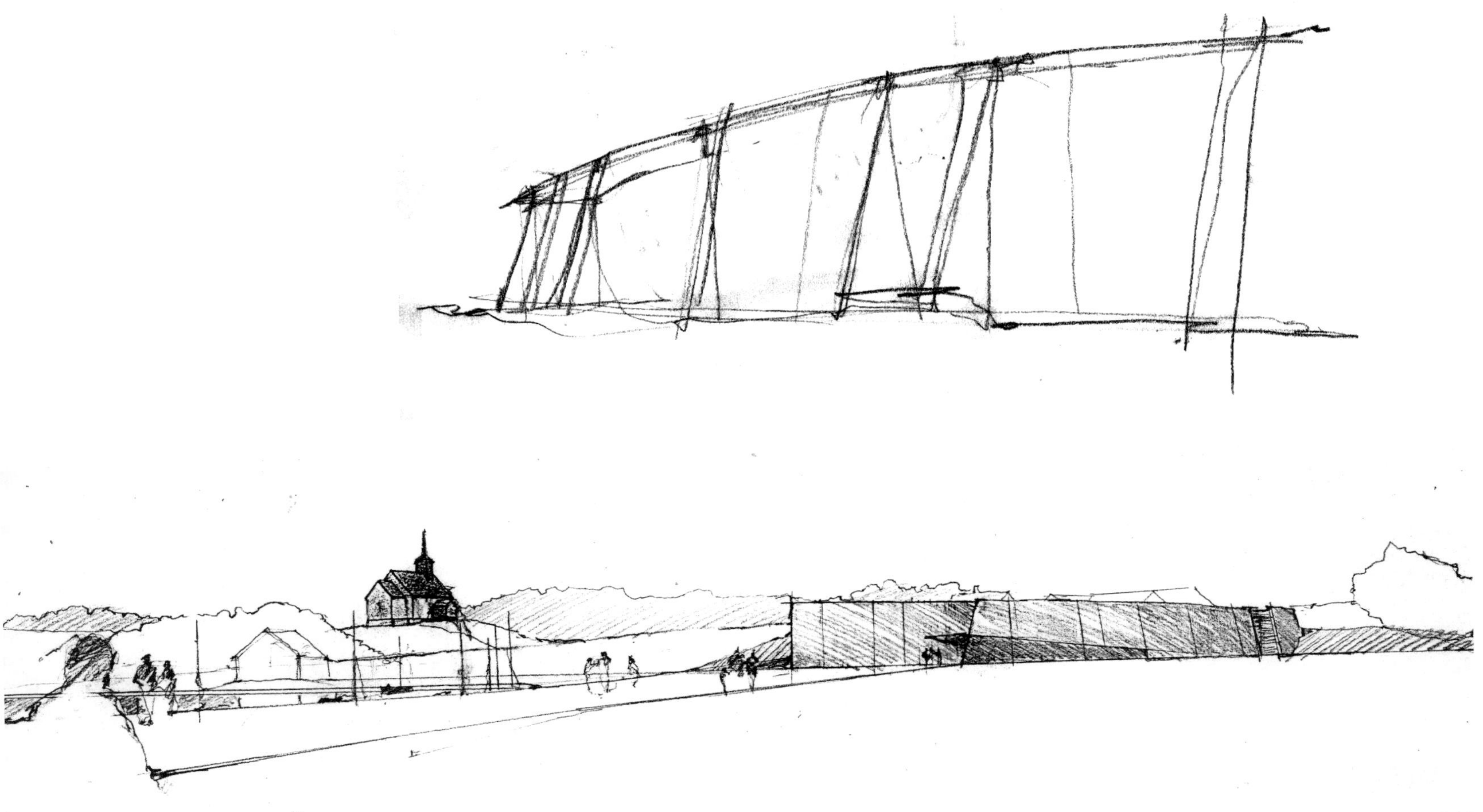

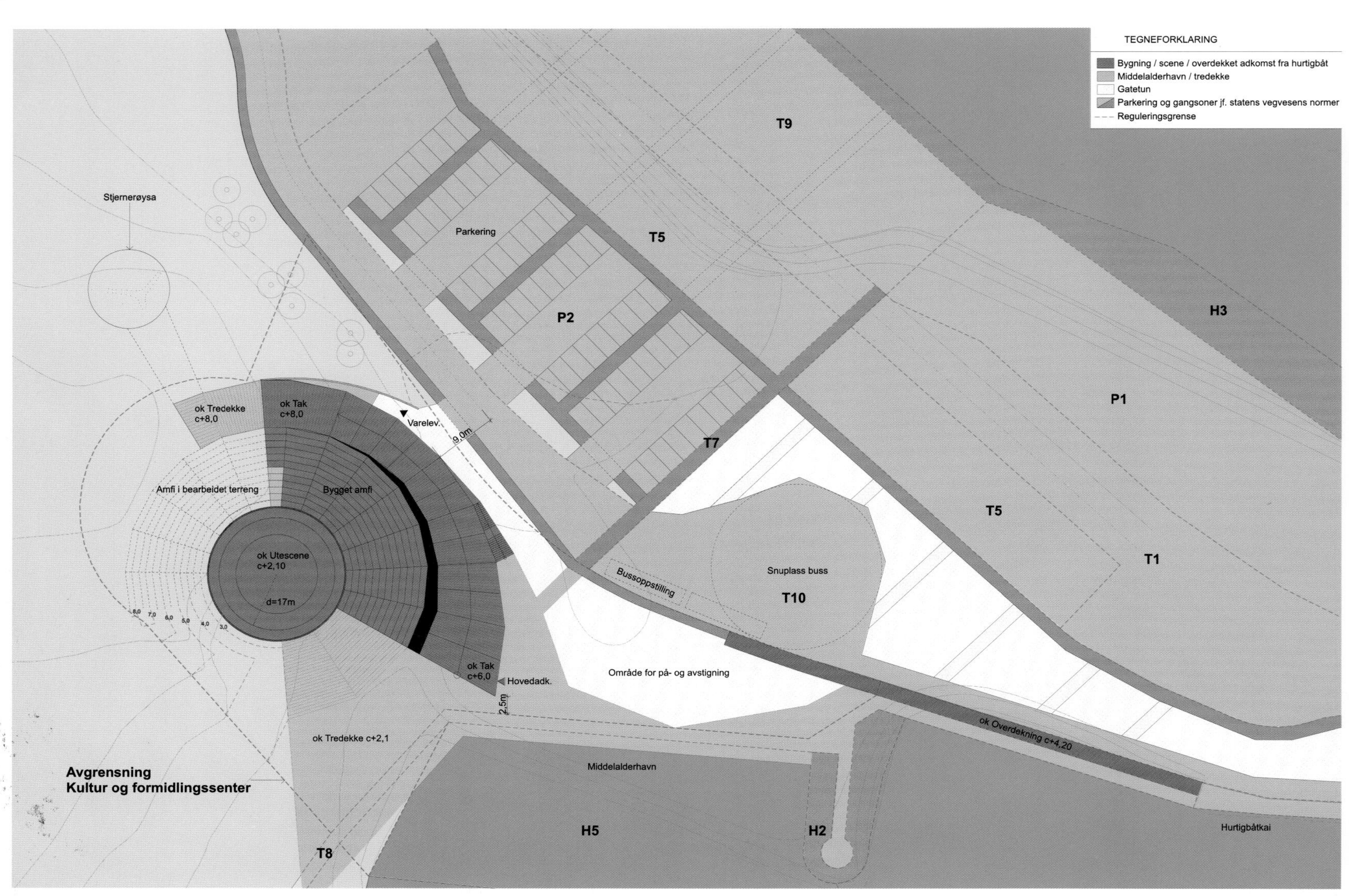

Site Plan 总平面

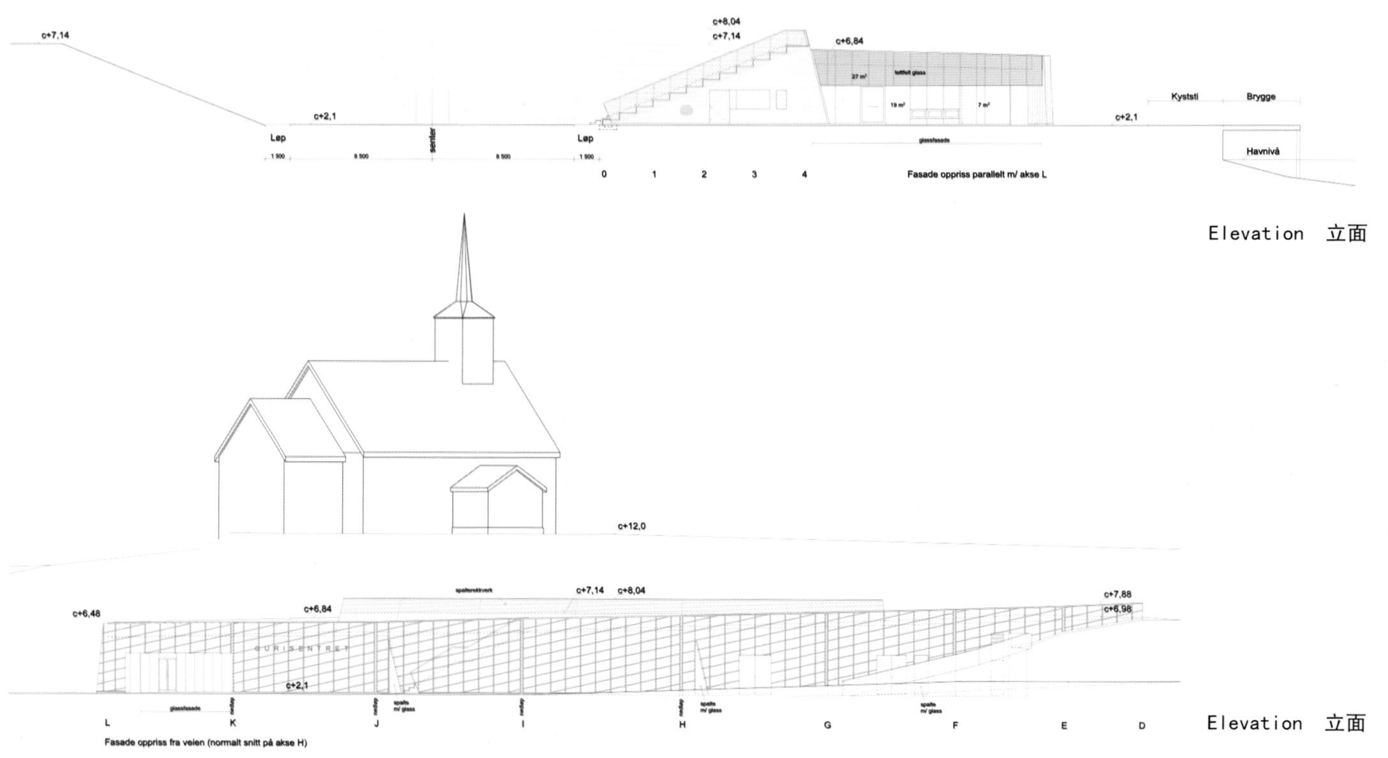

Elevation 立面

Elevation 立面

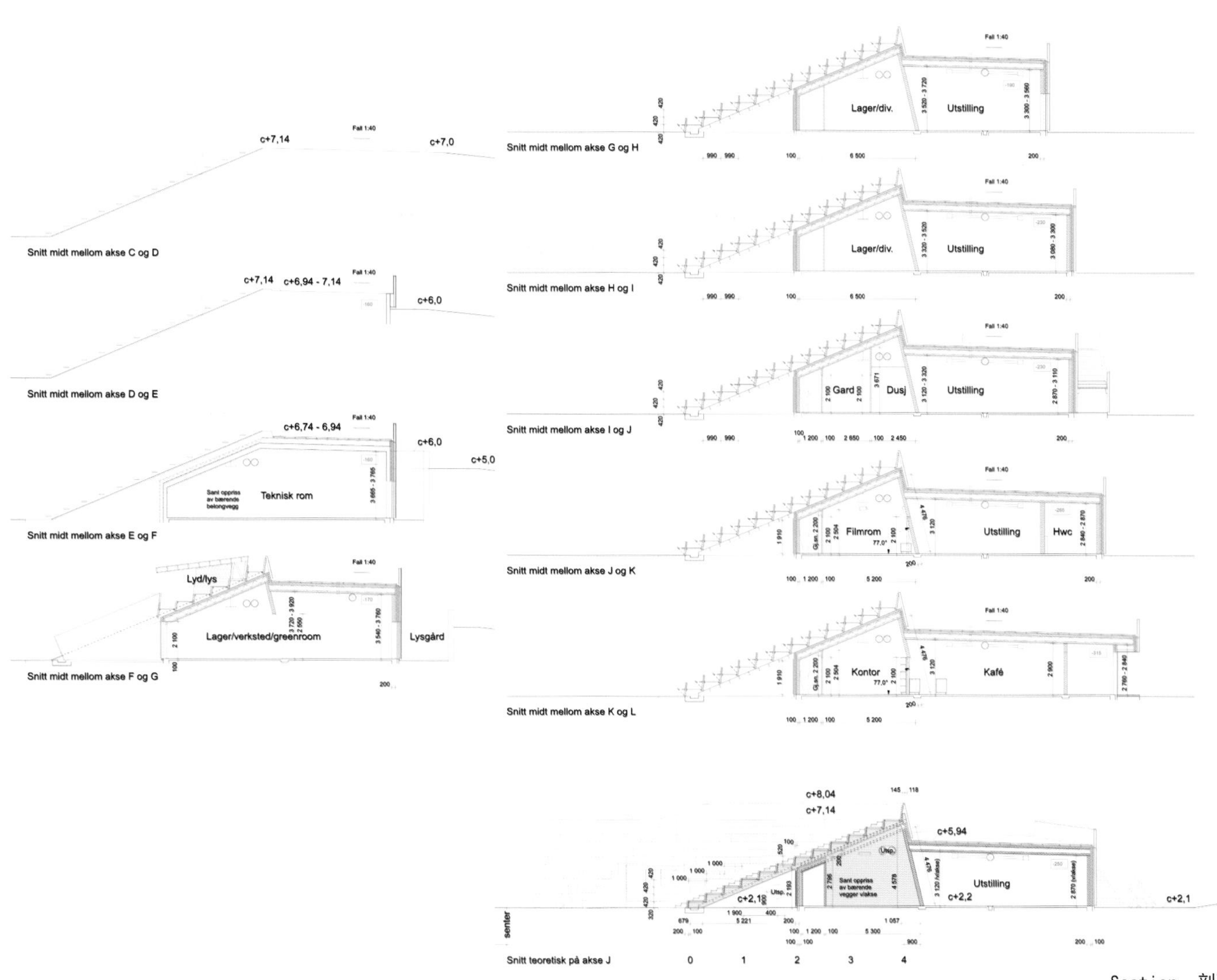
Snitt midt mellom akse C og D
c+7,14
c+7,0
Snitt midt mellom akse D og E
c+7,14
c+6,94 - 7,14
c+6,0
Snitt midt mellom akse E og F
c+6,74 - 6,94
c+6,0
c+5,0
Teknisk rom
Snitt midt mellom akse F og G
Lyd/lys
Lager/verksted/greenroom
Lysgård
Snitt midt mellom akse G og H
Lager/div.
Utstilling
Snitt midt mellom akse H og I
Lager/div.
Utstilling
Snitt midt mellom akse I og J
Gard
Dusj
Utstilling
Snitt midt mellom akse J og K
Filmrom
Utstilling
Hwc
Snitt midt mellom akse K og L
Kontor
Kafé
c+8,04
c+7,14
c+5,94
Utstilling
c+2,18
c+2,2
c+2,1
senter
Snitt teoretisk på akse J
0
1
2
3
4

Section 剖面

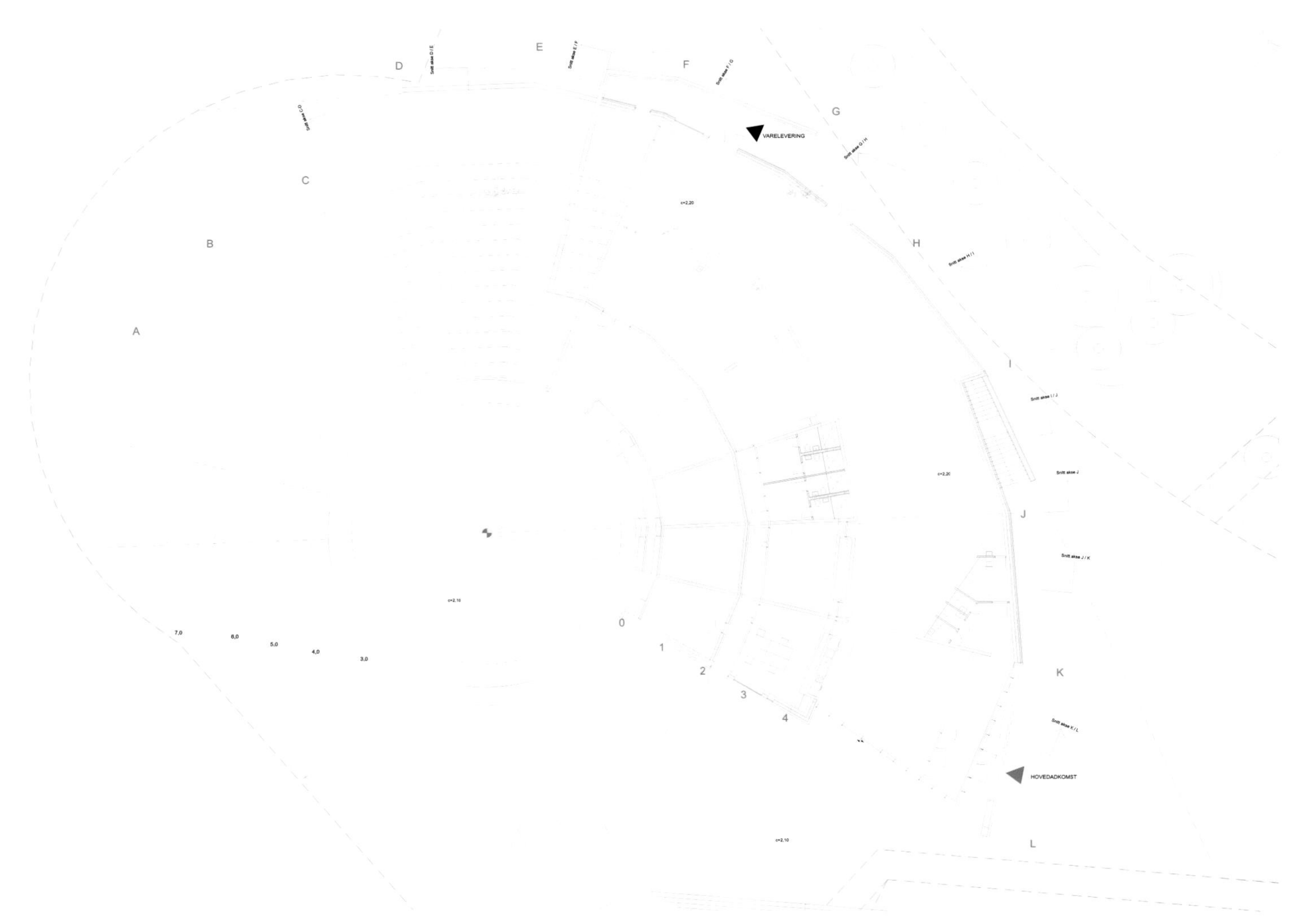

Plan Level 01　一层平面

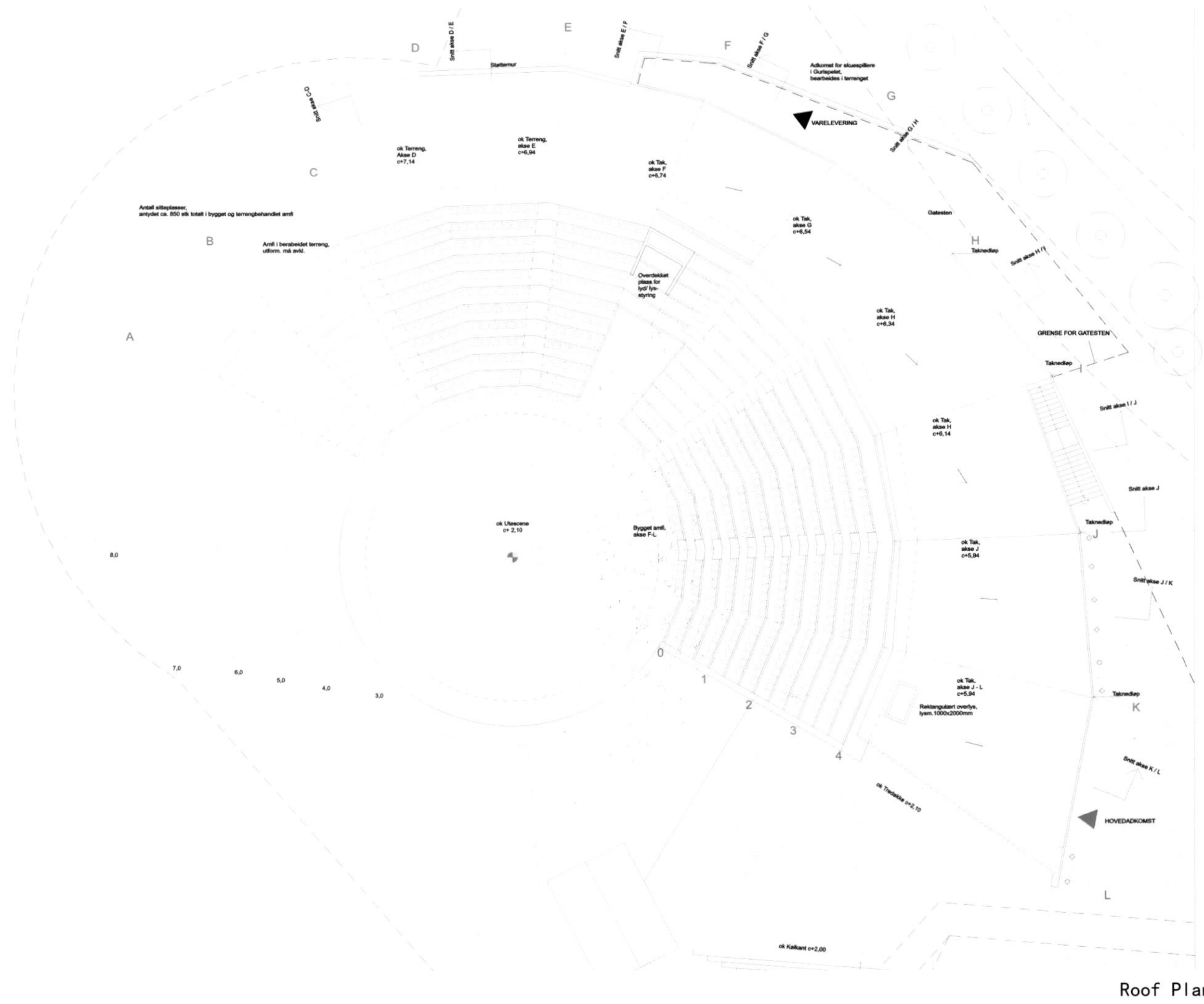

Roof Plan　屋顶平面

FLISAR
SYNNÅVIND

Jektvik Ferry Quay
捷克特维克轮渡码头

Carl-Viggo Hølmebakk AS　　Carl-Viggo Hølmebakk建筑事务所

Location: Helgeland, Norway
Typology: Public
Site Area: 500 m²
Gross Floor Area: 30 m²
Image Courtesy: Carl-Viggo Hølmebakk (CVH), Rickard Riesenfeld (CVH), Steinar Skaar

区位：挪威，海尔吉兰德
类型：公共
基地面积：500 m²
总建筑面积：30 m²
图片提供：Carl–Viggo Hølmebakk (CVH)，Rickard Riesenfeld (CVH)，Steinar Skaar

Despite the informal program, the small service building on Jektvik is very much an experiment. Besides of meeting some pragmatic functions – a waiting room and two rest rooms – the project is first and foremost about transparency and the architectural consequences of transparency. Vehicle for this study is built up as follows: A load-bearing, prefabricated, modular aluminium structure has a twisted facade glazing. ie. that a SG-facade system ("Structural Glazing") is assembled with the outside inwards. Both walls and ceiling have this structure and the rooms appear inside with plane glass surfaces. The glass units, which are composed of various combinations of frosted and coloured glass, give varying transparency and translucency to the side rooms and the surroundings.

On the outside of the supporting aluminium structure is mounted a lath work of 48x48 pine c/c approx. 250 mm. The lath work makes the basis for a seamless outer skin of reinforced polyester. The fibreglass skin is not cast against a smooth shape (as in for example a boat), but hung up on the timberwork as a wet cloth before it hardens. The fibreglass works was, like the main structure, made inside a production hall in Hamar City, before the house was taken 80 kilometers by car to Helgeland in the north of Norway. The main subcontractors were a glass contractor and a boat building company. All technical installations and lighting are positioned, half visible, in the zone between the glass and fibreglass. Over the flat glass ceiling is this zone functioning as a cold attic, which comprises ventilation aggregate, technical guidance and lighting. A working title for the project was "the shrimp" because the house's structure and internal organs were partially visible through the transparent layers of glass and fibreglass.

除了非正式的项目之外，这栋位于捷克特维克的小型服务建筑简直就是一项实验。除了满足些许实际功能的会议室和两间休息室之外，该工程首先就是关于透明度及透明建筑成果。这项研究配备的交通工具如下：一辆承重车、预制的铝质模块结构表面呈扭曲状的玻璃。也就是SG外观系统（"结构玻璃装配"），该系统由外至内进行安装。墙体和天花板均使用了这种结构，各个房间看起来就在玻璃平板表面的内部。玻璃单元是由有色磨砂玻璃的各类组合材料构成的，这种玻璃在边线外空地和周围环境中看来，变化出了透明和半透明的不同效果。

在发挥支撑作用的铝质结构外部安装有规格约为48 cmx48 cm的松木板条。该板条是增强聚酯无缝外表的基础。玻璃纤维表面并没有使建筑物的流畅造型（比如说好比一条船）显得突兀，但是在它变硬之前，它就像一块悬挂在木质结构之上的湿布一般。在该房屋移址于80 km车程外挪威北部的海尔吉兰德之前，玻璃纤维与主体结构一样，也是布局在哈马尔城生产大厅内部的。主要的分承包方包括一个玻璃承包商和一家造船公司。所有的技术装备和照明设备均安置在玻璃和玻璃纤维间的区域内，只能看见一半。位于平板玻璃天花板商贩的空间发挥着冷阁楼的功能，该阁楼是由通风设备聚集体、技术指导及照明设备构成的。该工程的暂定名是"虾"，取这个名称是因为透过透明的玻璃及玻璃纤维层，只能看见部分房屋结构及内部组件。

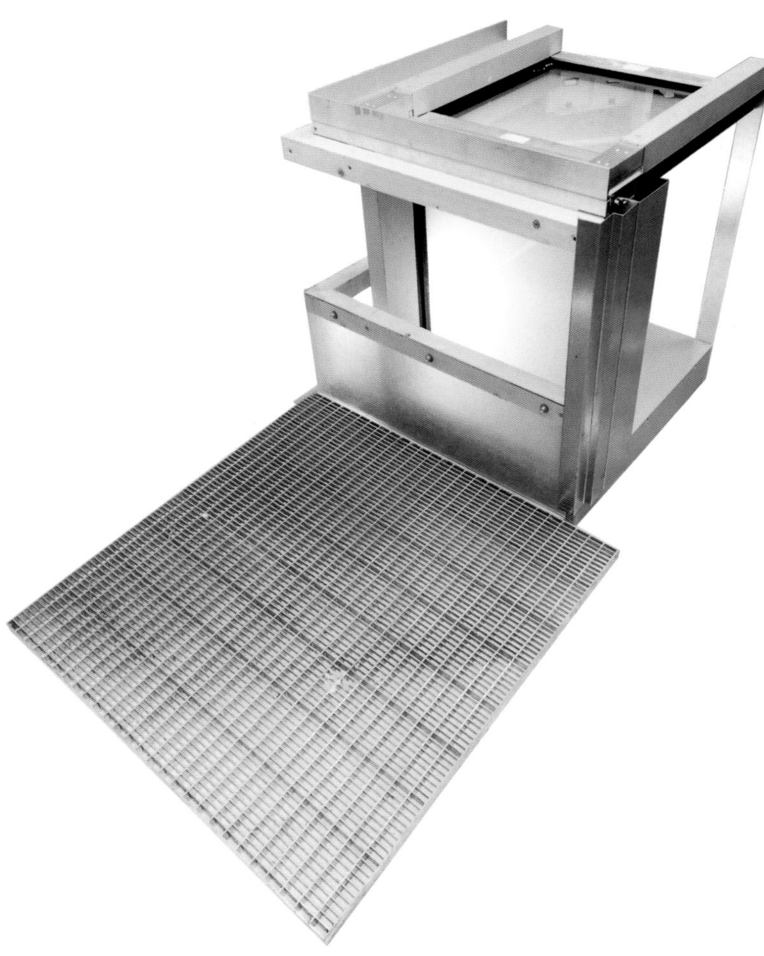

minikran.no
UNIC
CRAMO

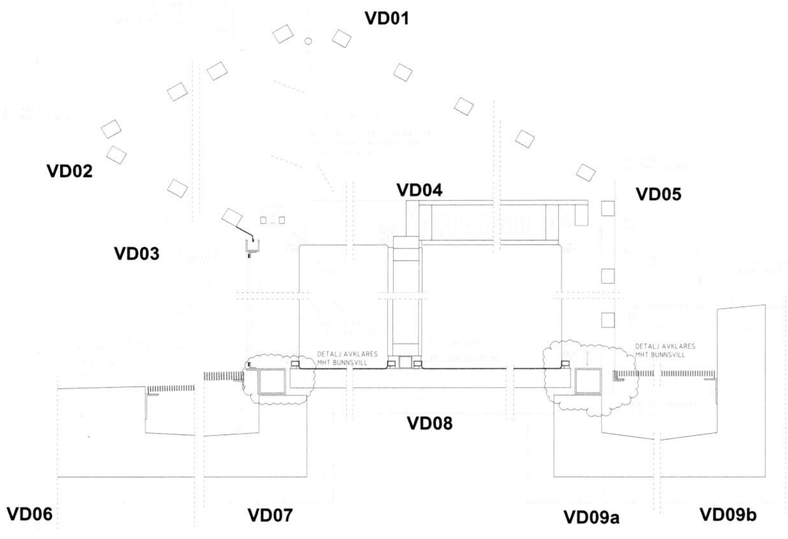
VD01
VD02
VD03
VD04
VD05
DETALJ AVKLARES MHT BUNNSVILL
DETALJ AVKLARES MHT BUNNSVILL
VD08
VD06
VD07
VD09a
VD09b

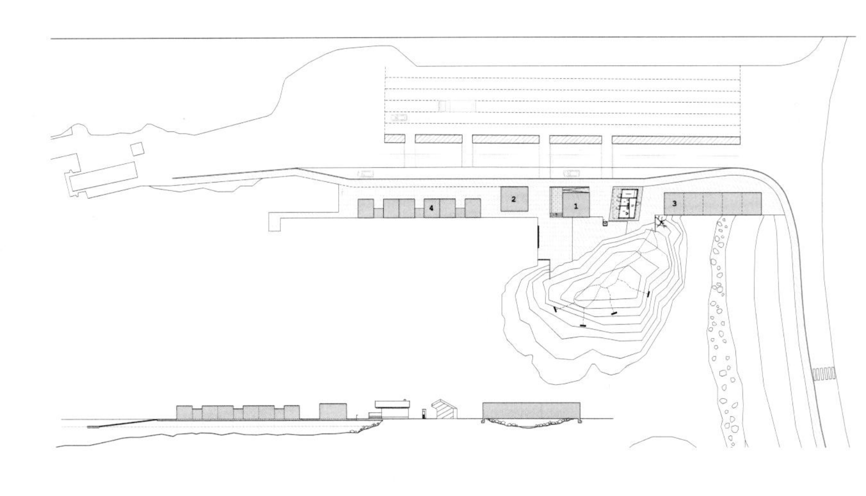

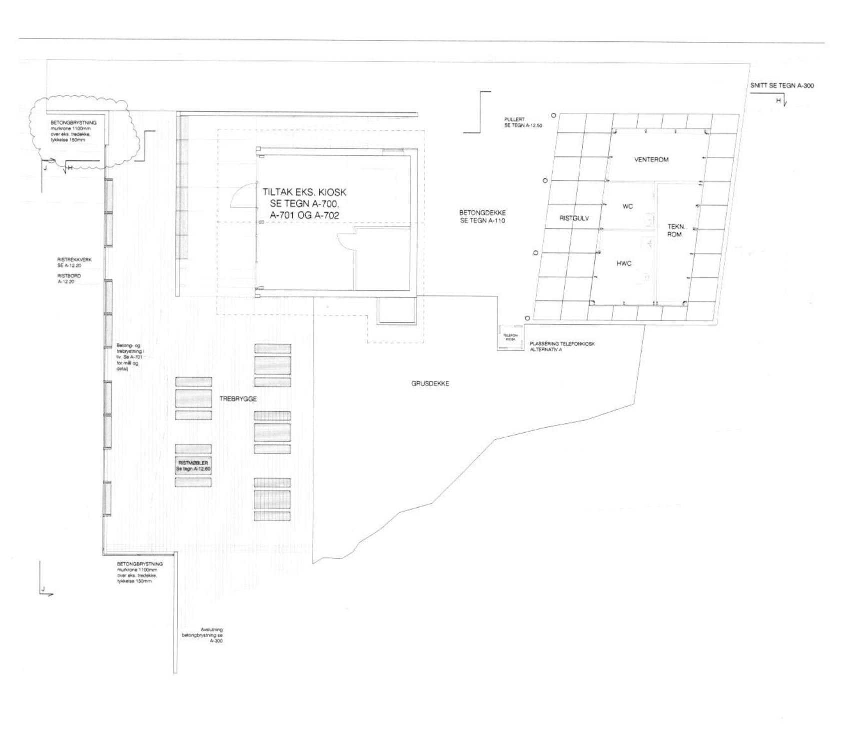
TILTAK EKS. KIOSK
SE TEGN A-700,
A-701 OG A-702
VENTEROM
WC
TEKN.
ROM
HWC
BETONGDEKKE
SE TEGN A-110
GRUSDEKKE
TREBRYGGE
SNITT SE TEGN A-300

HD01
HD02
HD03
VENTEROM
HD04
HD05
HD06
HD19
HD07
HD08
WC
HD10
TEKNISK ROM
HD09
HD11
HD20
HD12
HD13
HCWC
HD14
HD15
HD16
HD17
HD18
TELEFON-
KIOSK

Strømbu Service Centre and Rest Area

Strømbu服务中心和休息区

Carl-Viggo Hølmebakk AS Carl-Viggo Hølmebakk AS建筑事务所

Location: Folldal, Norway
Site Area: 8,000 m²
Gross Floor Area: 90 m²
Status: Completed in 2009
Image Courtesy: Carl-Viggo Hølmebakk (CVH); Rickard Riesenfeld (CVH); Jarle Wæhler ; Simon Skreddernes

区位：挪威，福尔达尔
基地面积：8000 m²
总建筑面积：90 m²
状态：2009年建成
图片提供：Carl-Viggo Hølmebakk (CVH); Rickard Riesenfeld (CVH); Jarle Wæhler ; Simon Skreddernes

The 42 km Rondane Tourist Road goes on the east side of the Rondane Mountains from Enden in the south to Folldal in the north. The centre is situated near the midpoint of this route, and the place also serves as a main starting point for mountain hikers. The placing together of the two user groups has been important for the layout. For the car tourist: How to be introduced to nature and the mountains during a few minutes stop. And for the hiker: How to leave or come back to civilisation after days in the wilderness. Strømbu is also an information centre for several activities in the district, and the place is basically run and operated by the local parish.

The main building has a section that creates quite different situations on the two sides. Towards the parking area, a planted slope with stairs and a ramp leads to a roof terrace. From the terrace there's a view towards the mountains and the passing mountain river. Towards north is a quiet room for rest, with fireplace and view to the wooded river terrain. In-between the two situations is a complex organisation of the kiosk and information area, serving both the outdoor and the indoor. It is also a practical aspect of the section, since the slope gives place for å large septic tank for the toilets, not possible to position below the ground water level. The toilet facilities are placed in a separate building.

The whole area is defined by a looped access with two connections to the main road. The parking, the footpaths and the buildings are placed along this loop and are all elevated to spring flood level, about one meter above the natural terrain. This condition establishes a distinct division between the cultivated and the given nature. In the middle of the loop there will be a large planted forest, giving the overall situation more intimacy and shelter (test plantation 2009). The plan is also to reopen a blocked river gully, passing close to the north glass facade of the building.

42 km的龙达纳观光公路位于龙达纳山东侧，从南端的安登延伸至北端的富达尔。本中心靠近路段中点，因此很多徒步登山旅行者都将此地作为主要的起始点。本中心整合了两类用户群组，这一点对其布局十分重要。对于乘车观光旅行者来说：在短短几分钟的逗留时间里，怎样给他们介绍自然风光以及山峦。而对于徒步旅行者来说：就牵涉到了怎样离开并且在几天的野外生活之后重返人类文明。Strømbu也是一个信息中心，它提供区域内的多项活动咨询，该区域是由当地教区来维持基本的经营和管理的。

主体建筑中有一个区域能够在建筑的两个面上呈现出不同的景观。朝向停车场的那一面，是一个带有楼梯的斜坡，通往屋顶平台。而从平台上望，就会有一个山峦以及越过山脉河流的景观。向北看，是一个安静的休息室，有壁炉，还可以看到森林河流的地形。在这两个景观的中间，是一个结构复杂的凉亭和信息区，提供户内外服务。这也是这一区域实用的方面，因为斜坡给卫生间不可能安装在低于地下水位置的大型化粪池提供了空间。卫生间的设施安放在另一座建筑里。

全部区域处于一个环状通道中，有两个连接点通向主要的公路。停车场、人行道以及各个建筑物都依环形通道构建，并且高于春汛线，大约高出天然地形1 m。这种环境造就了一个独立区，位于栽培区和自然区之间的独特区域。在环形通道的中间，是一片广袤的种植林，这使得整个情形更加使人亲近且有一种庇护感（见2009年种植测试）。这一规划也是对于空心河谷的再开发，更接近建筑北面的玻璃墙。

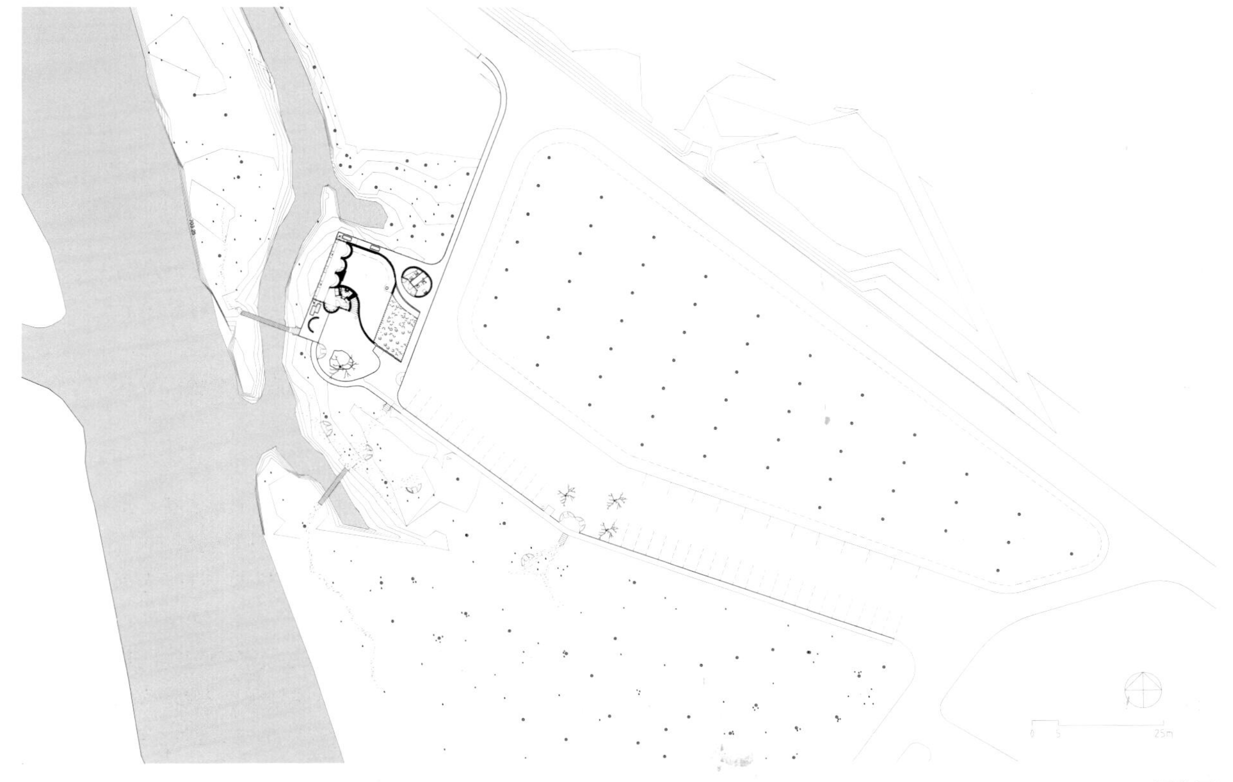

Situation 区位图

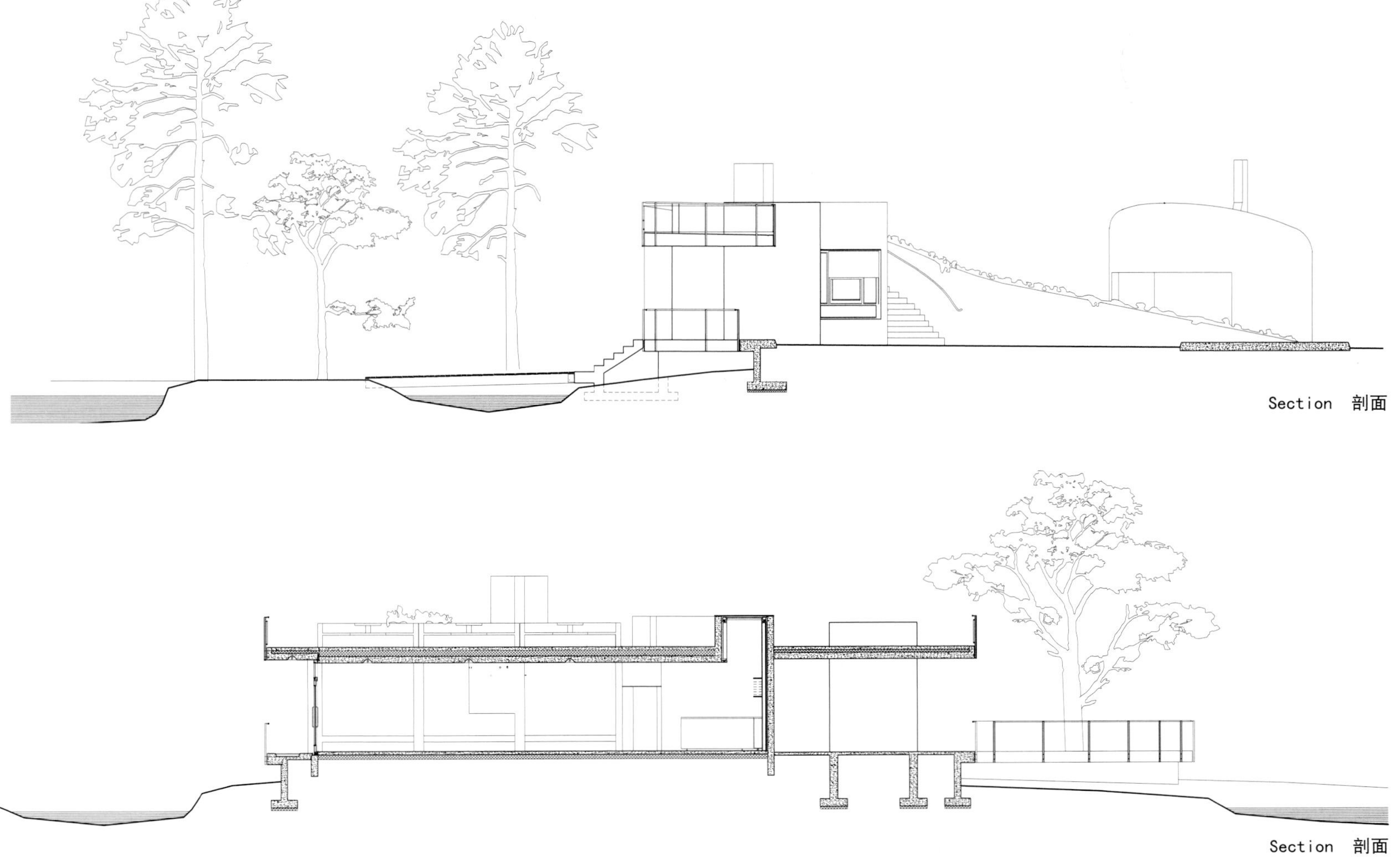

Section 剖面

Section 剖面

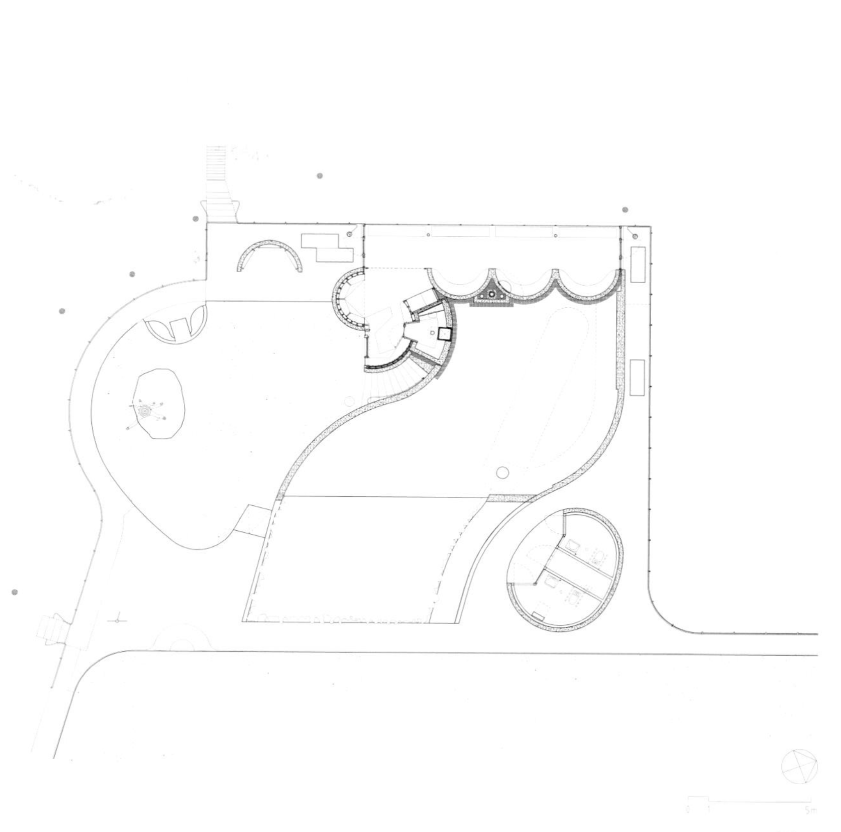

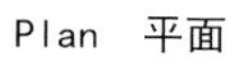

Plan 平面

Vøringsfossen Waterfall Area

韦奴弗森瀑布区

Carl-Viggo Hølmebakk AS   Carl-Viggo Hølmebakk建筑事务所

Location: Eidfjord, Norway
Typology: Cultural
Site Area: 150,000 m²
Gross Floor Area: 1,000 m²
Image Courtesy: Carl-Viggo Hølmebakk (CVH), Rickard Riesenfeld (CVH)

区位：挪威，埃菲杰
类型：文化
基地面积：150 000 m²
总建筑面积：1 000 m²
图片提供：Carl-Viggo Hølmebakk (CVH), Rickard Riesenfeld (CVH)

Vøringsfossen is the largest waterfall in Norway and the third most visited tourist attraction, next to the Vigeland Sculpture Park and Holmenkollen Ski Jump in Oslo. The project covers a large area, including a visitor centre, a footbridge over the waterfall, several look-out points and service facilities. The area also includes an existing hotel.

The edge between the mountain plateau and the canyon forms the most significant line on the site. This edge line defines the waterfall and explains the creation of the dramatic topography. The spatial impression of the canyon has an overwhelming and obscure power. The waterfall becomes the image to which these forces can be connected.

The ambition of the project is to make this event into one thing, one quality. The project embraces the canyon and the waterfall. The footpath establishes a continual experience, constantly pursuing new viewpoints, new sounds, spaces and moods.

The main challenge of the project is how to make structures on the very edge of the rock. Buildings and platforms have to be anchored about 1.5 m back from the cliff, both because of geological conditions and avoiding costly scaffold work. The 42-meter staircase footbridge is designed in seven pre-fabricated elements to be mounted by helicopter.

An important part of the plan is also to school local plant species and re-establish vulnerable vegetation destroyed by heavy foot traffic. The area will be drifted by grazing animals.

韦奴弗森瀑布是挪威最大的瀑布，也是奥斯陆第三大观光胜地，仅次于维吉兰雕塑公园和荷曼高兰山区滑雪场。该建筑项目覆盖一片广阔的区域，包括游客中心、瀑布上方的人行桥、若干瞭望点及服务设施。该区域也包括了一座宾馆。

山地和峡谷间的边缘形成了该景观最重要的线路。这条边界线界定了韦奴弗森瀑布，也解释了地貌的鬼斧神工。峡谷给人的空间印象让人感觉到了一种势不可挡却又说不清道不明的力量。瀑布成为了这些力量交汇处的景象。

该项目的追求目标是将整个建筑事件合为一体，展现一种品质。它拥抱了峡谷和瀑布。人行小道也建立起了一番绵延不断的体验经历，不断地追随新视角、新声音、新空间以及新心境。

该项目面临的主要挑战即为如何在岩石的边缘建造建筑结构。建筑物及平台必须固定在距离悬崖1.5 m处的后方，既是由于地理条件的限制，也是为了避免成本高昂的脚手架作业。七大由直升机完成安装任务的预制元素设计造就了长达42 m的阶梯人行桥。

建筑计划中的一大重要组成部分即将当地的植被物种分门别类，并重建被行人践踏导致破坏的脆弱植被。该区域将会成为食草动物的闲游场地。

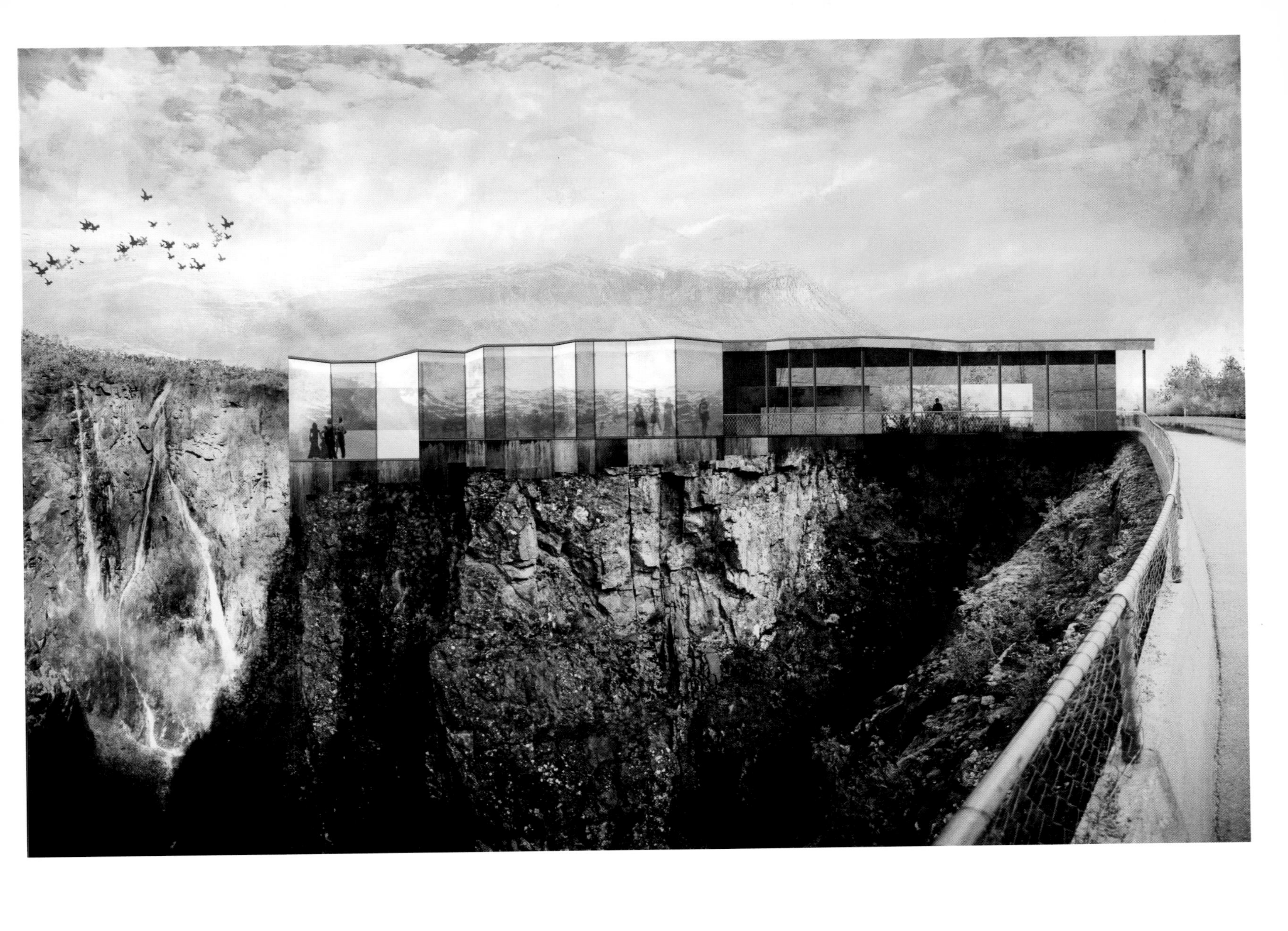

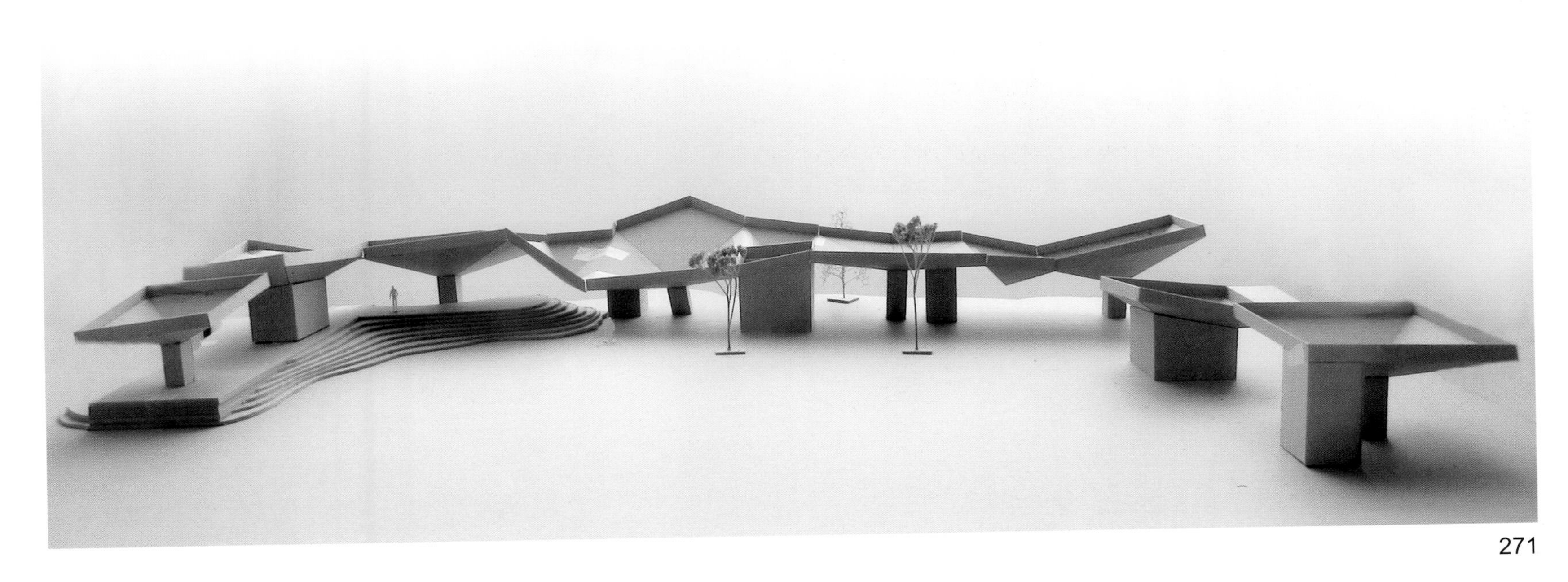

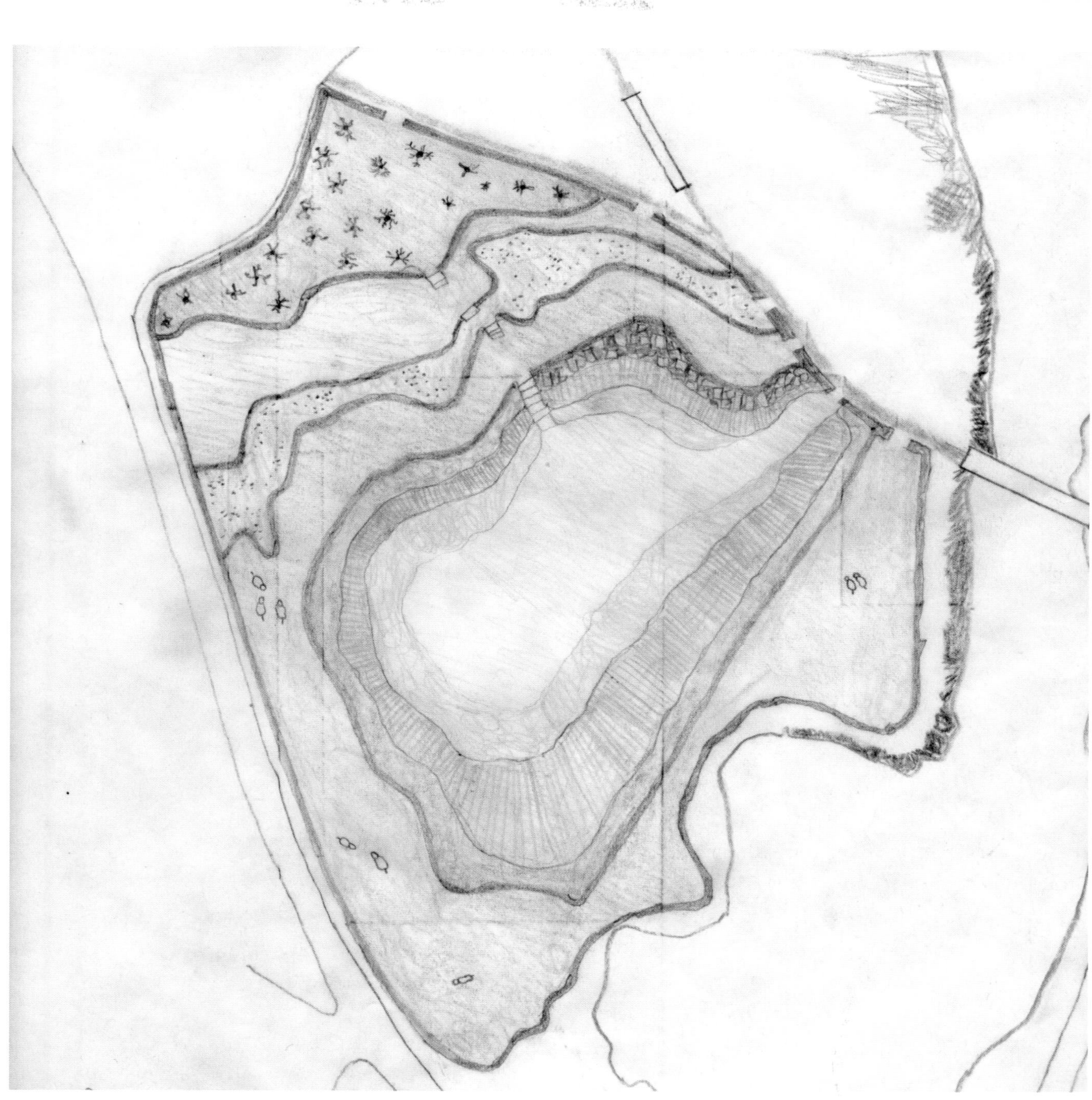

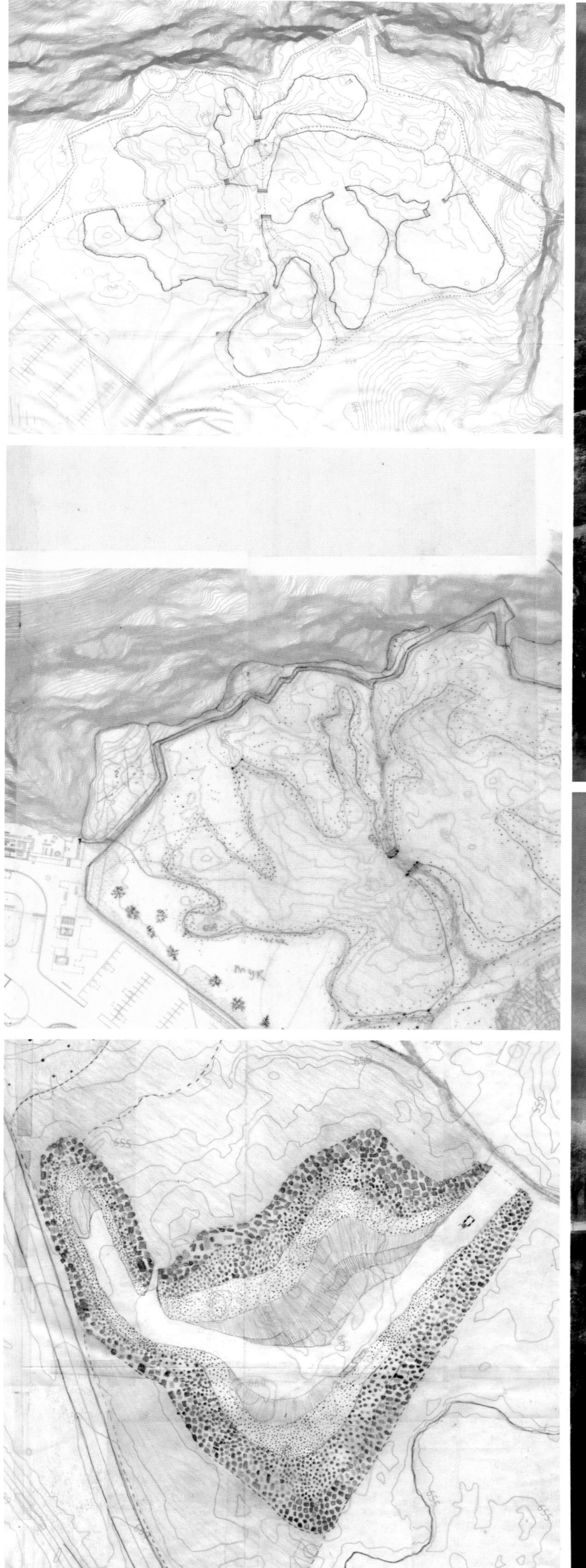